Springer Series in Materials Science

Volume 292

The Springer Series in Materials Science covers the complete spectrum of materials research and technology, including fundamental principles, physical properties, materials theory and design. Recognizing the increasing importance of materials science in future device technologies, the book titles in this series reflect the state-of-the-art in understanding and controlling the structure and properties of all important classes of materials.

More information about this series at http://www.springer.com/series/856

Gertz I. Likhtenshtein

Nitroxides

Brief History, Fundamentals, and Recent Developments

Springer

Gertz I. Likhtenshtein
Department of Chemistry
Ben-Gurion University of the Negev
Beersheba, Israel

Institute of Problems of Chemical Physics
Russian Academy of Science
Chernogolovka, Moscow Region, Russia

ISSN 0933-033X ISSN 2196-2812 (electronic)
Springer Series in Materials Science
ISBN 978-3-030-34821-2 ISBN 978-3-030-34822-9 (eBook)
https://doi.org/10.1007/978-3-030-34822-9

This Springer imprint is published by the registered company Springer Nature Switzerland AG
The registered company address is: Gewerbestrasse 11, 6330 Cham, Switzerland

Preface

It is known that there are delicate links and fine parallels between an art and science. Both these spheres of human endeavor involve a unique combination of professional skill and creative search. Sometimes, an intuitive line of a great poet or philosopher may be likened to the opening of a new horizon in science. Thus, the composer Maurice Ravel in his famous "Bolero" allegorically depicts the process of birth and development of an epochal discovery that gives rise to many advantages. Like that opening musical movement, the first publication on synthesis a nitroxide by Fremi in 1825 initiated, after a long lag period, the birth and rapid development of a novel class of stable nitroxide radicals which happened to be extremely important in both aspects, basic and applied.

At present, the science of nitroxides widely extended her "hands" to chemistry, physics, and biology. The application of nitroxides ranges from use as spin labels and antioxidants in biological studies, charge carriers for energy storage, basis for magnetic materials' mediators in polymerization reactions, functional spin probes for pH, oxygen, and thiol levels to catalysts in chemical and electrochemical oxidation reactions.

Classical and modern physical chemistry and chemical physics, chemical kinetics, organic, inorganic, and quantum chemistry provide an arsenal of physical methods and establish a basis for the investigation of structure and action mechanism of processes involving nitroxides.

This book embraces all principal aspects of structure and physicochemical action mechanisms of nitroxides. It is a view of nitroxides by a physicochemist with long-term experience in the area. The book is not intended to provide an exhaustive survey of each topic but rather a discussion of their theoretical and experimental background and recent developments. The literature of nitroxides is so vast, and many scientists have made important contribution in the area that it is impossible in the space allowed for this book to give a representative set of references. In fact, for each section in this area one can write several books. The author apologizes to those he has not been able to include.

Research on nitroxide which combines their fundamental importance for human welfare and intellectual fascination for investigation will promote solving exciting and complicated problems in chemistry, biology, and physics.

Chapter 1 of the monograph is a brief outline of 175 years history of nitroxides. Chapters 2 and 3 form the theoretical and experimental chemical background for nitroxide numerous application. Chapter 4 is a general survey of fundamentals of electron spin resonance and nuclear magnetic resonance, and main physical methods directly related to nitroxide. Advantages in design and use of nitroxide biradicals are reviewed in Chap. 5. Chapter 6 presents a review on the use of tethered nitroxide–fluorophore molecules as probes of redox status, antioxidant activity, oxidative stress, and free radical reaction. Nitroxide-mediated polymerization (NMP) is the subject of Chap. 7. Chapter 8 describes role nitroxides as the base for magnetic materials. Involving nitroxide radicals and their derivatives in biological processes is in focus of Chap. 9. Applications of nitroxides as spin labels and probes constitute the contents of Chap. 10. Chapter 11 is devoted to the use of nitroxyls to solve some of the physicochemical problems.

The monograph is intended for scientists and engineers working in the fields of chemistry, physics, and biology in which nitroxyls currently find or would find their application. Book, as a whole, and separate chapters can be used as a subsidiary manual for instructors, graduate and undergraduate students of university chemistry, physics, and biophysics departments.

Department of Chemistry, Ben-Gurion University of the Negev and Institute of Problem of Chemical Physics, Russian Academy of Science provided excellent conditions for writing this book for which the author is extremely grateful.

Moscow Region, Russia/Beersheba, Israel Gertz I. Likhtenshtein

Contents

About the Author

Prof. Gertz I. Likhtenshtein received his Ph.D. and his Doctor of Science from the Semenov Institute of Chemical Physics at the Russian Academy of Science in Moscow, where he was appointed to the position of Head of Laboratory of Chemical Physics of Enzyme Catalysis in 1966, becoming a full professor in 1976. In 1992, he moved to the Department of Chemistry at the Ben-Gurion University of Negev, Israel, as a full professor in charge of the Laboratory of Chemical Biophysics and has been an Emeritus since 2003. He has authored eleven scientific books and around 390 papers, and his awards include the Medal of the Exhibition of Economic Achievement, the USSR Diploma of Discovery, the USSR State Prize, the V. V. Voevodsky International Price for Electron Spin Resonance, the N. M. Emanuel Price for Biophysical Chemistry, and the Diploma of the Israel Chemical Society. Professor Likhtenshtein is a Foreign Member of the Academy of Science of Republic Tadjikistan and was a member of the International ESR Society, the American Biophysical Society, the Israel Chemical Society, and the Israel ESR Society. His recent main scientific interests focus on the analysis of biologically important molecules.

Chapter 1
Nitroxides: 170 Years of History

1.1 Long Preamble of a Tale (1845–1960)

As an each way, a long road of the 174 years history of stable radicals bearing a >N–O group has started with the first step. This step has been made in 1845 by Edmond Frémy (1814–1894), a director the Muséum national d'histoire naturelle, who was one of understanding chemists of nineteenth century [1]. Edmond Frémy published numerous articles in the *Annales de Chimie et de Physique*, seven volumes *of Traité de chimie générale,* and ten volumes *Encyclopédie Chimique*, in collaboration with several other scientists. But his main advantage, which left memories in the chemical history, was a synthesis and characterization of disodium nitrosodisulfonate (potassium nitrosodisulfonate, Frémy's salt (Fig. 1.1a).

A synthesis of potassium nitrodisulfonate,occurring by the following scheme:

$$HNO_2 + 2\,HSO_3^- \rightarrow HON(SO_3)_2{}^{2-} + H_2O$$
$$3\,HON(SO_3)_2{}^{2-} + MnO_4^- + H^+ \rightarrow 3\,ON(SO_3)_2{}^{2-} + MnO_2(s) + 2\,H_2O$$
$$2\,ON(SO_3)_2{}^{2-} + 4\,K^+ \rightarrow K_4\big[ON(SO_3)_2\big]_2(s)$$

now can be readily performed in a high school student chemical laboratory. Fremy's salt is long-lived radical in water and other solvents in unaerobic conditions. Nevertheless, a rapid and highly exothermic decomposition of this compound occurs spontaneously in air that is a serious limitation for its applications.

Next step in the area was done by O. Piloty and B. G. Schwerin in 1901 [2] who prepared the first organic radical containing >N–O. The radical was named as "porphyrexide" (Fig. 1.1b). Heinrich Wieland and Moriz Offenbacher reported in

G. I. Likhtenshtein, *Nitroxides*, Springer Series in Materials Science 292,
https://doi.org/10.1007/978-3-030-34822-9_1

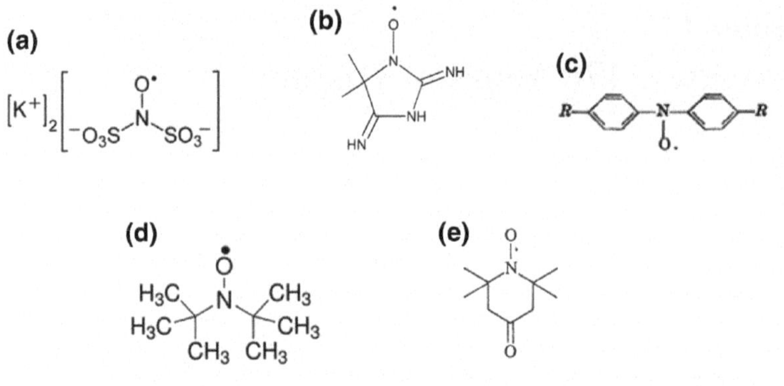

Fig. 1.1 Structures of pioneering nitroxides

1914 the synthesis of diaryl nitroxides in which the >N—O group is attached to two conjugated system (Fig. 1.1c) [3]. About 60 years after, stable di-*tert*-butyl nitroxide (Fig. 1.1d) was synthesized by Kentaro Hoffmann and Audrey T. Henderson [4]. In 1961, in the frame of G. A. Razuvaev chemical school, Lebedev et al. prepared 2,2,6,6-tetramethyl-1-piperidinyloxil from acetone and ammonia (Fig. 1.1e) [5]. Portraits of the first discovers of nitroxide chemistry are presented in Fig. 1.2. The synthesis of all nitroxides cited above was a remarkable achievement. However, the resulting compounds were not adapted to further modification.

Edmond Frémy Heinrich Otto Wieland Grigory A. Razuvaev

Fig. 1.2 Portraits of the first discovers of nitroxide chemistry

1.2 "Golden" Decade (1962–1972)

From period 1962–1972 practically all corner-stone ideas in the area of nitroxides were exposed and developed. These ideas were sustained by fundamental theoretical and experimental investigations in chemistry and physics of nitroxides. First nitroxides, presented in Fig. 1.1, though their principle importance, did not find wide application and were not be able to serve as a basis for synthesis of new paramagnetic compounds in a broad scale.

Until 1962, chemists adhered to the paradigm that the most chemically reactive portion of a radical can be a group bearing spin electron. This paradigm was broken by M. B. Neiman and E. G. Rozanzev (Figs. 1.3 and 1.4), who introduced nitroxide reactions with a nitroxide (Fig. 1.1e) without direct involvement of the spin center. A novel class of stable nitroxides radicals presented in publication [6, 7] was first met with skepticism, and even strong criticism, from qualified and very professional members of the scientific community. But, later, more and more young enthusiasts joined the ranks of scientists applying this new tool in their research, and ever increasing reports of nitroxides were published. These pioneering works have laid a chemical basis for the method of numerous nitroxide applications, spin labeling in particular. As a consequence, rapid and extensive progress in the area and a real burst of works on synthesis and application of nitroxide in chemistry, physics, biology, and even in medicine have been broken. The theoretical and experimental data presented in this book clearly demonstrate both history and the current progress within the nitroxide "empire" of about publications.

In parallel, the dependence of ESR spectra of nitroxides on their molecular dynamics in solutions was demonstrated. The radical motion leads to averaging of spin

Fig. 1.3 Professor Moisey B. Neiman (1889–1967)

Fig. 1.4 Gertz I. Likhtenshtein, Anatoly L. Buchachenko, Eduard G. Rozanzev and Nikolay N. Semenov (1977)

electron spin nuclear interaction and to a drastic change of the ESR spectra. Correlation time for isotropic and anisotropic rotation τ_c for a nitroxide on a homogeneous media in the area of fast ($\tau_c = 10^{-9}\ 10^{-10}$ s) and slow ($\tau_c = 10^{-7}\ 10^{-8}$ s) motion can be estimated in the frame of theory developed by Kivelson [8] and Freed [9].

At the end of 1960 new classes of stable radicals, α-Nitronyl nitroxides [10] and Doxyl nitroxides performed by the J. F. W., Keana group Fig. 1.5 [11], possessing great potential for further wide applications, were synthesized.

Fig. 1.5 Doxyl nitroxides [11]

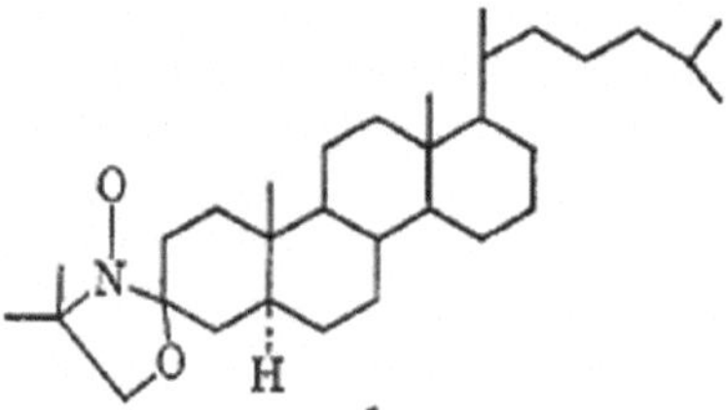

1

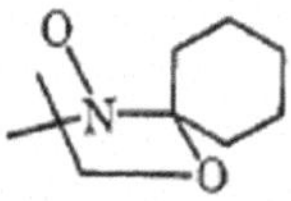

2

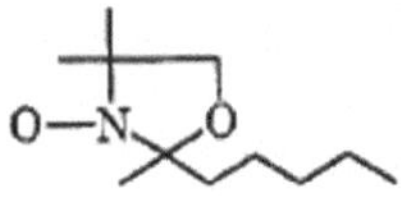

3

In the pioneering work of Golubev et al. [12], the first stoichiometric oxidation of alcohols to ketones mediated by an oxoammonium species with N^{+}-oxyl fragment was demonstrated.

Decisive breakthrough in the spin-labeling area was occurred owning a whole cascade of pioneering works of Howard McConnel and his collaborators. In the first publication in this series [13], a spin label, 2,2,6,5-tetramethyl-3-isocyanatopyrrolidine-1-oxyl, was covalently tethered to bovine serum albumin and to poly-L-lysine [13]. The nitroxide ESR spectra in solution indicated sensitivity of the nitroxide segment rotation to molecular motion of the labeled compound. Binding of a spin-labeled hapten, the 2,4-dinitrophenyl hydrazone of 2,2,6,6-tetramethyl 4-piperidone nitrogen oxide to antidinitrophenyl antibody was proved by the estimation of the rotational relaxation time of the bound hapten using analysis of the nitroxide ESR spectra [14] (Fig. 1.6).

Powerful potential capacity of spin-labeling method for study conformational transitions in proteins, which are necessary for its functional activity, was illustrated by the McConnell group on example of hemoglobin. Two reactive β-93SH groups in horse hemoglobin were modified with N-(1-oxyl-2,2,5,5-tetramethyl-3-pyrrolidinyl) iodoacetamide [15]. The ESR spectra of the labeled hemoglobin labeled were dependent on the degrees of oxygenation. The Hill constant n, a measure of the cooperativity of sigmoidal oxygen binding, was found to be $n = 2.3$ for the labeled hemoglobin, as compared to $n = 3$ for native hemoglobin. It was concluded that each of subunit of the protein tetramers undergoes a substantial conformational change when that subunit binds a molecule of oxygen. To tackle the problem of allosteric interactions in enzymes and proteins not having a quaternary structure, in works of G. I. Likhtenshtein group [16–18] the enzyme lysozyme was spin labeled by the hystidine-15

Fig. 1.6 Professor Harden McConnel (1927–2014)

group located at the distance 15 A from the substrate-binding center. Addition of specific inhibitors NAG and NAG–NAG induces distinct changes in ESR spectra, which were in a good quantitative agreement with the extent of the substrate binding. Similar transglobular effect was also detected in spin-labeled myoglobin [17].

The first work on the use spin labeling in enzyme catalysis has been published by L. J. Berliner, H. M. McConnell in 1966 [19]. It was shown that the nitroxide spin-labeled substrate, DL-2,2,5,5-tetramethyl-3-carboxypyrrolidine- *p*-nitrophenyl ester, can be used to study the activity of the proteolytic enzyme α-chymotrypsin. The most conclusive work in the investigation of active serine group in proteolytic enzymes based on phosphate and nitrobenzene derivatives was carried out by Hsia et al. [20]

One of remarkable achievement of the spin-labeling methods has been quantitative characterization of flexibility of model and biological membranes. Incorporation of hydrophobic nitroxide probe 2,2,4,4-tetramethyl-1,2,3,4-tetrahydro-'y-carboline-3-oxyl into the sodium dodecyl sulfate remarkably decreased in the rate of tumbling of the probe as compare with its motion on solutions, which is quantitatively described by a rotational correlation time, τ_c [21]. Progress in the study of biological and model membranes with the use of the steroid and lipid spin probes was demonstrated in the works of Keana [11] and MacConnell [15] groups. For example, in sonicated phospholipid dispersions of the walking leg nerve fibers of *Homarus americanus,* and in erythrocytes oriented by hydrodynamic shear, the nitroxide probes motion with rotational diffusion frequencies of the order of 10^7 to 10^8 s^{-1} was revealed. The first approach for quantitative investigation of lateral diffusion in membranes with the use of spin labels was developed by McConnell and McFarland [15], Hubbell and McConnell [22]. In a typical experiment, small drops of a lipid spin probe were inserted to films of oriented multilayers of lecithin. Because of radical diffusion, the probe ESR spectrum changes from a singlet, which is the characteristic of large local concentration, to a triplet for diluted radicals. Analysis of the process kinetics allowed to calculate the coefficient of translational lateral diffusion.

The application of nitroxide radicals to covalent modification and the study of the structure, dynamic and conformational changes of nucleic acids (poly rA, poly rU, and poly rG) were based on the principles established for proteins [23–25]. The possibilities of non-covalently bound nitroxide probes were first demonstrated in [24]. As early as in 1968, A. M. Vasserman, A. L. Buchachenko, A. L. Kovarskii, and M. B. Neiman have shown powerful potentiality of nitroxide spin-label methods in investigation molecular dynamics and microstructure of high molecular mass compounds [26]. The first observation of effect of nitroxide on radical polymerization, that is inhibition with nitroxide mono- and biradicals, was made in 1966 by the M. B. Neiman group [27]. The inhibiting effect of the nitroxides on styrene polymerization at 50 °C, initiated by azodiisobutyronitrile, was interpreted as a cross-recombination of nitroxides and polymer radicals which led to the process termination.

Synthesis and investigation of the chelate complex Cu^{+2}-Schiff bases ligand derivative of TEMPO, which appeared to be the first transition metal complexes with paramagnetic ligands, were carried out in 1969 [28].

Method of double spin labeling, invented by Likhtenshtein in 1968 [29], is based on specific modification of chosen groups of the object of interest by two or several spin labels, nitroxides, or complexes of paramagnetic metal followed by the analysis of effects of the spin–spin interactions on the label ESR spectra. This approach allows to estimate the distance between the paramagnetic centers up to 2.5 nm [29–32]. Later, the higher sensitivity of spin–lattice relaxation time of a radical to interactions between the radical and paramagnetic ions up to 100 nm was demonstrated [17, 18, 33]. In parallel, the effects of spin-exchange interaction on the labels ESR spectra were used for establishment of structure of systems under investigation, such as iron–sulfur clusters in nitrogenase, ferredoxins, and non-heme protein [34].

The first version of method of spin label–spin probe method (SLSPM) proposed by Likhtenshtein and coworkers [35] was based on a dynamic exchange spin–spin interaction of a stable radical, mostly nitroxide, attached to molecular object of interest with a spin probe which are chemically inert paramagnetic species capable of diffusing freely in solution. The value of the dynamic exchange rate constant k_{ex} depends on microviscosity, steric hindrances, and distribution of electrostatic charges, and the method was intensively employed for investigating microstructure of object under interest [17, 18, 36–38]. In parallel, a variant of the SLSP method based on direct measurement of electron spin–lattice relaxation times was developed in Hide group [39].

Principles of application of nitroxide spin-label method as a tool for experimental investigation of proteins molecular dynamics ("breathing") have been formulated at the end of 1960 by Likhtenshtein [40, 41] and then were implemented in collaborative works. Parameters of motion of a nitroxide in the labeled protein may serve as characteristics of surrounding media dynamics. For example, the monitoring motion of hydrophobic nitroxide spin probe in binding site of human serum albumen revealed low amplitude wobbling of the probe with the correlation time $\tau_c \approx 10^{-8}$ s modulated by the binding site dynamics [41]

Protective antitumor activity of nitroxide in animals was demonstrated in pioneering work of N. P. Konovalova, M. B. Neiman, E. G. Rozanzev, and Emanuel N. M. in 1964 [42]. Later in 1970, G. Sosnovsky and M. Konieczny have performed syntheses anticancer drugs belonging to the class of alkylating agents containing aminoxyl radicals [43].

Continuous wave electron–electron double resonance (CW ELDOR) independently reported in 1968 by the Hyde and Freed [44], Bendersky and Blumenfeld [45] groups allowed to resolve problems that are not accessible in the CW-ESR. The ELDOR spectra were shown to be sensitive to very slow rotations which may provide unique information on the details of molecular dynamics and can be used for distance estimation between centers bearing spin electron.

Thus, chemical and physical contributions within the "golden" decade have formed the basis for subsequent progress in the area. In subsequent decades, fundamentals and tendencies laid out in the "golden" period were intensively developed and number of publications in the area accelerated almost in a geometric progression [46–49] and references therein.

1.3 New Era

Starting from the pioneering works of Ya. S. Lebedev and his colleagues [50], problems of poor resolution of the 3 cm ESR have been solved by the use of the high-field–high-frequency (148 GHz), high-resolution 2-mm EPR spectroscopy. The comprehensive review on advanced biomolecular EPR spectroscopy addresses both the EPR and NMR communities has been recently published [51]. New contributions of K. Mebius and W. Lubiz groups to high-field–high-frequency EPR were summarized.

Synthesis of imidazoline and imidasolidine nitroxides (Fig. 1.7) markedly expanded ability of nitroxides for its applications as spin labels, ligands for materials with ferromagnetic properties, inhibitors in polymer processing, and initiators for "living" radical polymerization [52, 53] and references therein.

Figure 1.8 shows photography of leaders of groups involved in the nitroxide spin labeling till 1979.

Measurement of distance between spin labels by analysis of line shape of ESR spectra using new computation methods was carried out for double and multiple labeled proteins and enzymes [54]. Various aspects of spin-label magnetic resonance studies on lipid–lipid and lipid–protein interactions with integral proteins were

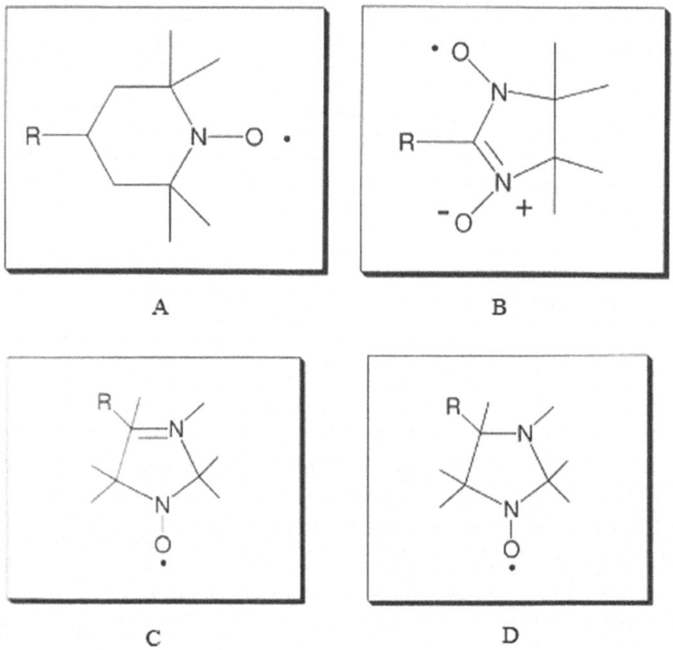

Fig. 1.7 Nitroxide derivatives: piperidine A, nitronyl B, imidasoline C, and imidasolidine D, and nitronyl D

Fig. 1.8 Leaders of groups involved in the nitrixide spin labeling till 1979. From left to right: rector of the Pech University, Leonid B. Volodarsky, Larry Berliner, Gertz Liktenshtein, R. Rassat, John Keena, George Sosnovsky, Kalman Hideg

reviewed [55–57]. Synthesis and determination of structural characteristics of magnetic all-organic radical liquid crystals were in focus of the program developed by R. Tamura with colleagues [58].

Works of 1970–1990 on chemical modification of DNA with nitroxide derivatives and ESR examination of structure and dynamics of the labeled biopolymer have laid the basis for further detail investigations in this extremely important area of molecular biology [59, 60]. After the first publication on labeling of linear polymers such as of cotton, silk, and wool by trichlorotriazine-based nitroxide [61], a series work on the investigation of cotton fibers and cellulose was reported by Marupov and Likhtenshein groups [62] and references therein.

Hundreds of complexes transition metals with nitroxide ligands were synthesized and investigated (for review see [63] and references therein).

The CW ESR saturation techniques were employed for the investigation of depth of immersions of nitroxide spin probes radicals up to 40 Å in lipid phases of biomembranes [64]. High sensitivity of spin–lattice relaxation parameters of nitroxide was also taken into consideration at development of theory of very slow motion ($\tau_c = 10^{-3}$–10^{-6} s) [65].

A new impact in nitroxide-mediated polymerization has started when P. G. Griffiths, E. Rizzardo, and D. H. Solomon showed that it was possible to prepare well-controlled and living (homo-, co) polymer by radical polymerization in the presence of nitroxyl radical as a controlling agent [66–68]. Since its discovery nitroxide-mediated radical polymerization (NMP) was proved to be a powerful method to synthesize well-defined macromolecular architectures with precisely controlled topologies, compositions, microstructures, and functionalities [69–73]. A significant contribution in the area has been brought by the groups of Solomon [69], Grishin [70], Fisher [71], and Bagryanskaya [72]. Kinetic aspects of the nitroxide-mediated radical polymerization were discussed in details in comprehensive review [73].

Significant progress in the high-field–high-frequency (HFHF) ESR has been achieved [74–76] especially owning to the design of millimeter-wave quasi-optic technique, permitting the construction of a 9-Tesla, 250-GHz (1.2 mm) [55]

After remarkable invention of echo-detected ELDOR (three-pulse ELDOR, PELDOR, double electron–electron resonance DEER) by Milov et al. [77] and its first application, a real burst of development of new ESR pulse methods has been broken [78–92]. Other methods used for the study of nitroxide labeling objects are four [80, 89, 90] and five-pulse techniques [91, 92], method to determine the effective saturation factor of nitroxide radicals for dynamic nuclear polarization (DNP) experiments in liquids [83], two-dimensional ELDOR [84]), high-frequency pulsed ENDOR/EPR [85], double and multiple quantum coherence pulsed ESR (DQC ESR) [79, 81], and ESR spectra hole burning [87].

The invention and use of site-directed mutagenesis in combination with modern ESR spectroscopy gave new breath to the nitroxide spin-labeling method. Site-directed spin-labeling (SDSL) designed by Hubbell group [93] is the substitution of a selected amino acid for cysteine via the site-directed mutagenesis technique following chemical modification with a sulfhydryl reactive nitroxide radical, *S*-(1-oxyl-2,2,5,5-tetramethyl-2,5-dihydro-1H-pyrrol-3-yl) methyl methanesulfonothioate) (MTSL). The main advantage of the SDSL is the possibility to overcome the limitations of a choice of amino acids suitable for the labeling in native proteins. The efficiency of combination of the site-directed spin-labeling with the advance pulse techniques can be illustrated by numerous works, for example in studies of T4 Lysozyme [94] and mutants maltose-binding protein (MBP) 09-11 [95].

Recently, new spin-labeling approaches have been shown to be an attractive alternative to the traditional method of nitroxide spin labels for pulse dipolar ESR (PD ESR). The first one, based on high-spin Gd^{3+} ($S = 7/2$) complexes, was designed and developed in Goldfarb group [96]. A combined method, utilizing NMR and EPR spectroscopies, was employed to compare different types of nitroxide-based and Gd(III)-based spin labels attached to isolated RBDs of the polypyrimidine tract-binding protein 1 (PTBP1) and to short RNA fragments [97] in complexes with short

RNAs. An idea of the use of carbon-centered triarylmethyl (trityl) radicals instead of nitroxides for nanometer distance measurements was first introduced and realized in 2012 [98]. Nowadays, triarylmethyl radicals TMA are successively used as spin labels for studies on the structure of proteins and nucleic acids utilizing site-directed spin labeling (SDSL) and pulse dipolar EPR spectroscopy [99, 100].

Spin oximetry, first reported by Subczynski and Hide [101], is a version of spin label–spin probe method [35, 39] in which molecular oxygen plays role of spin probe. A method invented for measurement of the oxygen diffusion–concentration product was based on the dependence of the spin–lattice relaxation time T_1 of the spin label, detected by using saturation recovery (SR), on the bimolecular collision rate with oxygen. Various aspects of the spin oximetry applications have been reviewed [102–104].

As it was pioneered in [105], EPR spectra of stable nitroxides of the imidazoline and imidazolidine types are sensitive to pH and can serve as spin pH probes. Data on synthesis and application of a wide set of pH-sensitive nitroxides of different sensitivity, stability to reduction, lipophilicity and its covalent-binding macromolecules have been reported [106, 107]. To quantitatively determine SH groups in high- and low-molecular weight compounds, a disulfide biradical (RS-SR), where R is imidazoline residue, has been used [108, 109]. The biradical is shown to participate in a thioldisulfide exchange reaction with compounds containing SH groups. In this case, the ESR spectra of the biradical RS-SR and the resulting monoradical R-SH are different.

Nitroxide radicals have found various applications in the field of materials science. The landmark was the discovery by M. Kinoshita group et al. in 1991 [110] who prepared the first purely organic ferromagnet with respect to a nitronyl nitroxide, 2-(4-nitrophenyl)-4,4,5,5-tetramethylimidazoline-1-oxy-3-oxide Since then, stable nitric oxide structures have been widely used as the spin source and building block for the elaboration of organic or molecule-based magnetic materials [58, 111, 112]. In the last two decades, π-conjugated superparamagnetic organic compounds including polymer magnets with stability at ambient temperature and/or higher magnetic ordering temperatures have been attracting attention as models of multispin systems and potential magnetic devices [113]. Novel molecular magnets Cu(hfac)2LR with nitronyl ligands were prepared and investigated in details in Ovcharenko group [114].

The spin redox probe techniques utilizing ability of nitroxides to be reacted with reducing agent to corresponding hydroxyl amine are widely used for quantitative characterization of redox processes and protection from radical damage by CW ESR spectroscopy [115–118] and references therein.

Novel methods of fast and sensitive analysis of antioxidant status of biological systems, spin redox probing and spin trapping, investigation of molecular dynamics, models for studies of photophysical and photochemical processes, and construction of new magnetic light-sensitive materials are based upon the use of dual fluorophore–

nitroxide compounds were designed and developed [119–126]. In pioneering work of Likhtenshtein and colleagues [120], three fundamental effects were first demonstrated, namely (1) The nitroxide fragment is a strong quencher of the fluorescence. (2) The radical photoreduction can lead to the decay of the EPR signal and the drastic increase of the fluorescence intensity. (3) The photoreduction kinetics strongly depends on molecular dynamics of environment. Next principle step was a series of excellent papers by Blough and Simpson [123] in which the potential of these tethered, optically switching molecules as potent redox and radical trapping probes was realized. A significant contribution to the synthesis and use of the dual compounds was made by the groups of Bottle [124] and Braslau [125]. Dual fluorescence nitroxide compounds are effectively used as convenient photochemical and photophysical models and form the basis for photoswitching magnetic materials [126] and references therein.

Stable nitroxide free radicals were utilized as antioxidants in animal models and human diseases (e.g., cancer) to protect processes of formation reactive oxygen species, ROS ($O_2^{-\cdot}$, H_2O_2, $\cdot OH$) involving oxidative stress [127–130]. The works, in which the principle possibility of effectiveness of the spin trapping was demonstrated, came to the light in 1968 [131], later numerous theoretical and experimental studies of the spin trapping of inorganic and organic radicals were carried out [132, 133]. After sited above pioneering works [42, 43], biologically active spin-label molecules have been the focus of biophysical, biochemical, and synthetic and medicinal chemical studies [134–136]. A noticeable contribution to the theory of spin relaxation and its application was made in the works of Eaton and Eaton [137].

1.4 Concluding Remarks

Author of this review wrote in the first book on spin labeling:

> *Likhtenshtein G. I.: Spin-Labeling Method in Molecular Biology.* Moscow, Nauka (In Russian). 1974: "It is thus our hope that spin labeling will continue to be an effective tool for solving various complicated problems in molecular biology". Now after 45 years, it is evident that present-day reality has surpassed all optimistic expectations. As far as concern outlook for further developments, there are all reasons to believe that slow but permanent progress in the area would continue in the next decades. Nevertheless, who knows, new unexpected bright ideas would be launched and implemented and ensure vigorous success, unexpected today.
>
> Portraits of scientists who significantly contributed in nitroxide chemistry, physics, and its miscellaneous applications are displayed in Fig. 1.9.

Fig. 1.9 Gallery of scientists who significantly contributed in nitroxide chemistry, physics and its miscellaneous applications

References

1. E. Fremi, Annales de Chimie et de Physique. Serie **3**(15), 408–488 (1845)
2. O. Piloty, B.G. Schwerin, Berichte den Deutschen Chemischen Gesselschaft **34**, 1870–1877 (1901)
3. H. Wieland, M. Offenbacher, Diphenylnitric oxide, a new organic radical with tetravalent nitrogen. Ber. Dtsch. Chem. Ges. **47**, 2111–2115 (1914)
4. A.K. Hoffmann, A.T. Henderson, A new stable radical: Di-t-Butyl nitroxide. Am. Chem. Soc. **83**, 4671–4672 (1961)
5. O.L. Lebedev, M.L. Khidekel, V.A. Razuvaev, Dokl. Acad. Nauk. **140**, 1327–1331 (1961)
6. M.B. Neiman, E.G. Rozantzev, Y.G. Mamedova, Free radical reactions involving no unpaired electrons. Nature **196**, 472–474 (1962)
7. E.G. Rozantzev, *Free Nitroxyl Radicals* (Plenum Press, New York, 1970)
8. D. Kivelson, Theory of EPR line widths of free radicals. J. Chem. Phys. **33**, 1094–1106 (1960)
9. J.H. Freed, Theory of the ESR spectra of nitroxids, in *Spin Labeling. Theory and Applications*, vol. 1, ed. by L. Berliner (Academic Press, New York, 1976)
10. J.H. Osiecki, E.F. Ullman, Studies of free radicals. I. alpha.-Nitronyl Nitroxides, a new class of stable radicals. J. Am. Chem. Soc. **90**, 1078–1079 (1968)
11. J.F.W. Keana, S.B. Keana, D. Beetham, New versatile ketone spin label. J. Am. Chem. Soc. **89**, 3055–3056 (1967)
12. V.A. Golubev, E.G. Rozantsev, M.B Neiman, Some reactions of free iminoxyl radicals with the participation of the unpaired electron. Bull. Acad. Sci. USSR, Div. Chem. Sci. **14**, 1898–1904 (1965)
13. T.J. Stone, T. Buckman, P.L. Nordio, H.M. McConnell, Spin-labeled biomolecules. Proc. Natl. Acad. Sci. **54**, 1010–1017 (1965)
14. L. Stryer, O.H. Griffith, A spin-labeled hapten. Proc. Natl. Acad. Sci. **54**, 1785–1791 (1965)
15. H.M. McConnell, B.G. McFarland, Physics and chemistry of spin labels. Q. Rev. Biophys. **3**, 91–136 (1970)
16. Y.D. Akhmedov, G.I. Likhtenshtein. L.V. Ivanov, Y.V. Kokhanov, Investigation of the lysozyme macromolecule by a spin-labeling method. Dokl. Acad. Nauk. SSSR **205**, 372–376 (1972)
17. G.I. Likhtenshtein, Spin labeling method in molecular biology (Moscow Nauka, 1974) (In Russian)
18. G.I. Likhtenshtein, Spin labeling method in molecular biology (N.Y., Wiley Interscience, 1976)
19. L.J. Berliner, H.M. McConnell, A spin-labeled substrate for α-chymotrypsin. Proc. Natl. Acad. Sci. **55**, 708–712 (1966)
20. J.S. Hsia, D.I. Kosman, J.S. Piette, Organophosphate spin-label studies of inhibited esterases, α-chymotrypsin and cholinesterase biochem. Biophys. Res. Commun. **36**, 75–78 (1969)
21. A.S. Waggoner, O.H. Griffith, C.R. Christensen, Magnetic resonance of nitroxide probes in micell-containing solutions, Proc. Natl. Acad. Sci. **57**, 1198–120 (1967)
22. W.L. Hubbell, H.M. McConnell, Orientation and motion of amfiphilic membranes. Proc. Natl. Acad. Sci. **64**, 20–27 (1969)
23. I.C.P. Smith, T. Yamane, Spin labeled nucleic acids. Proc. Natl. Acad. Sci. **58**, 884–887 (1967)
24. B.I. Sukhorukov, A.M. Wasserman, L.I. Kozlova, A.L. Buchachenko, Use the spin labeling method for study of the nucleic acid-water system. Dokl. Acad. Nauk SSSR. **177**, 454–457 (1967)
25. A.M. Bobst, Studies of spin-labeled polyrybo adenylic acids. Biopolymers **11**, 1421–1433 (1972)
26. A.M. Vasserman, A.L. Buchachenko, A.L. Kovarskii, M.B. Neiman, Study of molecular motion of polymers by method of paramagnetic probe. Vysok. Soedin., Seriya A, **10**, 1930–1936 (1968)
27. L.V. Ruban, A.L. Buchachenko, M.B. Neiman, Y.V. Kokhanov, Inhibition of radical polymerization with nitroxide mono-and biradicals. Vysok. Soedin. **8**, 1642–1646 (1966)

28. A.A. Medzhidov, L.N. Kirichenko, G.I. Likhtenshtein, Izvestiya akademii nauk SSSR, Seriya Khimicheskaya, 698–700 (1969)
29. G.I. Likhtenshtein, Determination of the topography of proteins groups using specific paramagnetic labels. Mol. Biol. (Moscow) **2**, 234–240 (1968)
30. J.C Taylor, S. Leigh, M. Cohn, The effect of dipole-dipole interaction between nitroxide radical and a paramagnetic ion on the line shape of the esr spectra of radical. Proc. Natl. Acad. Sci. **64**, 219–206 (1969
31. A.V. Kulikov, G.I. Likhtenshtein, E.G. Rozantzev, V. Suskina, A. Shapiro, Nitroxide Bi- and polyradicals as standard models for distance estimation between the nitroxide moieties. Biofizika **17**, 42–49 (1972)
32. A.I. Kokorin, K.I. Zamaraev, G.L. Grigoryan, V.P. Ivanov, E.G. Rozantsev, Distance estimation between nitroxyl radicals. Biofizika **17**, 34–41 (1972)
33. A.V. Kulikov, G.I. Likhtenshtein, The use of spin-relaxation phenomena in the investigation of the structure of model and biological systems by method of spin labels. Adv. Molecul. Relax. Proc. **10**, 47–78 (1977)
34. L.A. Syrtsova, L.A. Levchenko, E.N. Frolov, G.I. Likhtenshtein, T.N. Pisarscaya, L.V. Vorob'ev, V.A. Gromoglasova, Structure and function of the nitrogenase components from Azotobacter vinelandii. Mol. Biol. (Moscow) **5**, 726–734 (1971)
35. G.I. Likhtenshein, Y.B. Grebenshchikov, P. Bobodzhanov, Y.V. Kokhanov, Method of spin-labels spin probes. Molecul. Biol. (Moscow) **4**, 682–691 (1970)
36. G.I. Likhtenshtein, Study on the proteins microstructure by method of spin-label paramagnetic probe. Mol. Biol. (Moscow) **4**, 782–789 (1970)
37. G.I. Likhtenshtein, Y.B. Grebentchikov, E.G. Rosantev, V.P. Ivanov, Study on the electrostatic charges in proteins by method of paramagnetic probes. Mol. Biol. (Moscow) **6**, 498–507 (1972)
38. G.I. Likhtenshtein, *New Trends in Enzyme Catalysis and Mimicking Chemical Reactions* (Kluwer Academic/ Plenum Publishers, N.Y, 2003)
39. J.S. Hyde, H.M. Swartz, W.E. Antholin, The spin probe—spin label method, in *Biological Magnetic Resonance. Spin Labeling. Theory and Application*, vol. 8, ed. by L.J. Berliner J. Reubin (Plenum Press, New York, 1979), pp. 305–339
40. G.I. Likhtenshtein, A.P. Pivovarov, P.K. Bobodzhanov, E.G. Rozantsev, N.B. Smolina, Study of dynamic structure of proteins and enzymes by specific luminescent and paramagnetic labels. Biofizika **13**, 396–400 (1968)
41. E.N. Frolov, G.I. Likhtenshtein, N.V. Kharakhonycheva, Investigation of dynamic structure of human serum albumin by the method of spin-probe. Molek. Boil. (Moscow) **8**, 886–893 (1974)
42. N.P. Konovalova, G.N. Bogdanov, V.B. Mille, M.B. Neiman, E.G. Rozanzev, G. Emanuel, N.M. Doklady Akademii nauk SSSR **157**, 707–709 (1964)
43. G. Sosnovsky, M. Konieczny, Synthesis of phosphoorganic derivatives of nitroxide radicals, Naturforshung33b, 792–804 (1970)
44. J.S Hyde, J.C.W. Chien, J.H. Freed, Electron–electron double resonance of free radicals in solution. J. Chem. Phys. **48**, 4211–4226 (1968)
45. V.A. Benderskii, L.A. Blyumenfel'd, P.A. Stunzhas, F.A. Sokolov, Double electron-electron resonance of triplet excitons in ion-radical salts. Nature **220**, 365–367 (1968)
46. N. Kocherginsky, H.M. Swarts, *Nitroxide spin labels: Reactions in biology and chemistry* (CRC Press, 1995)
47. L.J. Berliner (ed.), *Spin Labeling: Next Millennium, Biological Magnetic Resonance* (Book 14) (Springer, 1998)
48. G.I. Likhtenshtein, *Biophysical Labeling Methods in Molecular Biology* (Cambridge University Press, Cambridge, N.Y., 1993)
49. G.I. Likhtenshtein, J. Yamauchi, S. Nakatsuji, A.I. Smirnov, R. Tamura, *Nitroxides: Application in Chemistry, Biomedicine, and Materials Science* (WILEY-VCH, Weinhem, 2008)

50. O.Y. Grinberg, A.A. Dubinskii, V.F. Shuvalov, L.G. Oranskii, V.I. Kurochkin, Y.S. Lebedev, EPR submillimeter spectroscopy of free radicals. Doklady Akademii Nauk SSSR **230**, 884–887 (1976)
51. K. Moebius, W. Lubitz, N. Cox, A. Savitsky, Biomolecular epr meets nmr at high magnetic fields. Magnetochemistry **4**, 50 (2018)
52. L.B. Volodarsky, V.A. Reznikov, V.I. Ovcharenko, *Synthetic Chemistry of Stable Nitroxides* (CRC Press Inc, Boca Raton, FL, USA, 1994)
53. L.B.Volodarsky, I.A. Grigor'ev, R.Z. Sagdeev, Stable imidazoline nitroxides, in *Biological Magnetic Resonance*, vol. 2, ed. by L. Berliner, J. Reuben (Plenum Press, 2000), p. 73
54. E.J. Hustedt, A.H. Beth, Structural information from CW-EPR spectra of dipolar-coupled nitroxide spin-labels, in *Biological Magnetic Resonance*, vol. 19, ed. by S.S. Eaton, G.R. Eaton, L.J. Berliner (2001)
55. J.H. Freed, New technologies in electron spin resonance. Ann. Rev. Phys. Chem. **51**, 655–689 (2000); D. Marsh, in *Handbook of Lipid Bilayers*, 2nd edn. (CRC Press, 2013)
56. A.K. Smith, J.H. Freed, Dynamics and ordering of lipid spin-labels along the coexistence curve of two membrane phases: an ESR study. Chem. Phys. Lipids **165**(3), 348–361 (2012)
57. M.A. Hemminga, L. Berliner, *ESR Spectroscopy in Membrane Biophysics* (Springer, 2007)
58. A.K. Vorobiev, N.A. Chumakova, D.A. Pomogailo, Y. Uchida, K. Suzuki, Y. Noda, R. Tamura, Determination of structural characteristics of all-organic radical liquid crystals based on analysis of the dipole—dipole broadened EPR spectra. J. Phys. Chem. B **118**, 1932–1942 (2014)
59. B.H. Robinson, C. Mailer, G. Drobny, Site-specific dynamics in experiments DNA. Annu. Rev. Biophys. Biomol. Struct. **26**, 629–635 (1997)
60. R.S. Keyes, Y.Y. Cao, E.V. Bobst, J.M. Rosenberg, A.M. Bobst, Spin-labeled nucleotide mobility in the boundary of the EcoRI endonuclease binding site. J. Biomol. Struct. Dyn. **14**, 163–172 (1996)
61. R.M. Marupov, P.K. Bobodzhanov, I.K. Usupov, E.N. Frolov, G.I. Likhtenshtein, Study of temperature stability of cotton fibers by spin labeling. Biofizika **24**, 519–523 (1979)
62. I.K. Ysupov, G.I. Likhtenshtein, Study of microstructure and molecular dynamics of cotton and cellulose fibers by methods of physical labels. Int. Res. J. Pure & Appl. Chemistry. **6**(3), 105–119 (2015)
63. S.S. Eaton, G.R. Eaton, Interaction of spin labels with transition metals. Coord. Chem. Rev. **83**, 29–72 (1988)
64. A.V. Kulikov, E.S. Cherepanova, V.R. Bogatyrenko, Determination of the closest distance between a radical and a paramagnetic Ion. Theor. Exper. Chem. **17**, 618–626 (1981)
65. J.S. Hyde, L. Dalton, Very slowly tumbling spin labels: adiabatic rapid passage. Chem. Phys. Lett. **16**, 568–572 (1972)
66. G.I. Likhtenshtein, Depth of immersion of paramagnetic centers. in *Magnetic Resonance in Biology*, vol. 19, ed. by L. Berliner, S. Eaton, G. Eaton (2000), pp. 309–347
67. D.H. Solomon, E. Rizzardo, P. Cacioli, Eur. Pat. Appl. (1985)
68. P.G. Griffiths, P.G.E. Rizzardo, D.H. Solomon, Initiation pathways in the polymerization of alkyl methacrylates with tert-butoxy radicals. J. Macromol. Sci., Chem. A **17**, 45–50 (1982)
69. D.H. Solomon, Genesis of the CSIRO polymer group and the discovery and significance of nitroxide-mediated living radical polymerization. J. Polym., Part A: Polym. Chem. **43**(23), 5748 (2005)
70. D.F. Grishin, I.D. Grishin, Controlled radical polymerization: prospects for application for industrial synthesis of polymers (Review) Russ. J. Appl. Chem. **84**(12), 2021–2028 (2011)
71. H. Fisher, M. Souaille, The Persistent radical effect in living radical polymerization –border cases and side-reactions. Macromol. Symp. **174**, 231–240 (2001)
72. M. Edeleva, G. Audran, S. Marque, E. Bagryanskaya, Smart control of nitroxide-mediated polymerization initiators' reactivity by pH, complexation with metals, and chemical transformations. Materials (Basel) **12**(5), 688 (2019)
73. E.G. Bagryanskaya, S.R.A. Marque, Kinetic aspects of nitroxide mediated polymerization. in *Nitroxide Mediated Polymerization: From Fundamentals to Applications in Materials Science*, ed. by Gigmes (RCS, 2015), pp. 45–113

74. V.I. Krinichny, *2-mm Wave band EPR Spectroscopy of Condensed Systems*. (CRC Press, Boca Raton, 1995)
75. S.K. Misra, J.H. Freed, Molecular motions, *Multifrequency Electron Paramagnetic Resonance*, ed. by S.K. Misra (2011), pp. 497–544
76. G.R. Eaton, S.S. Eaton, Multifrequency electron spin-relaxation times, in *Distance Measurements in Biological Systems by EPR Biological Magnetic Resonance*, vol. 19, ed. by L.J. Berliner, S. Eaton, G.R. Eaton (Springer, 2000), pp. 719–753
77. A.D. Milov, K.M. Salikhov, Y.D. Tsvetkov, Fiz. Tverd. Tela (Leningrad) **15**, 1187 (1973)
78. A.D. Milov, Y.D. Tsvetkov, A.G. Maryasov, M. Gobbo, C. Prinzivalli C.M. De Zotti, F. Formaggio, C. Toniolo, Conformational properties of the spin-labeled tylopeptin B and heptaibin peptaibiotics based on PELDOR spectroscopy. Appl. Magn. Res. **44**, 495–508 (2013)
79. P.P. Borbat, J.H. Freed, Multiple-quantum ESR and distance measurements. Chem. Phys. Lett. **313**, 145–154 (1999)
80. A. Schweiger, G. Jeschke, *Principles of Pulse Electron Paramagnetic Resonance* (Oxford University Press, Oxford, 2001)
81. P.P. Borbat, J.H. Freed, Double-quantum ESR and distance measurement. in *Biological Magnetic Resonance,* vol. 19, ed. by L.J. Berliner, S. Eaton, G.R. Eaton (Springer, 2001), pp. 383–460
82. A.D. Milov, R.P. Samoilova, Y.D. Tsvetkov, V.A. Gusev, F.M. Formaggio, M. Crisma, C. Toniolo, J. Raap, Spatial distribution of spin-labeled trichogin GA IV in the gram-positive bacterial cell membrane determined from PELDOR data. Appl. Magn. Reson. **23**, 81–95 (2002)
83. M.-T. Türke, M. Bennati, Saturation factor of nitroxide radicals in liquid DNP by pulsed ELDOR experiments. Phys. Chem. Chem. Phys. **13**, 3630–3633 (2011)
84. Y.-W. Chiang, A.J. Costa-Filho, B. Baird, J.H. Freed, 2D-ELDOR study of heterogeneity and domain structure changes in plasma membrane vesicles upon cross-linking of receptors. J. Phys. Chem. B **115**, 10462–10469 (2011)
85. T.I. Smirnova, A.I. Smirnov, S. Pachtchenko, O.G. Poluektov, Geometry of hydrogen bonds formed by lipid bilayer nitroxide probes: a high frequency pulsed ENDOR/EPR study. J. Am. Chem. Soc. **129**, 3476–3477 (2007)
86. S. Saxena, J.H. Freed, Theory of double quantum two-dimensional electron spin resonance with application to distance measurements. J. Chem. Phys. **107**, 1317–1134 (1997)
87. S.A. Dzuba, A. Kawamori, Selective hole burning in EPR: spectral diffusion and dipolar broadening. Concepts Magn. Reson. **8**, 49–61 (1996)
88. A.D. Milov, Y.D. Tsvetkov, A.G. Maryasov, M. Gobbo, C. Prinzivalli C, M. De Zotti, F. Formaggio, C. Toniolo, Conformational properties of the spin-labeled tylopeptin B and heptaibin peptaibiotics based on PELDOR spectroscopy. Appl. Magn. Res. **44**, 495–508 (2013)
89. G. Jeschke, M. Panier, H.W. Spies, Double electron-electron resonance, in *Distance Measurements in Biological Systems by EPR, in Biological Magnetic Resonance*, vol. 19, ed. by L.J. Berliner, S. Eaton, G.R. Eaton (Springer Verlag, 2001), pp. 493– 512
90. G. Jeschke, DEER distance measurements on proteins. Ann. Rev. Phys. Chem. **63**, 419–446 (2012)
91. I. Krstić, R. Hänsel, O. Romainczyk, J.W. Engels, V. Dötsch, T.F. Prisner, Long-range distance measurements on nucleic acids in cells by pulsed EPR spectroscopy. Angew. Chem. Int. Ed. Engl. **50**, 5070 (2011)
92. P.P. Borbat, E.R. Georgieva, J.H. Freed, Improved sensitivity for long-distance measurements in biomolecules: five-pulse double electron-electron resonance. J. Phys. Chem. Lett. **4**, 170–175 (2013)
93. W.L. Hubbell, R.L. Gross, M.A. Lietzow, Curr Opin Struct Biol **8,** 649–656 (1988)
94. P.P. Borbat, H. Mchaourab, J.H. Freed, Protein structure determination using long-distance constraints from double-quantum coherence ESR: study of T4 lysozyme. J. Am. Chem. Soc. **124**, 5304–5314 (2002)
95. B. Selmke, P.P. Borbat, N. Chen, V. Raghavan, J.H. Freed, W.E. Trommer, Open and closed form of maltose binding protein in its native and molten globule state as studied by electron paramagnetic resonance spectroscopy. Biochemistry **57**(38), 5507–5512 (2018)

96. A. Feintuch, G. Otting, D. Goldfarb, Gd3+ spin labeling for measuring. Distances in biomacromolecules: why and how? in *Methods in Enzymology*, vol. 563 (2015), pp. 416–457
97. C. Gmeiner, G. Dorn, F.H.T. Allain, G. Jeschke, M. Yulikov, Spin labelling for integrative structure modelling: a case study of the polypyrimidine-tract binding protein 1 domains in complexes with short RNAs. Phys. Chem. Chem. Phys. **19**, 28360 (2017)
98. G.W. Reginsson, N.C. Kunjir, S.T. Sigurdsson, O. Schiemann, Trityl radicals: spin labels for nanometer-distance measurements. Chemistry **18**, 13580–13584 (2012)
99. M.K. Bowman, C. Mailer, H.J. Halpern, The solution conformation of triarylmethyl radicals. J. Magn. Reson. **172**, 254 (2005)
100. O. Krumkacheva, E. Bagryanskaya, EPR-based distance measurements at ambient temperature. J. Magn. Reson. **280**, 117–126 (2017)
101. W.K. Subczynski, J.S. Hyde, The diffusion-concentration product of oxygen in lipid bilayers using the spin-label T_1 method. Biochim. Bioph. Acta **643**, 283–291 (1981)
102. R. Springett, H.M. Swartz, Measurements of oxygen in vivo: overview and perspectives on methods to measure oxygen within cells and tissues. Antioxid. Redox Signal. **9**, 1295–1301 (2007)
103. G. Ilangovan, J.L. Zweier, P. Kuppusamy, Oxygen sensing. Methods Enzym **381**, 747–762 (2004)
104. W.K. Subczynski, L. Mainali, T.G. Camenisch, W. Froncisz, J.S. Hyde, Spin-label oximetry at Q-and W-band. J. Magn. Reson. **209**, 142–148 (2011)
105. V.V. Khramtsov, L.M. Weiner, I.A. Grigor'ev, L.B. Volodarsky, Proton exchange in stable nitroxyl radicals. ESR study of the pH of aqueous solutions. Chem. Phys. Lett. **91**, 69–72 (1982)
106. I.A. Kirilyuk, A.A. Bobko, V.V. Khramtsov, I.A. Grigor'ev, Nitroxides with two p*K* values—useful spin probes for pH monitoring within a broad range". Org. Biomol. Chem. **3**, 1269 (2005)
107. S. Koda, J. Goodwin, V.V. Khramtsov, H. Fujii, H. Hirata, Electron paramagnetic resonance-based pH Mapping using spectral-spatial imaging of sequentially scanned spectra. Anal. Chem. **84**, 3833–3837 (2012)
108. V.V. Khramtsov, V.I. Yelinova, L.M. Weiner, T.A. Berezina, V.V. Martin, L.B. Volodarsky, Quantitative determination of SH groups in low- and high-molecular-weight compounds by an electron spin resonanc emethod. Anal. Biochem. **182**(1), 58–63 (1989)
109. L.M. Weiner, Quantitative determination of thiol groups in low- and high-molecular-weight compounds by an electron paramagnetic resonance. Methods Enzymol **251**, 87–105 (1995)
110. M. Tamura, Y. Nakazawa, D. Shiomi, K. Nozawa, Y. Hosokoshi, M. Ishikawa, M. Takahashi, M. Kinoshita, Bulk ferromagnetism in the β-phase crystal of the *p*-Nitrophenyl nitronyl nitroxide radical. Chem. Phys. Lett. **186**, 401–404 (1991)
111. D.B. Amabilino, J. Veciana, Nitroxide-based organic magnet, in *Magnetism: Molecules to Materials II*, ed. by J.S. Miller, M. Drillon (Wiley-VCH, Weinheim, Germany, 2001), pp. 1–60
112. R. Tamura, Organic functional materials containing chiral nitroxide radical units, in *Nitroxides: Application in Chemistry, Biomedicine, and Materials Science*, ed. by G.I. Likhtenshtein, J. Yamauchi, S. Nakatsuji, A. Smirnov, R. Tamura (Weinhem, WILEY-VCH, 2008), pp. 303–331
113. A. Rajca, From high-spin organic molecules to organic polymers with magnetic ordering. Chem. Eur. J. **8**, 4834–4841 (2002)
114. S.L. Veber, M.V. Fedin, K.Y. Maryunina, A. Potapov, D. Goldfarb, E. Reijerse, W. Lubitz, R.Z. Sagdeev, V.I. Ovcharenko, E.G. Bagryanskaya, Temperature-dependent exchange interaction in molecular magnets Cu (hfac) 2LR studied by EPR: methodology and interpretations. Inorg. Chem. **50**, 10204–10212 (2011)
115. H.M. Swartz, N. Khan, V.V. Khramtsov, Use of electron paramagnetic resonance spectroscopy to evaluate the redox state in vivo. Antioxid. Redox Signal. **9**, 1757–1771 (2007)
116. N. Kocherginsky, H.M. Swarts, in *Nitroxide Spin Labels. Reactions in Biology and Chemistry* (CRC Press, 1995)

117. P. Kuppusamy, M.C. Krishna, EPR imaging of tissue redox status. Curr. Top. Biophys. **26**, 29–34 (2002)
118. G.I. Likhtenshtein, Nitroxide redox probes and traps, nitron spin traps, in *Nitroxides: Application in Chemistry, Biomedicine, and Materials Science*, ed. by G.I. Likhtenshtein, J. Yamauchi, S. Nakatsuji, A. Smirnov, R. Tamura (WILEY-VCH, Weinhem, 2008), pp. 303–331
119. G.I. Likhtenstein, V.R. Bogatyrenko, A.V. Kulikov, K. Hideg, H.O. Hankovsky, N.V. Lukoianov, A.I. Kotelnikov, B.S. Tanaschelchuk, Study of superflow motion in solid solutions by the method of physical probes. Dokl. Akad. Nauk SSSR **253**, 481–484 (1980)
120. I.M. Bystryak, G.I. Likhtenshtein, A.I. Kotelnikov, O.H. Hankovsky, K. Hideg, The influence of the molecular dynamics of the solvent on the photoreduction of nitroxy-radicals. Russ. J. Phys. Chem. **60**, 1679–1983 (1986)
121. V.R. Vogel, E.T. Rubtsova, G.I. Likhtenshtein, K. Hideg, Factors affecting photoinduced electron transfer in a donor- acceptor pair (D-A) incorporated into bovine serum albumin. J. Photochem. Photobiol. A, Chem. **83**, 229–236 (1994)
122. E. Lozinsky, V.V. Martin, T.A. Berezina, A. Shames, A.L. Weis, G.I. Likhtenshtein, Dual fluorophore-nitroxide probes for analysis of vitamin C in biological liquids. J. Biochem. Biophys. Meth. **38**, 29–42 (1999)
123. N.V. Blough, D.J. Simpson, Chemically mediated fluorescence yield switching in nitroxide-fluorophore adducts, optical sensors of radical/redox reactions. J. Am. Chem. Soc. **110**, 1915–1917 (1988)
124. J.P. Allen, M.C. Pfrunder, J.C. McMurtrie, S.T. Bottle, J.P. Blinco, K.E. Fairfull-Smith, BODIPY-based profluorescent probes containing Meso- and β-Substituted Isoindoline Nitroxides European. J. Org. Chem. **3**, 476–483 (2017)
125. R. Braslau, F. Rivera III, E. Lilie, M. Cottman, Urushiol detection using a profluorescent nitroxide. J. Org. Chem. **78**, 238–245 (2013)
126. G.I. Likhtenstein, K. Ishii, S. Nakatsuji, Dual chromophore-nitroxides: novel molecular probes, photochemical and photophysical models and magnetic materials. Photochem. Photobiol. **83**, 871–881 (2007)
127. N. Khan, B.B. Williams, H.M. Swartz, Clinical applications of in vivo EPR: rationale and initial results. Appl. Magn. Reson. **30**, 185–199 (2006)
128. J.B. Mitchell, M.C. Krishna, P. Kuppusamy, J.A. Cook, A. Russo, Protection against oxidative stress by Nitroxides. Exp. Biol. Med. **226**, 620–621 (2001)
129. V.V. Khramtsov, In vivo spectroscopy and imaging of nitroxide probes. in *Nitroxides—Theory, Experiment and Applications*, ed. by A.I. Kokorin (September, 2012), pp. 317–346
130. G.I. Likhtenshtein, Biomedical and medical application of nitroxides, in *Nitroxides: Application in Chemistry, Biomedicine, and Materials Science*, ed. by G.I. Likhtenshtein, J. Yamauchi, S. Nakatsuji, A. Smirnov, R. Tamura (WILEY-VCH, Weinhem, 2008), pp. 371–401
131. E.G. Janzen, Y.Y. Wang, R.V. Shetty, Spin trapping with alpha-pyridyl 1-oxide N-tert-butyl nitrones in aqueous solutions. A unique electron spin resonance spectrum for the hydroxyl radical adduct. J. Am. Chem. Soc. **100**, 2923–2925 (1978)
132. F.A. Villamena, E.J. Locigno, A. Rockenbauer, C.M. Hadad, J.L. Zweier, Theoretical and experimental studies of the spin trapping of inorganic radicals by 5, 5-dimethyl-1-pyrroline N-oxide (DMPO). 2. Carbonate radical anion. J. Phys. Chem. A **111**, 384–391
133. R.P. Mason, Free using anti-5, 5-dimethyl-1-pyrroline N-oxide (anti-DMPO) to detect protein radicals in time and space with immuno-spin trapping. Radic. Biol. & Med. **36**, 1214–1223 (2004)
134. S. Schocha, V. Senb, S. Gajewskia, V. Golubevb, B. Straucha, A. Hartwiga, B. Köberlea, Activity profile of the cisplatin analogue PN149 in different tumor cell lines. Biochem. Pharmacol. **156**, 109–119 (2018)
135. M. Balog, T. Kálai, J. Jekő, H-J. Steinhoff, M. Engelhard, K. Hideg, Synthesis of New 2, 2, 5, 5-Tetramethyl-2, 5-dihydro-1H-pyrrol-1-yloxyl radicals and 2-Substituted-2, 5, 5-trimethylpyrrolidin-1-yloxyl radicals based a-amino acids. Synlett **14**, 2591 (2004)
136. O.D. Zakharova, T.S. Frolova, Y.V. Yushkova, E.I. Chernyak, A.G. Pokrovsky, M.A. Pokrovsky, S.V. Morozov, O.I. Sinitsina, I.A. Grigor'ev, G.A. Nevinsky, Antioxidant and

antitumor activity of trolox, trolox succinate, and α-tocopheryl succinate conjugates with nitroxides. Eur. J. Med. Chem. **122**, 127–137 (2016)
137. S. Eaton, G. Eaton, Determination of distance based on T_1 and T_2 effects, in *Magnetic Resonance in Biology*, ed. by L. Berliner, S. Eaton, G. Eaton (Kluwer Academic Publishers. Dordrecht), pp. 348–382

Chapter 2
Nitroxide Basic Physical Properties

2.1 Introduction

Nitroxides belong to the class of heterocyclic compounds with typical structures of five- or six-membered heterocyclic derivatives of piperidine, pyrrolidine, imidasoline, imidasolidine, nitronyl, in particular (Figs. 1.2 and 1.7). Ninety-seven nitroxide structures were listened in a table presented in comprehensive review [1].

In a non-conjugated nitroxide

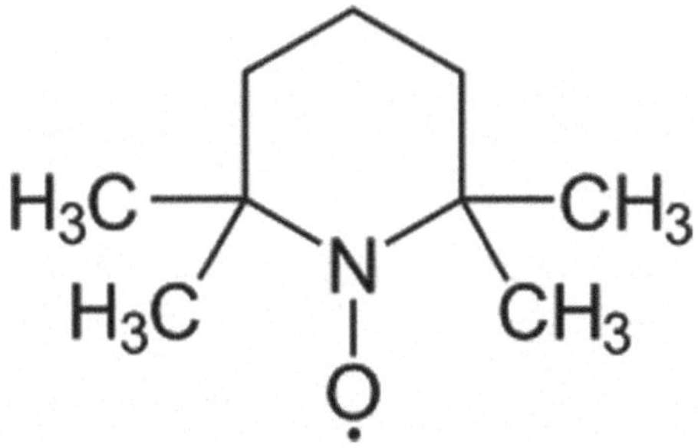

the unpaired electron is located mainly on the nitrogen and oxygen atoms and only slight delocalized for adjacent alkyl groups. The unpaired electron is considered to reside in the p orbital. According to a nitroxide molecular orbitals (MO) diagram, which is in the first approximation close to nitric oxide diagram, N–O·moiety has five fully bonding orbitals with unpaired electron resulting in one on of the fourth antibonding π*π*-orbitals (single occupied molecular orbital (SOMO) (Fig. 2.1). The nitrogen and oxygen are held together by 2.5 bonds. Nitroxide radicals feature a $(\pi)2(\pi^*)1$ three-electron bond between a nitrogen atom and an oxygen one.

The unpaired electron in the antibonding orbitals weakens over bonding of the nitrogen to oxygen (N–O·). Therefore, the transfer of the electron away from the antibonding orbital significantly stiffens the $N–O·^+$ bond. Conversely, the transfer of

G. I. Likhtenshtein, *Nitroxides*, Springer Series in Materials Science 292,
https://doi.org/10.1007/978-3-030-34822-9_2

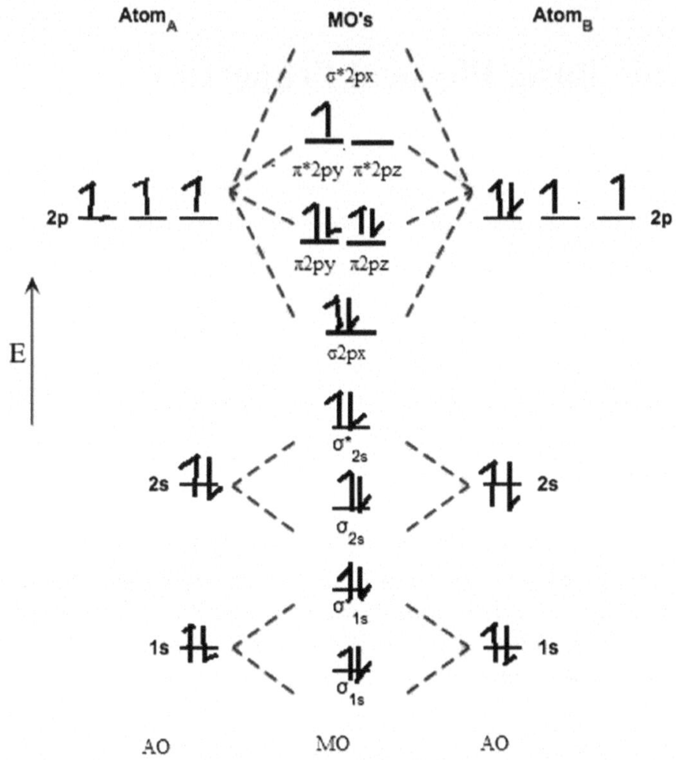

Fig. 2.1 Molecular orbitals diagram of NO fragment. https://en.wikipedia.org/wiki/Molecular_orbital_diagram

additional electron density into the antibonding additionally weakens over the N–O$^-$ bond. The aforementioned features of the NO molecule substantially determine its physicochemical properties and redox potential, in particular. The resulting magnetic moment is determined by the electronic spin only. The value of the intrinsic electron spin moment is a product of g-factor (free electron g-value is 2.002319) and Bohr magneton ($\beta_B = 9.274009994(57) \times 10^{-24}$ J T^{-1}).

The nitroxide radical fragments are stabilized by the steric screening imparted by its four adjacent methyl groups and by quantum mechanical interaction of the unpaired electron on oxygen and electron pair on nitrogen (three electron bond). In certain conditions, these structural features protect the radicals from reduction or other processes.

The preference for a planar (sp2 hybridization) or pyramidal (sp3 hybridization) geometry around the N atom and the partitioning of the spin density between the N and O atoms are strongly dependent on the nature of the substituents. The change in hybridization of the nitrogen atom from sp2 to sp3 that occurs on reduction causes

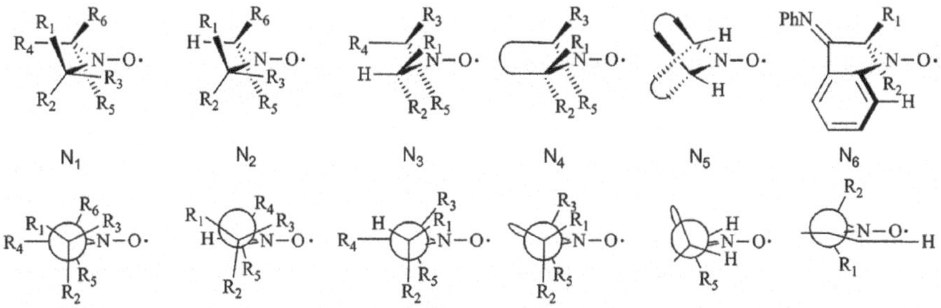

Fig. 2.2 Six deduced conformations (N1–N6) around the nitroxide center and their corresponding Newman projections: N1 (has a syn and an anti b-R group), N2 (with anti b-hydrogen atom and a syn b-R group), N3 (with an anti-R group and an anti-hydrogen atom), N4 (with an anti-ring), N5 (with two syn hydrogen atoms), and N6 (with the radical centre fused to an aromatic ring). Reprint from reference [1], Copyright 2012, American Chemical Society

change torsional strain which is small for six-membered rings, but increases for five-membered rings. Conformation N1–N6 radicals around the nitroxide center and their corresponding Newman projections are presented in Fig. 2.2 [1].

Unique chemical stability of nitroxide is partly due to the strong delocalized three-electron bond within the radical center between the nitrogen and the oxygen atoms conferring its substantial thermodynamic stability, and as a result, prohibition on dimerization between the radicals with the free energy of about 146 kJ/mol Insignificant weight of the ionic resonant structure 95.6 kJ/mol is actually extinguished nucleophylic ability of нитроксилов [2].

2.2 Spin Electron–Spin Nuclear Interactions in Nitroxides

The total spin Hamiltonian of the system of an electron and many nuclear spins are given by [3–5]

$$\widehat{H} = \widehat{H}_e + \widehat{H}_N + \widehat{H}_{eN} + \widehat{H}_{NN}, \tag{2.1}$$

and composed of the single-spin Zeeman energies, Hamiltonian for interaction in the applied magnetic field along the *z*-axis, the hyperfine interaction and the intrinsic nuclear–nuclear interaction, respectively. Generally, both the spin and orbital angular momentums of the electron contribute to nuclear magnetic moment. The following equation shows the total energy related to electron transitions in EPR:

$$\Delta E = g_e \mu_e S_s B + \sum g_{Ni} \mu_{Ni} I_{Ni} (1 - \sigma_i) + \sum a_i S_s I_{Ii} \tag{2.2}$$

The first two terms correspond to the Zeeman energy of the electron and the nucleus of the system, respectively. The third term is the hyperfine coupling between the electron and nucleus where a_i is the hyperfine coupling constant.

Two mechanisms with the different physical origins can contribute in coupling between electron and nuclear spins. The hyperfine interaction between the electron and the nuclear spins consists of the isotropic Fermi contact interaction and the anisotropic dipole–dipole interaction. The Fermi contact interaction is the magnetic interaction between an electron and an atomic nucleus when the electron is inside that nucleus. The electron–nuclear spin interactions for an electron in an *s*-symmetry orbital *non-s-symmetry orbital* are expressed as follows:

$$H_{\mathrm{HF}} = \frac{16\pi}{3}\gamma_I \mu_{\mathrm{B}} \mu_{\mathrm{N}} \delta(\mathbf{r})[\mathbf{S}\cdot\mathbf{I}], \quad l = 0 \tag{2.3}$$

where γ_I, μ_{B}, and μ_{N} are the gyromagnetic factor of the nuclear spin, the Bohr magneton, and the nuclear magneton, respectively. S and I are the spin operators for the electron and the nucleus, and l is the angular momentum operator for the electron. The nuclear magnetic moment m_I originating from the nuclear spin quantum number I is $m_I = 2I + 1$. Therefore, for nitrogen isotopes ^{14}N ($I = 1$), $m_I = 1, 0, -1$ and for ^{15}N ($I = 1/2$), $m_I = 1/2, -1/2$. Selection rules for ESR-transitions: $\Delta m_S = \pm 1$, $\Delta m_I = 0$.

The contact interaction is given by

$$\widehat{H}_{eN} = \sum_n a_n \widehat{\mathrm{S}}_{\mathrm{e}} \cdot \widehat{\mathbf{J}}_n, \tag{2.4}$$

where $\hat{\mathrm{S}}_{\mathrm{e}}$ and $\hat{\mathrm{J}}_n$ are the spin and nucleus operators, respectively, and

$$a_n = \frac{\mu_0}{4\pi}\gamma_{\mathrm{e}}\gamma_n \frac{8\pi}{3}|\Psi(\mathbf{R}_n)|^2, \tag{2.5}$$

where μ_0 is the vacuum magnetic permeability, $\mathbf{R}n$ denotes the coordinates of the nth nucleus, γ_n and γ_{e} are the nuclear gyromagnetic ratio and the electron gyromagnetic ratio, respectively; $|\psi(R_n)|^2$ is the spin density of s-electron on the nucleus ρ. The intrinsic interaction between nuclear spins includes the direct dipole–dipole interaction, the indirect interactions mediated by virtual excitation of electron–hole pairs, and the intranuclear quadrupole interaction.

The second mechanism is dipole–dipole interaction between electron and nuclear spins and is given by the term

$$\widehat{H}^{d}_{NN} = \sum_{n<m} \frac{\mu_0}{4\pi}\frac{\gamma_n\gamma_m}{\mathrm{R}^3_{n;m}}\left(\widehat{\mathbf{J}}_n \cdot \widehat{\mathbf{J}}_m - \frac{3\widehat{\mathbf{J}}_n \cdot \mathbf{R}_{n;m}\mathbf{R}_{n;m} \cdot \widehat{\mathbf{J}}_m}{R^2_{n;m}}\right), \tag{2.6}$$

$$\mathcal{H}_{\text{dip}} = -g\mu_B g_n \mu_n [\mathbf{S}_x\ \mathbf{S}_y\ \mathbf{S}_z] \cdot \begin{bmatrix} \left\langle \frac{r^2-3x^2}{r^5} \right\rangle & -\left\langle \frac{3xy}{r^5} \right\rangle & -\left\langle \frac{3xz}{r^5} \right\rangle \\ -\left\langle \frac{3xy}{r^5} \right\rangle & \left\langle \frac{r^2-3y^2}{r^5} \right\rangle & -\left\langle \frac{3yz}{r^5} \right\rangle \\ -\left\langle \frac{3xz}{r^5} \right\rangle & -\left\langle \frac{3yz}{r^5} \right\rangle & \left\langle \frac{r^2-3z^2}{r^5} \right\rangle \end{bmatrix} \cdot \begin{bmatrix} \mathbf{I}_x \\ \mathbf{I}_y \\ \mathbf{I}_z \end{bmatrix}$$
$$= \mathbf{S} \cdot \mathbf{A}_{\text{dip}} \cdot \mathbf{I}$$

Fig. 2.3 Hamiltonian for anisotropic hyperfine dipolar of anisotropic electron spin with anisotropic nuclear spin [4]

where $\mathbf{R}_{n;m}$ is the distance between electron and nucleus. $\hat{J}_m$ (S_e) and $\hat{J}_n$ are the spin and nucleus operators.

For asymmetric molecules, the dipolar interaction depends on the electron spin–electron nuclei distances and on their orientations, which links interacting spins referred to the direction of the applied magnetic field. This interaction can be described by a matrix of parameters A_{xx}, A_{yy}, and A_{zz}, corresponding to the principal axes of the radical. The appearance of the spectrum will change depending on the orientation of the molecule relative to the magnetic field.

For anisotropic hyperfine dipolar of anisotropic electron spin with anisotropic nuclear spin, Hamiltonian is shown in Fig. 2.3 [4].

$$\mathcal{H}_{\text{dip}} = -g\mu_B g_n \mu_n [S_x S_y S_z] \cdot \begin{bmatrix} \left\langle \frac{r^2-3x^2}{r^5} \right\rangle & -\left\langle \frac{3xy}{r^5} \right\rangle & -\left\langle \frac{3xz}{r^5} \right\rangle \\ -\left\langle \frac{3xy}{r^5} \right\rangle & \left\langle \frac{r^2-3y^2}{r^5} \right\rangle & -\left\langle \frac{3yz}{r^5} \right\rangle \\ -\left(\frac{3xz}{r^5} \right) & -\left\langle \frac{3yz}{r^5} \right\rangle & \left\langle \frac{r^2-3z^2}{r^5} \right\rangle \end{bmatrix} \cdot \begin{bmatrix} \mathbf{I}_x \\ \mathbf{I}_y \\ \mathbf{I}_z \end{bmatrix}$$
$$= \mathbf{S} \cdot \mathbf{A}_{\text{dip}} \cdot \mathbf{I}$$

2.3 Relationships Between the Structure and Properties of Nitroxide

Review [6] described the relationships between the structure and properties of nitroxide spin labels, methods for their synthesis, advances in methods for their incorporation into biomolecules, and selected examples of applications in biomolecule structural investigations. Various aspects in this area are covered in the original articles.

Density functional and ab initio (MP2) coupled cluster methods indicated that the spin density is distributed roughly evenly between the oxygen and nitrogen atoms except for $(CF_3)_2$.

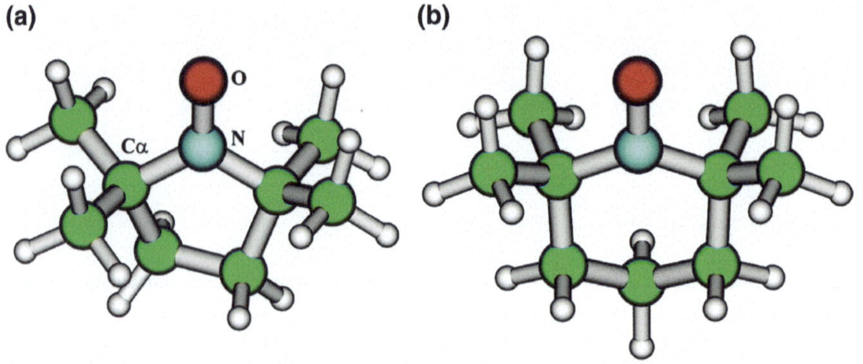

Fig. 2.4 Structure and label of the Proxyl **a** and Tempo **b** spin-probe molecules Michele Pavone [8]. Reprint from reference [8], Copyright 2010, American Chemical Society

- Synthesis and properties of diradicals in which nitronyl nitroxide fragments and a fused-thiophene bridge lie in the range 2.6–8.5° were described [7]. The planarity of **nitroxides** makes the Kekulé-type coupler units effective bridges to transmit magnetic interactions by a spin polarization mechanism. According to the magnetic susceptibility measurements, supported by density functional theory calculations, the singlet–triplet energy splitting ($2J/k_B$) ranges from −130 to −80 K. The singlet diradicals **D** demonstrated two one-electron reversible electrochemical oxidation peaks and two-electron quasi-reversible reduction peaks resulting in the formation of the corresponding bis(oxoammonium) dication or bis(aminooxidanide) dianion.

Stereoelectronic, environmental, and dynamical effects in the proxyl and TEMPO (Fig. 2.4) were investigated in [8]. The parameter-free PBE023 density functional, which includes an amount of exact exchange, and the polarized double-basis set N07D tailored for DFT-based calculations of magnetic parameters were employed.

According to the calculation, in the planar NO moiety of Proxyl, both N and O atoms exhibit a sp2 hybridization, whereas in the non-planar case of TEMPO, the N hybridization is close to sp3. It was shown that pyramidalization at the nitrogen center allows for the direct involvement of the nitrogen s-type orbital in the SOMO (Fig. 2.5), leading to a large direct contribution to the Fermi contact term.

The following findings were also pointed out: (1) Vibrational effects are relevant in the case of Proxyl, with a shift of –2 G, whereas they are negligible in the case of TEMPO. (2) Dynamical effects are not markedly affected by the presence of the solvent; i.e., the hccN shifts from the minimum energy structure are –2 G in the case of Proxyl and smaller than 1 G for TEMPO. (3) The isotropic nitrogen hyperfine coupling constant (hccN) experimental values in nonpolar solvents were found to be 14.2 and 15.28 G for Proxyl and TEMPO, respectively. (4) The hccN value increases when going from a planar to a pyramidal nitroxide conformation. (5) The planar structure is the minimum energy configuration for Proxyl, while in the case of TEMPO it corresponds to a transition state governing inversion between two equivalent pyramidal energy minima. (6) Calculating the hccN and Δg_{iso} values

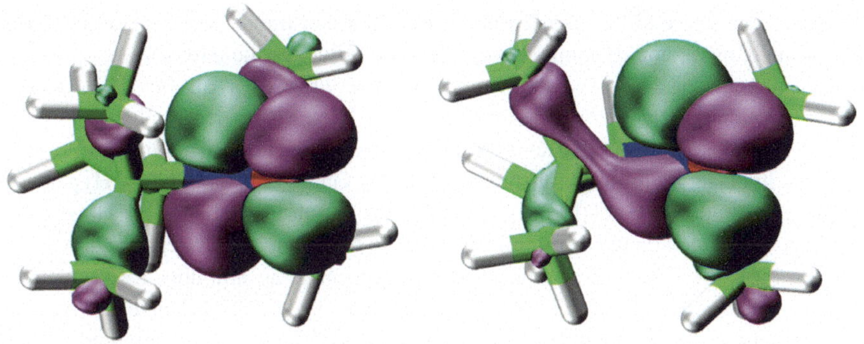

Fig. 2.5 Isodensity surface of Proxyl's SOMO from a planar configuration (left) to a pyramidal one (right) [8]. Reprint from reference [8], Copyright 2010, American Chemical Society

for the Proxyl and TEMPO is close. (7) The hccN shift, when going from gas phase to nonpolar solution, is in qualitative agreement with the experimental observations regarding a similar nitroxide (–1 G). Three tables summarized data on structural and magnetic parameters of proxyl and TEMPO in the gas phase, structural and magnetic parameters of proxyl and TEMPO in DMSO and isotropic nitrogen hyperfine coupling constant.

The calculation of large nitroxides in condensed phases based on geometries, vibrational frequencies, and potential energy surfaces computed at the DFT level was carried out employing AMBER force-field program [9]. Environmental and short-time dynamic effects on the hyperfine and gyromagnetic *g*-factor tensors of Proxyl, TEMPO, and (2-methyl-3-oxo-2-phenyl-2,3-dihydro-1H-indol-1-yl)oxidanyl molecule (INDCO) were taking in consideration. The partial atomic charges were calculated with the restrained electrostatic potential *(RESP)* procedure, by fitting an electrostatic potential grid computed for the nitroxide molecules. The HF/6-31 G (d) level of theory was used for study the partial atomic charges. The suggested model is based on a least squares fitting of the electrostatic potential with the addition of hyperbolic restraints on the charges of non-hydrogen atoms, which reduce the charges on buried carbons. Molecular dynamic calculations for Proxyl led to the following sequence in the values of hcc (in G): 13.4 (vacuum), 13.9 (toluene), 15.3 (water), while experimental a_N are 14.2 and 16.4 for toluene and water, respectively. For TEMPO, hcc (in G) are 14.6, (vacuum), 15.0 (toluene), 16.4 (water), and experimental values are 15.5 and 17.3 for toluene and water, respectively. The minima and transition state together with their relative energies for structures of DTBN, Proxyl, and TEMPO were also calculated. For PROXYL, transition state energy calculated at AMBER level was found to be 2.56 kcal mol^{-1}, and for TEMPO, the correspondent value is 5.11 kcal mol^{-1}. The average number of solvent molecules coordinating the NO moiety is two in aqueous solution and one in methanol. For Proxyl, both water molecules lie in the molecular plane approximately along the directions of the two lone pairs. The calculation showed that the behavior of the *g* tensor is dominated by component g_{xx}, and that Δg_{iso} and g_{xx} have parallel trends while g_{yy} and g_{zz} are almost unchanged, and dynamical effects do not influence the g tensor.

A systematic study of structure and physical properties of the INDCO has been undertaken [10]. Geometry optimization of INDCO performed by the QM (PBE0/N07D) or MM (AMBER) level produces an almost flat structure. A nitrogen isotropic hyperfine coupling constant (a_N) of 7.9 Gauss was calculated. Hyperfine coupling constants (in G) of the nitrogen and hydrogen atoms in the INDCO molecule were also calculated on the minimum structure in gas phase and in benzene and were listed in the table. The overall vibrational correction of a_N amounts 1.2 G. Calculated a_N values were found to be 9.1 G in gas phase and 9.3 G in benzene which is in agreement with experiment (9.25 G). The values of the g tensor computed on the energy minimum and those averaged along the MD trajectories (2.00593) are close to the experimental value of 2.00598.

Density functional and ab initio (MP2, coupled cluster) methods were used to obtain insight into the molecular structures, harmonic vibrational frequencies, inversion barriers, and hyperfine coupling constants of nitroxides: (i) acyclic: dimethylaminoxyl (Me_2NO), bis(trifluoromethyl)aminoxyl $(CF_3)_2NO$, and di-*tert*-butyl nitroxyl[$(Me_3C)_2NO$]; (ii) cyclic: aziridine-N-oxyl, azetidine-N-oxyl, pyrrolidine-N-oxyl and piperidine-N-oxyl; and (iii) imino nitroxides [11]. Scheme of the optimized structures of several nitroxides studied in this work is shown in Fig. 2.6. The cyclic aziridine-N-oxyl exhibits an inversion barrier of ~hc 3500 cm^{-1} compared to

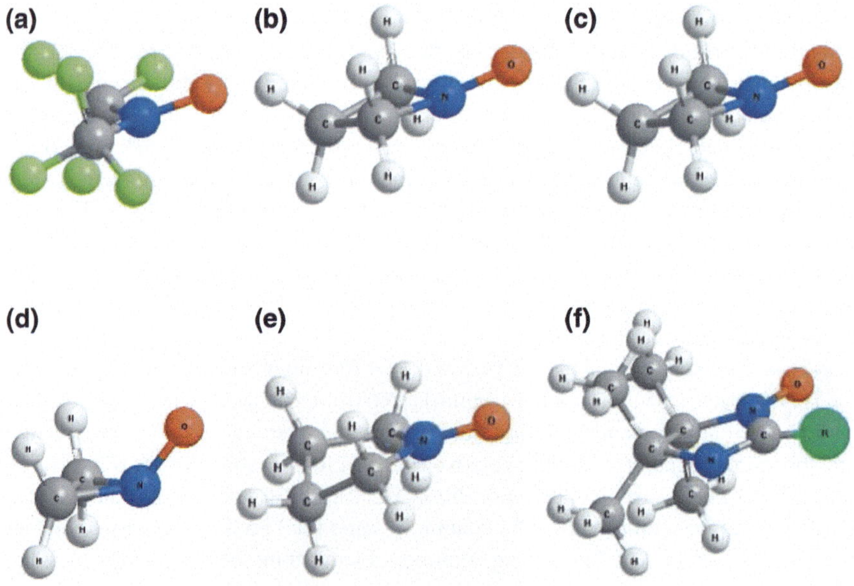

Fig. 2.6 Scheme of the optimized structures of **a** bis(trifluoromethyl)aminoxyl (CF3)2NO, **b** aziridine-N-oxyl, **c** azetidine-N-oxyl, **d** pyrrolidine-N-oxyl, **e** piperidine-N-oxyl, **f** imino nitroxide [11]. Reprint from reference [8], Copyright 2010, Schweizerische Chemische Gesellschaft

only ~hc 500 cm^{-1} for the other nitroxides. The possibilities that some of the chiral derivatives may be dominated by molecular parity violation in their dynamics were discussed.

Effects of solvent on structural and electronic characteristics of 4-amino-2,2,5,5-tetramethyl-3-imidazoline-N-oxyl (ATP)

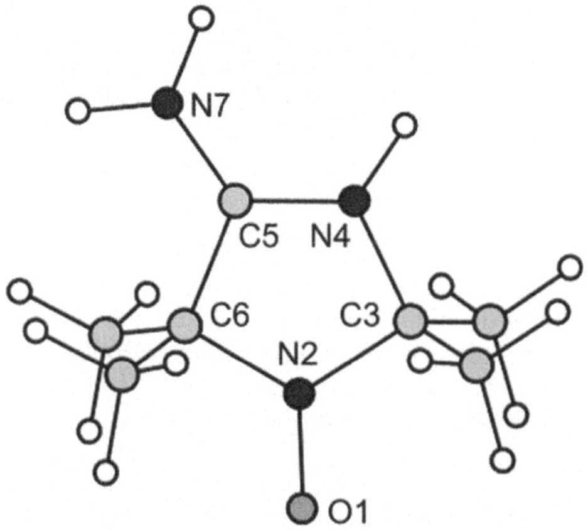

were evaluated in detail, employing the density functional theory (DFT) [12]. All calculations were carried out with the use of the original program Priroda. The calculation an aqueous sphere around a spin label protonated at the position of the ring nitrogen N4, ATI(H^+), was started with a small system (ATI(H^+) + $2H_2O$) and then expanded to 41 water molecules when a complete hydrogen-bonded water network surrounding ATI(H^+) was formed (Fig. 2.7). Calculations showed that (1) ATI protonation occurs at the nitrogen atom of the imidazoline ring rather than at the amino group. (2) Direct proton transfer between the amino group and N4 atom is hindered kinetically. (3) Spin density on the nitrogen atom of the nitroxide fragment increased with an extension of a water shell around ATI. (4) Protonation of ATI leads to a decrease in a spin density on the nitrogen atom of the nitroxide fragment > N–O•. (5) The ring nitrogen N4 is characterized by higher affinity to proton, as compared to the amino group. (6) The energy of the transition state [ATI-H^+] of the interconversion process ATI(H^+) T ATI(-NH_3^+) is higher than the energies of ATI(H^+) and ATI(-NH_3^+) by 55.2 and 23.8 kcal/mol, respectively (Fig. 2.8). (7) The energy difference between EATI(-NH_3^+)-(H_2O)41 and EATI(H^+)-(H_2O)41 is 19.1 kcal/mol, which is smaller than Egas. (8) Both protonation and hydration of ATI caused certain changes in calculated bond lengths and valence angles. The authors concluded that a hydrogen-bonded cluster of HO $n \geq 41$ water molecules could be considered as an appropriate model for simulation of ATI hydration effects.

The calculation also revealed the following peculiarities of the hydration process (Fig. 2.9): (1) The most significant changes in charge distribution occur in the nitroxide fragment N2–O1 and in the amino group and N4–H bond. (2) Hydration of ATI and ATI-(H+) molecules causes a rise in the positive charges on their N2 atoms.

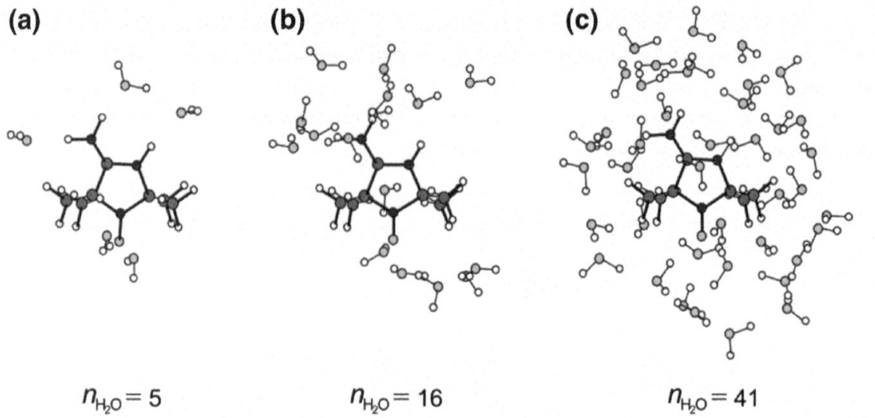

Fig. 2.7 Chemical structures of protonated 4-amino-2,2,5,5-tetramethyl-3-imidazoline-*N*-oxyl, ATI(H^+), surrounded by water molecules: **a** ATI(H +) surrounded by five H_2O molecules; **b** ATI(H^+) surrounded by 16 H_2O molecules; **c** ATI(H^+) surrounded by 41 H_2O molecules [12]. Reprint from reference [12], Copyright 2004, American Chemical Society

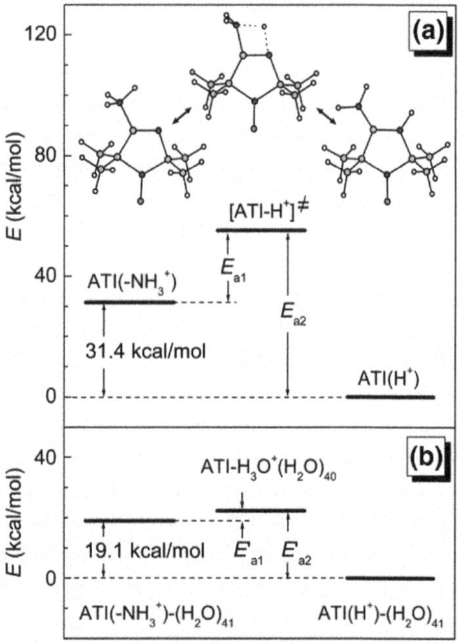

Fig. 2.8 Energy diagrams characterizing proton transfer between the amino group and the ring nitrogen N4 of ATI molecule: **a** ATI in the gas phase; top figures show calculated structures of ATI molecule protonated either at the amino group (left) or at the ring nitrogen atom N4 (right) [12]; **b** ATI in water. Reprint from reference [12], Copyright 2004, American Chemical Society

(3) The negative charges on O1 atoms become smaller. (4) Effects of water-induced redistribution of charges on the N4-H bond and amino group are different in cases of protonated and deprotonated ATI.

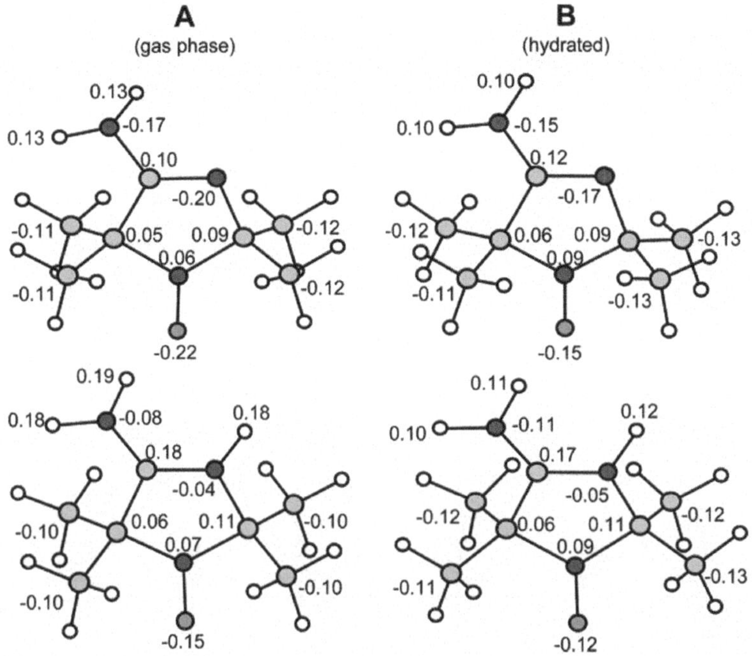

Fig. 2.9 Electric charges on atoms of deprotonated (top pictures) and protonated (bottom pictures) ATI molecules: **a** spin-labels in the gas phase; **b** spin-labels surrounded by 41 water molecules. For deprotonated form, negative charges of the nitrogen atoms N4 and N7 decrease after a spin-label hydration, while the hydration of pe ATI(H^+) results to a certain increase in the values of negative charges of N4 and N7 atoms. Calculation also revealed positive charges of hydrogens in the amino group and the positive charge of the hydrogen atom bound to the N4 atoms [12]. Reprint from reference [12], Copyright 2004, American Chemical Society

Figure 2.10 presents the distribution of spin densities over the atoms of protonated and deprotonated ATI molecules in the gas phase or surrounded by 41 water molecules.

The electronic structure of nitronyl nitroxide 2-phenyl-4,4,5,5-tetramethyl-4,5-dihydro 1H-imidazole-1-oxyl-3-oxide (NitPh) was investigated by conventional and polarized neutron diffraction and compared to those obtained by various theoretical ab initio calculations [13]. It was found that in the 0-N-C-N-0 fragment, most of the spin density the 0-N-C-N-0 fragment is equally shared between the four atoms of the two NO groups: 0.27 (O) and 0.27 (N). In addition, the bridging sp2 carbon atom carries a significant negative spin density (−011), while delocalization of the unpaired spin onto the phenyl fragment is weak.

Modulating spin delocalization in conjugated nitroxides, 2-(N-aminoxyl-N-tert-butyl)-benzothiazols, was carried out using UB3LYP/cc-pVDZ//UB3LYP/6-31G [14]. In solvent CH_2Cl_2, this radical exhibits UV–Vis absorbances at 476, 512, and 558 nm and solution ESR hyperfine coupling constants (hfc) of a_N (NO) = 9.16, (ring N) 2.67 G, a_H = 1.05, 1.31, 0.33, 0.43 G. In the same conditions, the nitroxide

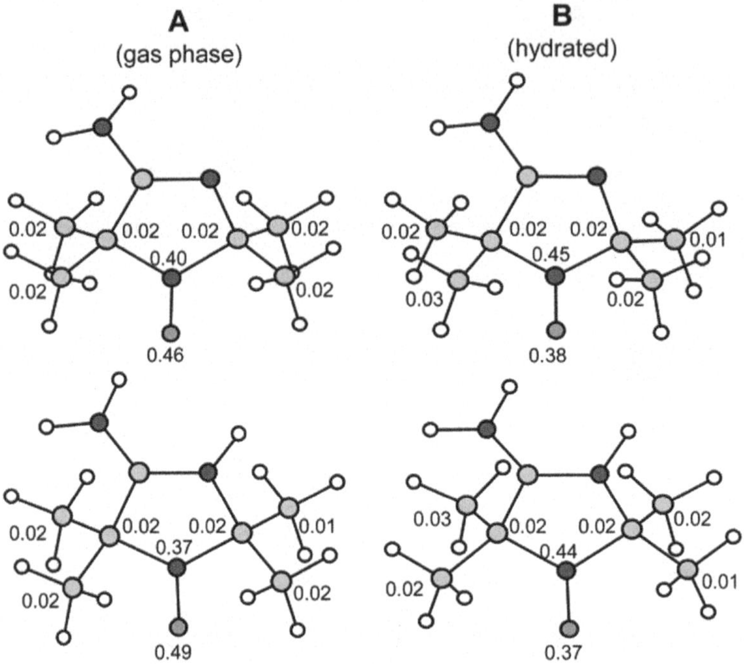

Fig. 2.10 Spin densities on atoms of ATI (top pictures) and ATI(H +) (bottom pictures) molecules: **a** spin-labels in the gas phase; **b** spin-labels surrounded by 41 water molecules [12]. Reprint from reference [12], Copyright 2004, American Chemical Society

hcc observed for isoelectronic 2-(N-aminoxyl-N-tert-butyl)-benzimidazole is 10.0–10.2 G. Distribution of spin densities was estimated from ESR hyperfine coupling (a) using the formulae $a_{(H-[p-C])} = (-22\ G)\rho(\pi\text{-}[C])$ and $a_{(N)} = 30\ G)\rho[N])$, where ρ are the estimated spin densities for p-carbons and nitrogens. The hyperfine constants showed that spin density throughout the benzothiazole ring is about 10% from the total value and the sp3 carbons attached to the nitroxide nitrogen has ≤1–2% spin density. The optimized structure with the N–O group syn to the thiazole nitrogen was found to be lower in energy than the anticonformer by 7.5 kcal/mol.

The analysis of the high-resolution W-band (95 GHz) pulsed ELDOR-detected NMR (EDNMR) and 244 GHz cw EPR spectra yielded precise electron Zeeman parameter, g_{xx}, ^{14}N hyperfine, A_{zz}, and ^{14}N quadrupole, P_{zz}, tensor components for the perdeuterated nitroxide radical 3-hydroxymethyl-2,2,5,5-tetramethylpyrrolin-1-oxyl dissolved in deuterated frozen solvents of polar or unpolar, protic, or aprotic character (*ortho*-terphenyl, methanol, propanol, glycerol, aniline, phenol, and water) [15]. The results, obtained on the basis of semi-empirical and DFT quantum chemical calculations, indicated that the principal values of all the magnetic interaction parameters primarily depend on the nitroxide hydrogen-bond situation and the type of hydrogen-bond donor group of the solvent. To characterize the heterogeneity of the nitroxide local environment A_{zz} hyperfine and P_{zz} quadrupole values of the

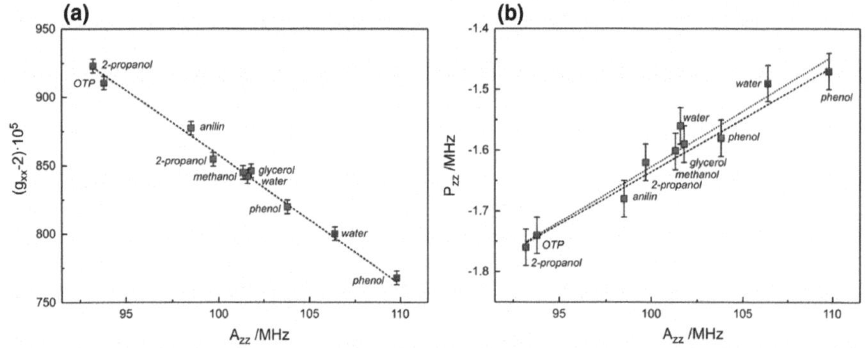

Fig. 2.11 **a** The g_{xx} versus A_{zz} plot showing g_{xx} values extracted from the analysis of the 244 GHz cw EPR spectra and A_{zz} values obtained from W-band EDNMR recordings. The dashed line shows the best linear least squared fit with a slope of $(-9.5 \pm 0.3) \times 10^{-5}$ MHz^{-1} (or -2.59 ± 0.07 T^{-1}). **b** The P_{zz} versus A_{zz} plot using values obtained from W-band EDNMR experiments. The dotted and dashed lines show best linear least-square fits (with the slopes of $+ (18 \pm 2) \times 10^{-3}$ and $+ (17 \pm 1) \times 10^{-3}$) including and excluding data for water, respectively. The red-, green- and blue-colored squares denote nitroxide fractions assigned to nitroxide forming zero, one, or two H-bonds with the solvent molecule, respectively (color figure online) [15]

perdeuterated nitroxide radical in different matrices, W-band EDNMR spectra were recorded. Figure 2.11 demonstrates perfect linear correlation between the hyperfine A_{zz} and g-tensor g_{xx} components with an R-squared coefficient $R^2 = 0.994$.

According to the author's discussion, A_{zz} is expected to react to polarity changes in non-bonding as well as H-bonding situations through changes in the spin density distribution of the N–O bond as a consequence of charge displacements between N and O in the N–O bond. The effective g_{xx} value of nitroxide radicals is, among other perturbations, significantly affected by additional matrix perturbations of the n-π energy gap of the O-atom in H-bonding situations.

References

1. E.G. Bagryanskaya, S.R. Marque, Y.P. Tsentalovich, Alkoxyamine re-formation reaction. Effects of the nitroxide fragment: a multiparameter analysis. J. Org. Chem. **77**, 4996–5005 (2012)
2. E.G. Bagryanskaya, S.R.A. Marque, Scavenging of organic C-centered radicals by nitroxides. Chem. Rev. **114**, 5011–5056 (2014)
3. G.R. Eaton, S.S. Eaton, D.P. Barr, R.T. Weber, *Quantitative EPR* (Springer, 2010)
4. G.I. Likhtenshtein, J. Yamauchi, S. Nakatsuji, A. Smirnov, R. Tamura, *Nitroxides: Application in Chemistry, Biomedicine, and Materials Science* (WILEY-VCH, Weinhem, 2008)
5. W.R. Hagen, *Biomolecular EPR Spectroscopy* (CRC, 2008)
6. M.M. Haugland, E.A. Andersona, J.E. Lovet, Tuning the properties of nitroxide spin labels for use in electron paramagnetic resonance spectroscopy through chemical modification of the nitroxide framework, in *Electron Paramagnetic Resonance*, vol. 25 (Royal Chemical Society, 2017), pp. 1–34

7. E. Tretyakov, K. Okada, S. Suzuki, M. Baumgarten, G. Romanenko, A. Bogomyakov, V. Ovcharenko, Synthesis, structure and properties of nitronyl nitroxide diradicals with fused-thiophene couplers. J. Phys. Org. Chem. **29**(12) (2016). https://doi.org/10.1002/poc.3561
8. M. Pavone, M. Biczysko, N. Rega, V. Barone, Magnetic properties of nitroxide spin probes: reliable account of molecular motions and nonspecific solvent effects by time-dependent and time-independent approaches. J. Phys. Chem. B, **114**, 11509–11514 (2010)
9. E. Stendardo, A. Pedone, P. Cimino, M.C. Menziani, O. Crescenzi, V. Barone, Extension of the AMBER force-field for the study of large nitroxides in condensed phases: an ab initio parameterization. Phys. Chem. Chem. Phys. **12**(37), 11697–11709 (2010)
10. P. Cimino, A. Pedone, E. Stendardo, V. Barone, Interplay of stereo-electronic, environmental, and dynamical effects in determining the EPR parameters of aromatic spin-probes: INDCO as a test case. Phys. Chem. Chem. Phys. **12**, 3741 (2010)
11. L. Horný, F. Mariotti, M. Quack, Ab initio study of some persistent nitroxide radicals. CHIMIA **62**, 256–259 (2008)
12. L.N. Ikryannikova1, L.Y. Ustynyuk, A.N. Tikhonov, DFT study of nitroxide radical1. Effects of solvent on structural and electronic characteristics o 4-amino-2,2,5,5-tetramethyl-3-imidazoline-N-oxyl. J Phys. Chem. A, **108**(21), 4759–4768 (2004)
13. A. Zheludev, V. Barone, M. Bonnet, B. Deuey, A. Grand, E. Ressouche, P. Rey, R. Subra, J. Schweizer, Spin density in a nitronyl nitroxide free radical. Polarized neutron diffraction investigation and ab initio calculations. J. Am. Chem. Soc. **116**, 2019–2027 (1994)
14. P.S. Taylor, P. Ghalsasi, P.M. Lahti, Modulating spin delocalization inconjugated nitroxides: 2-(N-aminoxyl-N-tert-butyl)-benzothiazole. Tetrahedron Lett. **45**, 6295–6298 (2004)
15. A. Nalepa, K. Möbius, M. Plato, W. Lubitz, A. Savitsky, Nitroxide spin labels—magnetic parameters and hydrogen-bond formation: a high-field EPR and EDNMR study. Appl. Magn. Reson. **50**, 1–16 (2019)

Chapter 3
Nitroxide Chemical Reactions

3.1 Introduction

After the pioneering work of Neiman et al. [1], the possibilities of modifying nitroxides without affecting the free valence appeared to be practically unlimited. Modern synthetic chemistry represents a rich and wide arsenal for carrying out almost any projects in these fields.

Nitroxides possess unique properties and chemical reactivity. A radical power of the N-oxyl moiety is partially redeemed due to the formation of a three-electron bond. This property provides chemical inertness with respect to many biological and other molecules under ambient conditions. The chemical activity increases with increasing temperature, which in particular is used in syntheses and living polymerization. On the other hand, nitroxides readily react with active radicals, which allows them to be used as antioxidants and intermediate agents in the living polymerization. Another avenue is the use of nitroxyls for analyzing antioxidants and establishing the antioxidant status in biological systems objects. Nitroxides are very weak nucleophyl reagent.

Nitroxyl radicals, oxoammonium cations, and hydroxylamines form a stable organic redox triad with efficient one- and two-electron transfer (Scheme 3.1).

In the pioneering work of Golubev et al. [3], the first stoichiometric oxidation of alcohols to ketones mediated by an oxoammonium species with N^+-oxyl fragment was performed. Since then, numerous applications of aminoxyl radicals in organic

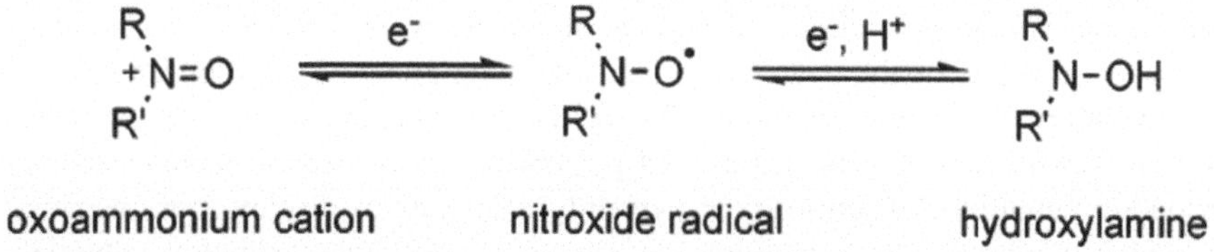

Scheme 3.1 Redox triad: oxoammonium cations, nitroxyl radicals, and hydroxylamine anion [2]. Reprinted from [2], Copyright 2007 American Chemical Society

G. I. Likhtenshtein, *Nitroxides*, Springer Series in Materials Science 292,
https://doi.org/10.1007/978-3-030-34822-9_3

oxidation reactions have focused on catalytic methods in which the oxoammonium species is (re)generated by a stoichiometric secondary oxidant. The chemical activity of the N-oxyl and N^+-oxyl moieties is mainly determined by their redox potential.

3.2 Nitroxide Redox Potential

The standard redox potential of nitroxides dictates the thermodynamic possibility and, to a large extent, the kinetics of the reactions with the participation of these compounds. Tables presented in review [4] listened the following data on: (1) experimental aminoxyl/oxoammonium and N-hydroxyimide/imidoxyl redox potentials, (2) effect of ring structure on the calculated one-electron potentials for oxidation of cyclic aminoxyl radicals, (3) redox potential of the aminoxyl to oxoammonium redox process for various N-oxyl radicals, (4) redox potentials (V vs. NHE) of N-aryl hydroxylamines, hydroxyamic acids, and other acyclic N-hydroxy derivatives as determined by cyclic voltammetry (CV) or differential pulse voltammetry, and (5) second-order rate constants of the reaction of phthalimide N-oxyl (PINO) with substrates containing activated C–H bonds.

The value of the midpoint potentials ($E_{1/2}$), for 48 nitroxyl/oxoammonium pairs varies in the range from 0.67 to 1.31 eV, for 16 N-hydroxyimide/imidoxyl pairs from 0.57 to 1.12 eV, and for 18 redox potential of the nitroxyl to oxoammonium redox process for various N-oxyl radicals in a range 0.57–1.20 eV. The following general conclusions were made regarding the redox potential data presented: (1) The change of potentials for oxidation of cyclic nitroxyl radicals was explained by the change in hybridization of the nitrogen atom from sp2 to sp3 that occurs on reduction and causes change torsional strain; (2) this change is small for six-membered rings, but increases for five-membered rings; (3) an attachment of aromatic groups to the radical rings leads to the significant increase of the potential; and (4) replacing the group t-Bu by SO_3^- leads to a substantial decrease in the nitroxide redox potential.

The oxidation and reduction potentials of 54 piperidine, pyrrolidine, isoindoline, and azaphenalene nitroxides, substituted with COOH, NH_2, NH_3^+, OCH_3, OH, and NO_2 groups nitroxides were calculated using G3(MP2)-RAD//B3-LYP/6-31G(d) gas-phase energies and PCM solvation calculations at the B3-LYP/6-31G(d) level [2]. This approach gave theoretical values of oxidation and reduction potentials within 40 mV of experimental values. Effects of level of the theory used for geometry optimization on the adiabatic ionization energies, on the adiabatic electron affinities, and on the oxidation potentials of various nitroxides were listen in corresponding tables. As a result of the calculation, the following conclusions were drawn: (1) The overall ring structure has more effect on the electrode potentials than the inclusion of substituents; (2) piperidine and pyrrolidine derivatives have intermediate oxidation potentials but on average pyrrolidine derivatives display more negative reduction potentials; (3) isoindoline derivatives show higher oxidation potentials and more negative reduction potentials; (4) within a ring, the substituents have a relatively small effect with electron donating groups such as amino and hydroxy groups; (5) the oxidized species and electron-withdrawing groups such as carboxy groups stabilize the reduced species; and (6) azaphenalene derivatives display the lowest oxidation potentials and negative reduction potentials. In addition, the calculations revealed the following patterns: (1) the five-membered cyclic nitroxides (pyrrolidine and isoindoline derivatives) have reduction potentials centered around −1570 mV; (2) the six-membered cyclic nitroxides (piperidine and azaphenalene derivatives) have reduction potentials centered at −1370 mV; (3) piperidine and pyrrolidine derivatives show oxidation potentials centered around 830 mV; (4) the highest for the isoindoline derivatives centered around 1020 mV; and (5) the lowest oxidation potentials are seen for the azaphenalene. Derivatives centered around 510 mV. It was suggested that azaphenalene derivatives show even lower oxidation potentials than piperidines due to the fact that the positive charge in the cation species can be delocalized onto the ring by hyperconjugation.

The redox potential of piperidine nitroxides and *N*-oxoammonium cation couples modified at the C4 position was measured, and the nitroxide structure–reactivity relationship was evaluated [5]. It was concluded that the superoxide dismutase activity of a nitroxide can be controlled based on its redox potential.

Redox potentials of a wide group of nitronyl (NN) and imino nitroxides with methyl, phenyl, iodo, and cyanosubstituents (Fig. 3.1) were determined by CV [6]. Nitronyls studied underwent reversible one-electron reduction, which formed the oxoammonium cation. The oxidation potentials E_{ox} of NNs were calculated as the mean potentials of the anode (E_p^{ai}) and cathode (E_p^{ci}) peaks. The reduction potentials E_{red} of NNs were calculated by the equation $E_{\text{red}} = (E_p^{a2} + E_p^{c2})/2$. The experiments revealed the following trends in the values E_{red} depending on the structure of the nitroxides: (1) The highest potential, $E_{\text{red}} = 1.41$ V, is required for the reduction of pyrazol-4-yl-substituted **1a**, (2) the alkyl derivatives **1b–1e** and **1i** of this nitroxide

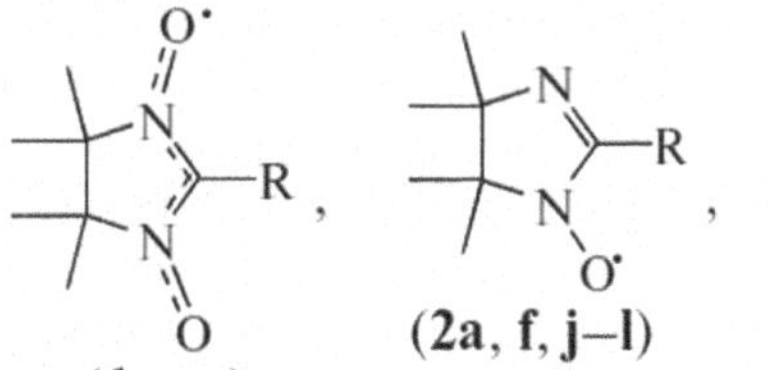

(1a–p)

(2a, f, j–l)

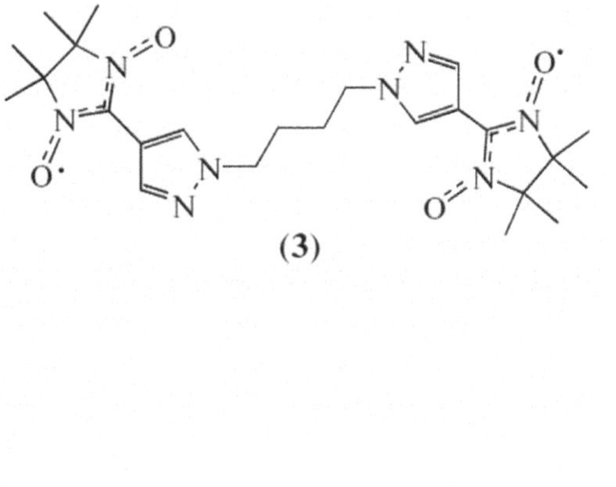

(3)

Fig. 3.1 Nitroxides used in [6]

were reduced at slightly lower potentials E_{red} ~ 1.23–1.26, (3) the stronger electron_accepting substituent in **1l**, **1n**, and **1o** led to a decrease in E_{red} to ~1.2 V, and (4) the lowest E_{red} ~ 0.9 V is inherent in **1f**, **1j**, **1k**, and **1p**.

In the case of the oxidation potential, the tendency is as follows: (1) Pyrazol-4-yl-substituted NNs **1a–1e** and **1i**, pyrazol-3-yl-substituted NN **1g**, imidazole-4-yl-substituted NNs **1j** and **1k**, and methyl-substituted NN **1 m** are characterized by the lowest values of *E*ox ~ 0.35–0.40 V, (2) as the electron accepting properties of substituents increased in the series **1n**, **1o** < **1l** < **1f**, **1h**, *E*ox gradually increased 0.55 and 0.81 V on passing to **1p**, (3) changes in *E*ox correlate with the ability of the N–O group to act as a Lewis base in reactions with transition metal ions, and (4) imino nitroxides are oxidized irreversibly.

The redox potentials of seven aromatic nitroxides derived from *tert*-butyl phenyl nitroxide

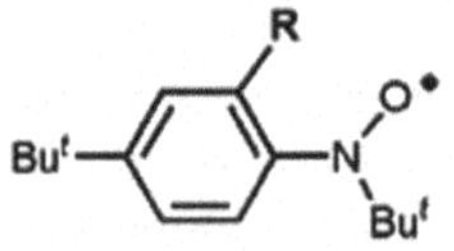

II

were determined by CV in non-aqueous solution, and the first oxidation potential to the formation of the corresponding N-oxo ammonium cation was assigned [7]. The electrochemical experiments were conducted in a three-electrode glass cell. A platinum wire and a saturated calomel electrode (SCE) were used as a counter and a reference electrode, respectively. The following general conclusions were made: (1) The electrochemical reduction of the aromatic nitroxides reveals an irreversible wave, and (2) the first oxidation potential as well as the reduction potential strongly depend on the electron donating and withdrawing substituents of the compounds. The experimental values of oxidation and reduction potentials of the studied nitroxides were tabulated. The following experimental results and their explanation were reported: (1) The reduction potentials of the aromatic nitroxides (1.13–140 eV) are relatively low compared to TEMPO (1.61 eV) probably due to the mesomeric effect of the aromatic cycle, stabilizing the negative charge on the oxygen; (2) for the neutral nitroxides having the ester and sulfonate ester electron-withdrawing groups a shift of 150 mV toward less negative potentials took place; (3) the negatively charged nitroxides are more prone to reduction than their corresponding neutral nitroxides with potential shifts of 150 and 220 mV, respectively; (4) the observed high-potential shift from 0.918 to 0.358 mV in the case of the carboxylate was explained by charge effect; and (5) the proximity of the carboxylate group in the ortho position and the rigid structure of compound allow for a cyclic conformation of the corresponding N-oxo ammonium via an intramolecular electrostatic interaction, which can stabilize the nitroxide structure.

The redox potentials of 25 cyclic nitroxides from four different structural classes (pyrrolidine, piperidine, isoindoline, and azaphenalene) (Fig. 3.2) determined experimentally by cyclic voltammetry in acetonitrile were found to be influenced by the

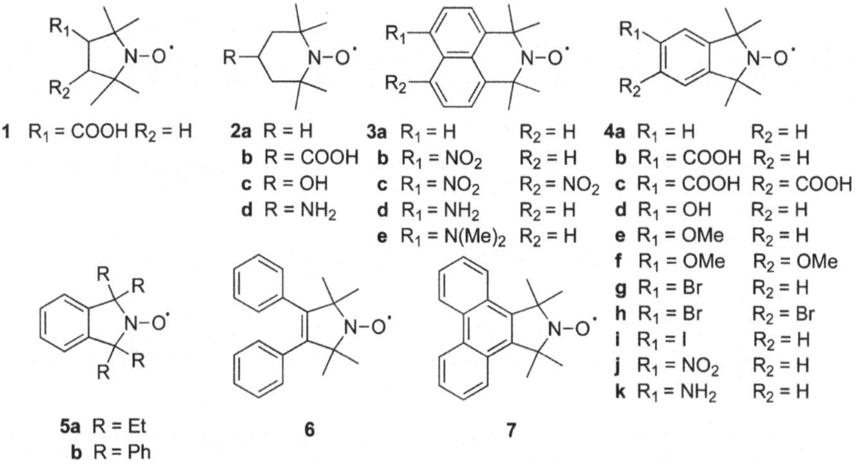

Fig. 3.2 Nitroxides investigated in cyclic voltammetry and theoretical studies [8]. Reprint from [8], Copyright 2008, American Chemical Society

type of ring system, ring substituents and/or groups surrounding the radical moiety [8]. The redox potentials, E_o of the oxidation of the nitroxide, were estimated as half the sum of the anodic (*E*pa) and cathodic (*E*pc) peak potentials. In addition, the potential values were calculated by high-level ab initio molecular orbital calculations with agreements with experimental data. Exceptions are the azaphenalenes, for which is an extraordinary discrepancy (mean absolute deviation of 0.60 V). It was found that for investigated nitroxides, the peak separation, *E*pa − *E*pc, ranged from 65 to 109 mV and for the unsubstituted ring systems, the order of oxidation from most easily oxidized to least easily oxidized is: piperidine (**2a**, 0.577 V), azaphenalene (**3a**, 0.736 V) > isoindoline (**4a**, 0.771 V). Azaphenalenes were more easily oxidized than the piperidine derivatives, which was attributed to the ability of the aromatic ring of the azaphenalene to stabilize the cation.

Orbital diagrams showing the singly occupied molecular orbital (SOMO) of the parent nitroxide radicals and the highest occupied molecular orbital (HOMO) of the oxidized species are presented in Fig. 3.3. Oxidation potentials of monosubstituted isoindoline nitroxides plotted with respect to the Hammett constant of the functional group (σ_p), an empirical constant based on the electronic effect of the subsistent, showed the linear correlation.

The kinetics and thermodynamics of comproportionation of eight pairs of oxoammonium and hydroxyammonium salts *3H+* were investigated, and the effect of substituents and heterocyclic ring structure on the comproportionation equilibrium constant K4 was revealed [9]. The following finding was shown: (1) Experimental values of the reduction potentials $E_{2/1}$, dissociation constants of R_2NO^+ and R_2NOH to $R_2NO\cdot$ hydroxyammonium cations pK_{3H+}, the equilibrium constant K4; (2) the values of bond dissociation energies D(O–H) for hydroxylamines in water (0.972–0.734 V vs. NHE), (3) pH dependences of reduction potentials of nitroxyl radicals; (4) linear

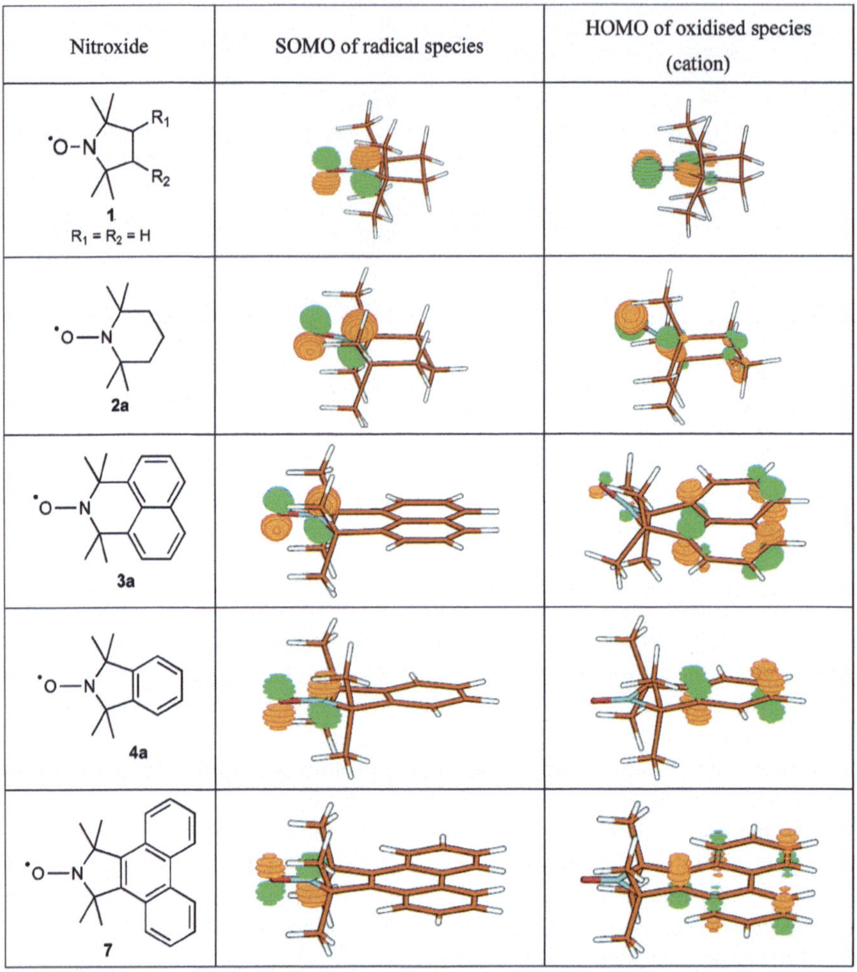

Fig. 3.3 B3LYP/6-311+G(3df,2p) orbital diagrams showing the singly occupied molecular orbital (SOMO) of the parent nitroxide radicals and the highest occupied molecular orbital (HOMO) of the oxidized species [8]. Reprint from [8], Copyright 2008, American Chemical Society

free energy relationships between redox potentials ($E_{1/2}$), dissociation constants of hydroxyammonium cations K_{3H+}, and the equilibrium constant K4; (5) correlations with the inductive substituent constants σ_I for the radicals of piperidine and pyrrolidine series, values of $pK3H_+$ and $E_{1/2}$, (6) increasing in –I-effects of substituents which reduces the overall equilibrium constant of the reaction K4; (7) dissociation energies for hydroxylamines in water; and (8) correlation between the Pauling Group Electronegativity (Eg) of substituents versus redox potentials ($E_{1/2}$) for the 5-membered ring hydroxylamine/aminoxyl radicals in MeOH (0.1 M $Bu4NClO_4$) and phosphate buffer (0.1 M, pH 7.4) was also established [10].

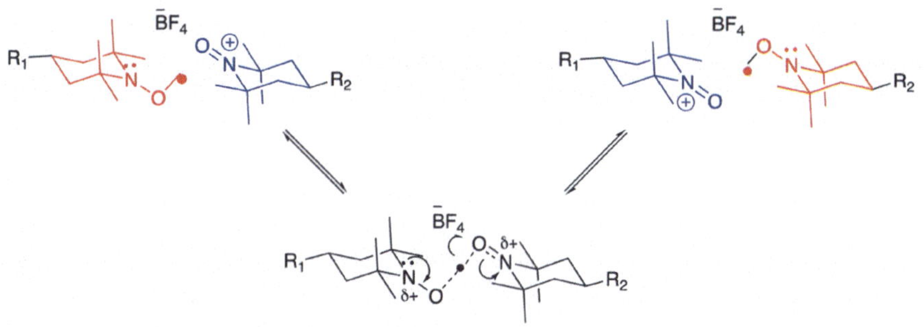

Fig. 3.4 A redox equilibrium between oxoammonium salts and trace amounts of corresponding nitroxide [12]. Reprint from [12], Copyright 2017, American Chemical Society

The influence of the structure of 21 nitroxyl radicals on the thermodynamics and the kinetics of the reverse comproportionation reaction of R_2NO^+ and R_2NOH to $R_2NO^{\cdot}$ in aqueous H_2SO_4 was investigated [11]. pH dependences of reduction potentials of nitroxyl radicals $E_{1/2}$ and bond dissociation energies D (O–H) for hydroxylamines in water were determined. The experiments indicated: (1) the linear free energy relationships between the reduction potentials $E_{2/1}$, dissociation constants of hydroxyammonium cations K3H+, and the equilibrium constant K4; (2) for the radicals of piperidine and pyrrolidine series, values of $pK3H^+$ and $E_{2/1}$ correlate with the inductive substituent constants σ_I; and (3) a correlation between the basicity of nitroxyl groups and reduction potentials of oxoammonium cations takes place.

Three new homologous TEMPO oxoammonium salts and three homologous nitroxide radicals have been prepared, and their oxidation properties of the salts have been explored [12]. Overall paramagnetism in the oxoammonium salt solutions revealed by EPR spectroscopy was explained by a redox equilibrium between oxoammonium salts and trace amounts of corresponding nitroxide (Fig. 3.4). This equilibrium was confirmed by electron interchange reactions between nitroxides with an N-acetyl substituent and oxoammonium salts with longer acyl side chains.

Nitronyl nitroxides (NNs) are capable of scavenging physiologically relevant reactive oxygen (ROS) and nitrogen (RNS) species, namely superoxide, nitric oxide (NO), and nitroxyl (HNO). The redox properties of 2-(4-carboxyphenyl),-4,5-dihydro-4,4,5,5-tetramethyl-1H-imidazolyl-1-oxy-3-oxide monopotassium salt, cPTIO, 4,4,5,5-tetramethyl-2-[4-trimethylammoniophenyl]-2-imidazoline-3-oxide-1-yloxy methyl sulfate (NN^+), and 4,4,5,5-Tetramethyl-2-[4-trimethylammoniophenyl]-2-imidazoline-1-yloxy methyl sulfate (IN^+), including determination of the equilibrium and rate constants of their reduction by HNO and ferrocyanide, and reduction potential of the couple NN/hydroxylamine of nitronyl nitroxide (hNN), were investigated [13]. The reduction potential of the NNs and iminonitroxides (Ins), product of NNs reaction with NO were calculated based on their reaction constants with ferrocyanide. The values of the reduction potential were found to be for NN/hNN ($E'_0 \approx 285$ mV) and IN/hIN ($E' \approx 495$ mV). Obtained data on equilibrium and bimolecular rate and constants for the redox reactions of

NNs and Ins were tabulated. A correlation between the reduction potential of cyclic oxammonium derivatives and substituent Hammett σ^* constants was established. In addition, the rate constants of the reaction of the NNs with HNO were measured to be equal to (1–2) × 10^4 M^{-1} s^{-1}.

N-tert-Butyl-*N*-2-pyridylhydroxylamines were characterized using NMR, electrochemistry, and density functional theory [14]. Substitution of the pyridyl ring in the 3-, 4-, and 5-positions was used to vary the reduction potential (E_{red}) of the nitroxyl/oxoammonium redox couple by 0.95 V. The values were found to follow Hammond's correlation. DFT computations of the electrochemical properties agreed with experiment. Electrochemical studies of the reduction and oxidation reactions of five different organic nitroxyls have been performed across a wide pH range (0–13) [15]. The hydroxylammonium pK_a was determined by NMR. The resulting Pourbaix diagrams illustrate structural effects on their various redox potentials and on pH (Fig. 3.5). Evidence was also given for the reversible formation of a hydroxylamine *N*-oxide when nitroxyls are oxidized in alkaline media. Several conclusions have been drawn from the Pourbaix diagram: (1) the oxoammonium reduction potential is pH insensitive, (2) the potential is insensitive to the identity of the buffering electrolyte, and (3) the nitroxyl reduction potential is pH-dependent and has a break in slope, corresponding to the hydroxylammonium pK_a.

3.3 Oxoammonium Cation Reactions

The oxoammonium species can be commonly generated by a stoichiometric secondary oxidant such as NaOCl, bromine, NO_x/O_2, hypervalent iodide species, or electrochemically. The species may be isolated as a stable salt, and it serves as two-electron oxidant in a number of organic oxidation reactions. The first reaction of the oxidation of alcohols by oxoammonium cation was demonstrated by Golubev et al. [2]. Primary alcohols were oxidized to corresponding aldehydes, while secondary alcohols were oxidized to ketone. This approach has been commonly used for alcohol oxidation including in industrial scale what attracts particular interest [16–19].

Oxidation of alcohols containing a β-oxygen atom, using 4-acetylamino-2,2,6,6-tetramethylpiperidine-1-oxoammonium tetrafluoroborate, in the presence of pyridine, yielded dimeric esters, while in the presence of 2,6-lutidin, the product is aldehyde [20]. The betaine is oxidized by the oxoammonium salt to give an *N*-acylpyridinium ion that serves as an acylating agent for ester formation. A suggested mechanism for oxoammonium cation oxidations premised on nucleophilic additions to the oxygen atom of the positively charged nitrogen–oxygen double bond. The rates of oxidation of a series of primary, secondary, and benzylic cycle of alcohols oxidation oxoammonium salt with CF_3 moiety have been found to oxidize alcohols more rapidly than does with CH_3 (Fig. 3.6). The rate of oxidation of meta- and para-substituted benzylic alcohols displays a strong linear correlation to Hammett parameters ($r > 0.99$) with slopes (ρ) of −2.7 and −2.8, respectively (Fig. 3.7). It was

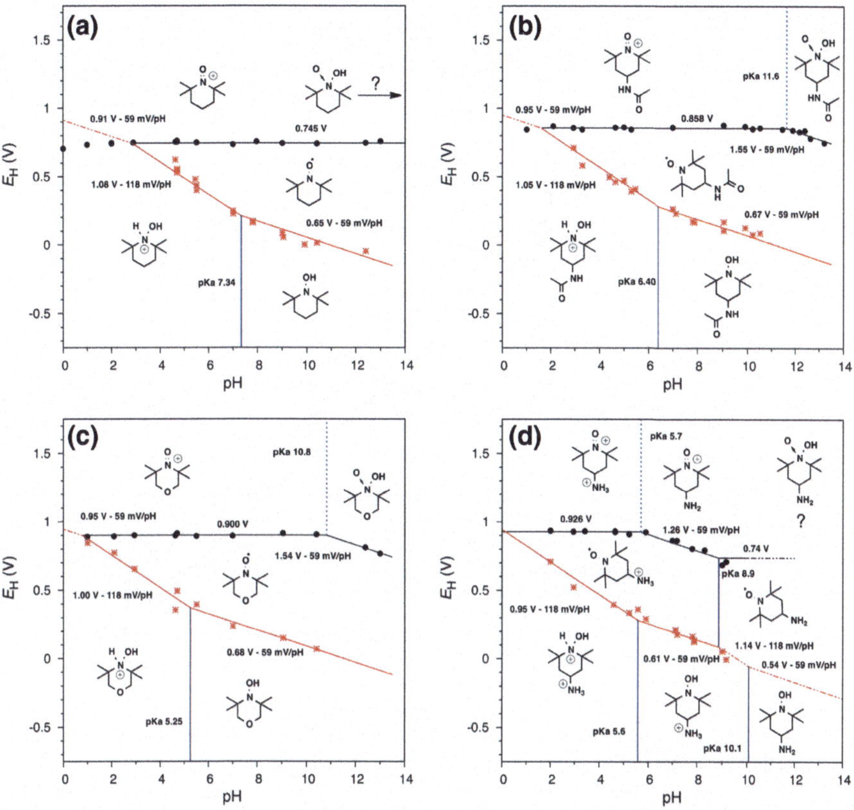

Fig. 3.5 Line fits have been constrained to have slopes corresponding to 0, 1, or 2 H^+/e^- and to intersect at points. Black circles and lines correspond to oxoammonium reduction to the nitroxyl, red stars and solid lines correspond to aminoxyl reduction, and the dashed red line corresponds to the theoretical $2e^-/2H^+$ oxidation of hydroxylammonium to oxoammonium. EH denotes redox potential referenced to NHE. Reprint from [15], Copyright 2018, American Chemical Society

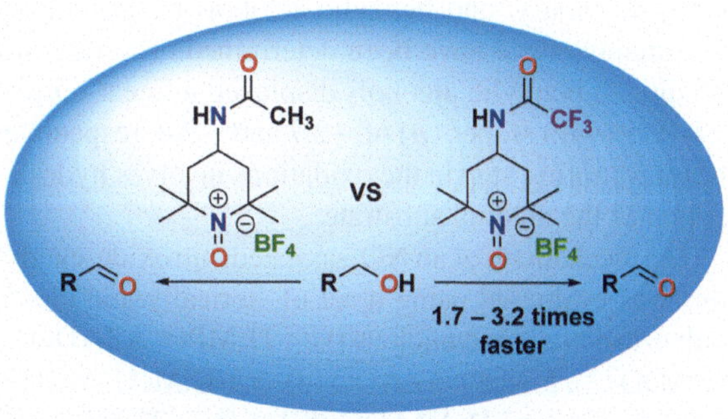

Fig. 3.6 Illustration of effect of structure of oxoammonium cation on an alcohol oxidation [21]. Reprint from [21], Copyright 2017, American Chemical Society

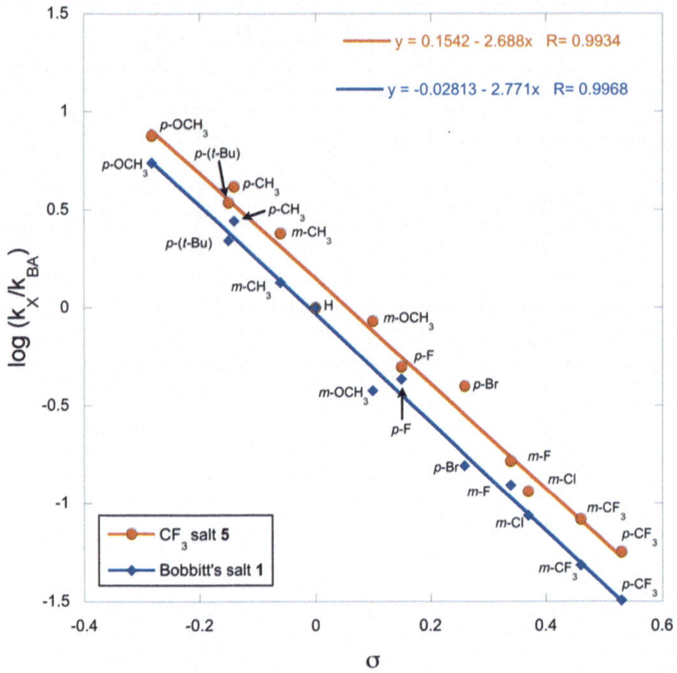

Fig. 3.7 Hammett plot of the rates of oxidation of substituted benzyl alcohols by 1 and 5; k_{BA} is the rate constant for oxidation of benzyl alcohol, and k_X is the rate constant for oxidation of substituted benzyl alcohol [21]. Reprint from [21], Copyright 2017, American Chemical Society

suggested that the rate-limiting step in the oxidations involves hydride abstraction from the carbinol carbon of the alcohol substrates.

A novel TEMPO-based oxoammonium salt, 2,2,6,6-tetramethyl-4-(2,2,2-trifluoroacetamido)-1-oxopiperidinium tetrafluoroborate, and its corresponding nitroxide were synthesized and characterized [21]. The rates of oxidation of primary, secondary, and benzylic alcohols by the nitroxide in acetonitrile solvent at room temperature have been determined. The rate of oxidation of meta- and para-substituted benzylic alcohols displayed a linear correlation to Hammett parameters ($r > 0.99$) with slopes (ρ) of -2.7 and -2.8, respectively. This finding indicated that the rate-limiting step in the oxidations involves hydride abstraction from the carbinol carbon of the alcohol substrate.

The mechanism of an NO_x-assisted, nitroxide-catalyzed aerobic oxidation of alcohols was evaluated using a set of sterically and electronically modified nitroxides with azaadamantane nucleus [i.e., TEMPO, AZADO (1), 5-F-AZADO (2), 5,7-DiF-AZADO (3), 5-MeO-AZADO (4), 5,7-DiMeO-AZADO (5), oxa-AZADO (6), TsN-AZADO (7), and DiAZADO (8)]. s [i.e., TEMPO, AZADO (1), 5-F-AZADO (2), 5,7-DiF-AZADO (3), 5-MeO-AZADO (4), 5,7-DiMeO-AZADO (5), oxa-AZADO (6), TsN-AZADO (7), and DiAZADO (8)] [22]. The introduction of an F atom at a remote position from the nitroxyl radical moiety on the azaadamantane nucleus enhanced the catalytic activity under typical NO_x-mediated aerobic oxidation conditions. The kinetic profiles of the azaadamantane-*N*-oxyl-[AZADO -, 5-F-AZADO -, and 5,7-DiF-AZADO]-catalyzed aerobic oxidations were investigated, revealing that: (1) AZADO showed a high initial reaction rate compared to 5-F-AZADO and 5,7-DiF-AZADO (Fig. 3.8), (2) AZADO-catalyzed oxidation exhibited a marked slowdown, resulting in ~90% conversion, (3) 5-F-AZADO-catalyzed oxidation smoothly reached completion. Experiments with Oxa-AZADO, Ts*N*-AZADO, and DiAZADO supported evidence for the electronic effect on the catalytic efficiency of F atom at a remote position from the nitroxyl radical moiety on the azaadamantane nucleus to enhance the catalytic activity under NO_x-mediated aerobic oxidation conditions.

4-acetamido-2,2,6,6-tetramethylpiperidine-1-oxoammonium tetrafluoroborate (1) in wet CH_3CN at room temperature oxidatively cleaved benzylic ethers and related $ArCH_2OR$ substrates to give the corresponding aromatic aldehyde and alcohol in high yield (Fig. 3.9) [20]. Primary or secondary alcohol products are further oxidized by 1 to give carboxylic acids and ketones, respectively. The oxidation likely involves a hydride abstraction from the benzylic carbon. In addition, the following findings were reported: (1) oxidation of alcohols containing a β-oxygen atom in the presence of pyridine by 1 yields dimeric esters; (2) in the presence of 2,6-lutidine the product is a simple aldehyde; (3) the betaine is oxidized by the oxoammonium salt to give an *N*-acylpyridinium ion; and (4) a series of alcohols containing a β-oxygen

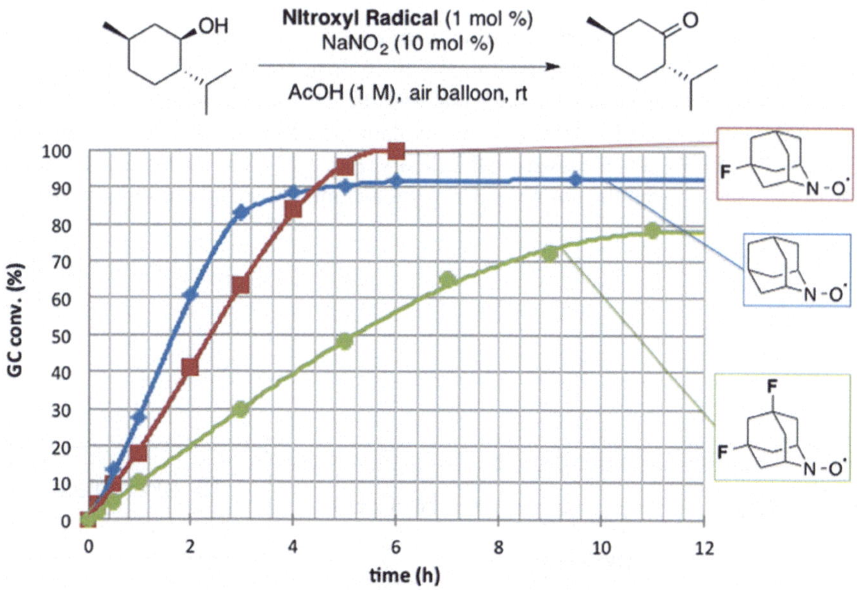

Fig. 3.8 Illustration of the electronic effect on the catalytic efficiency of the heteroatoms under NO_x-assisted aerobic oxidation conditions [22]. Reprint from [22], Copyright 2014, American Chemical Society

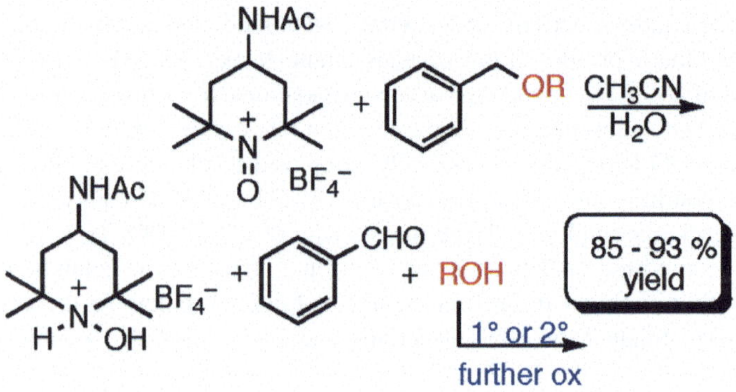

Fig. 3.9 Benzylic ethers and related $ArCH_2OR$ substrates are oxidative cleavage of enzylic ethers [20]. Reprint from [21], Copyright 2014, American Chemical

substituent were oxidized to aldehydes in the presence of 2,6-lutidine. Suggested mechanism for oxoammonium cation oxidations premised on nucleophilic additions to the oxygen atom of the positively charged nitrogen–oxygen double bond (Fig. 3.9).

Facile oxidation of primary amines to nitriles using a stoichiometric quantity of 4-acetamido-2,2,6,6-tetramethylpiperidine-1-oxoammonium tetrafluoroborate I was performed in CH_2Cl_2 – pyridine solvent at room temperature with yield of 75–95% (Fig. 3.10) [23].

At room temperature, benzylic amines bearing electron-withdrawing substituents were oxidized in 12 h, while aliphatic amines in 24-36 h. The detailed oxidation mechanism, taking ethylamine as the substrate and the 2,2,6,6-tetramethylpiperidine-1-oxoammonium cation (TEMP·O+) as the oxidant, was computationally investigated using the B3LYP/6-311+G* level of DFT theory (Fig. 3.11). The mechanism involves a hydride transfer from the amine to the oxygen atom of I as the rate-limiting step.

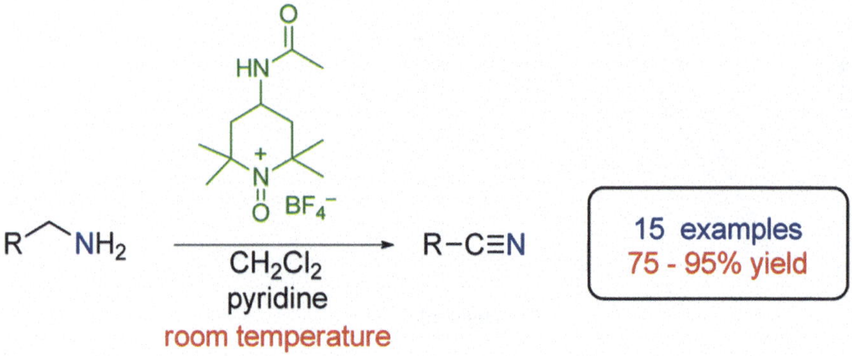

Fig. 3.10 Oxidation of primary amines to nitriles (Scheme 3.1, 1) a% [23]. Reprint from [23], Copyright 2017, American Chemical Society

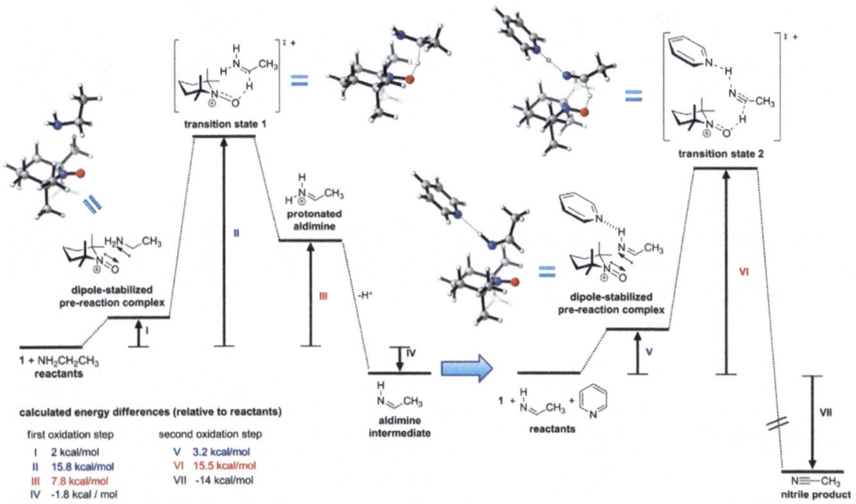

Fig. 3.11 Computed profile (B3LYP/6-311+G*) for the oxidation of primary amines to nitriles [23]. Reprint from [23], Copyright 2017, American Chemical Society

Efficient reduction of O_2 to water is a challenge in energy conversion and many aerobic oxidation reactions. It was demonstrated that the electrochemical oxygen reduction reaction (ORR) can be achieved at high potentials by using soluble organic nitroxyl and nitrogen oxide (NO_x) mediators [24]. The combination of nitroxyl/NOx species, such as 2,2,6,6-tetramethyl-1-piperidinyl-N-oxyl (TEMPO), or sodium nitrite, mediated sustained O_2 reduction with overpotentials as low as 300 mV in acetonitrile containing trifluoroacetic acid. The nitrogen oxide catalyst drives aerobic oxidation of a nitroxyl mediator to an oxoammonium species, and the electrolysis potential is dictated by the oxoammonium/nitroxyl reduction potential. The oxoammonium species are reduced by cathode. The overpotentials accessible to this ORR system were found to be significantly lower than widely studied molecular metal-macrocycle ORR catalysts.

3.4 Nitroxide Cross-Coupling Reactions

Cross-coupling reactions of nitroxides with organic C–centered radicals (OCCR) (Fig. 3.12) comprehensively reviewed by Bagryanskaya and Marque [25] are attracted special interest. The review discussed the data on 75 carbon- and heteroatoms-centered active radicals and 100 nitroxides. Numerous examples of the involving of the cross-coupling stages in various chemical processes and methods of the processes investigation were described. General trends and specific effects in the frame of the multiparameter approach were also discussed in detail.

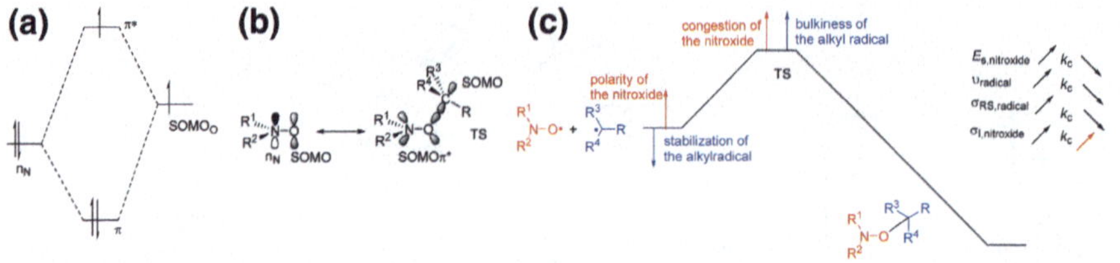

Fig. 3.12 Orbital diagram for the SOMO interactions and the energy profile of the cross-coupling reaction between a nitroxide and a C-centered radical [26]. Reprinted from [26] Copyright 2012 American Chemical Society

Orbital diagram for the SOMO interactions and the energy profile of the cross-coupling reaction between a nitroxide and a C-centered radical are presented in Fig. 3.12 [26].

In Fig. 3.12, the SOMO of the nitroxide is described as a three-electron bond orbital with the odd electron located in the π* orbital. Factors affected on the efficiency of the cross-coupling reaction are: (1) the energy gap, (2) the overlapping between the SOMOs of the nitroxide (SOMO π*) and the alkyl radical, (3) the spin density in the SOMOs, (4) strength of stabilization of the alkyl sRS (low electron density on the coupling sites), (5) bulkiness of the alkyl radical υ (large steric hindrance) that increases the activation energy of the process, and (6) the polarity (electron-withdrawing groups modifying the stabilization, matching of electrophilic/nucleophilic character) of the radicals. Quantitative experimental data on reactions in which the cross-coupling plays a key role were summarized in tables: (1) values of rate constant kc at room temperature for various active radicals and nitroxides and in various solvent ranging from 9.5 to 520 $M^{-1}S^{-1}$; (2) data on Arrhenius parameters, pre-exponent factor A_c, energy activation Eac and values of bond dissociation energies [BDEs(C–H)] for the cross-coupling reaction of nitroxides and C-centered radicals; (3) values of bond dissociation energies [BDEs(C–H)] and stabilization (σRS), polarity (σ_I), and steric (υ); (4) descriptors for selected C-centered radicals corresponding re-evaluated rate constants k_c for the cross-coupling with nitroxides; (5) correlation of log kc at room temperature with a linear combination of the molecular descriptors; (6) values of polarity (σ_I) and steric (υ) molecular descriptors for selected nitroxides; and (7) steric effect accounted for by υ takes into account both the hampered approach of the nitroxide by the congested radical center and the rehybridization sp2 → sp3 impeded by the bulkiness of the groups attached to the radical center [26].

To determine the cross-coupling rate constant (k_c), several pathways for the constant determination were suggested [25–28]. The reaction rate of cross-coupling after photo homolysis can be measured by methods of time domain absorption spectroscopy or EPR [25]. Radical clock reactions involve a competition between a unimolecular radical reaction with a known rate constant and a bimolecular radical reaction with an unknown rate constant to produce unrearranged and rearranged products [27]. In the C-centered radical nitroxide trapping (competitive method), scavenging of the C-centered radicals in the presence of two nitroxides, Nα (k_c unknown) and Nβ (k_c known) provides indirect measurement of k_c [28]. Time-Resolved Chemically Induced Dynamic Nuclear Polarization (TR CIDNP) has been proved to be a powerful method for investigation of radical processes [29, 30] In TR CIDNP, the NMR spectra of emission or absorption of reactants and products are detected. The analysis of the spectra yielded information on the intermediate radical pairs and mechanism of the reaction studied. For example, the measurements of the CIDNP intensity of the alkoxyamine in the absence and in the presence of nitroxide and at different time delays between the laser pulse and radio frequency pulse τ_0 provide the observed alkoxyamine formation rate constant k_{obsd}. The dependence of k_{obsd} on the nitroxide concentration gives the value of cross-coupling rate constant k_c.

A key role of the radical cross-coupling in a chemical reaction can be illustrated by an example of reaction of hindered amine light stabilizers (HALSs), which includes the generation of nitroxides from amines [31]. Thermodynamically favored proposed cycle for the nitroxide regeneration and competitive amine formation via H-atom abstraction from an alkoxyamine and subsequent aminyl radical formation are schematically illustrated in Fig. 3.13.

Along with the effects of the electron structure of nitroxides in various solvent, values of the dipole moment (μ), dielectric constant (ε_r), cohesive pressure (c), normalized Reichardt solvent polarity constant (E_T^N), EPR nitrogen hyperfine coupling constant (a_N) (spin density on the NO moiety) can also affect on values of the cross-coupling rate constant k_c. [25]. For example, linear trends were observed between k_c the cohesive pressure c, the normalized Reichardt solvent polarity constant E_T^N, [32] and the nitrogen hyperfine coupling constant a_N.

Effects polarity and sterics on the rate constant of cross-coupling of alkyl radicals were investigated in detail [33]. The analysis of the experimental data led to the following conclusions: (1) congestion around the nitroxide moiety and the bulkiness of the alkyl radical destabilize the transition state (TS) and cause the decrease in k_c,

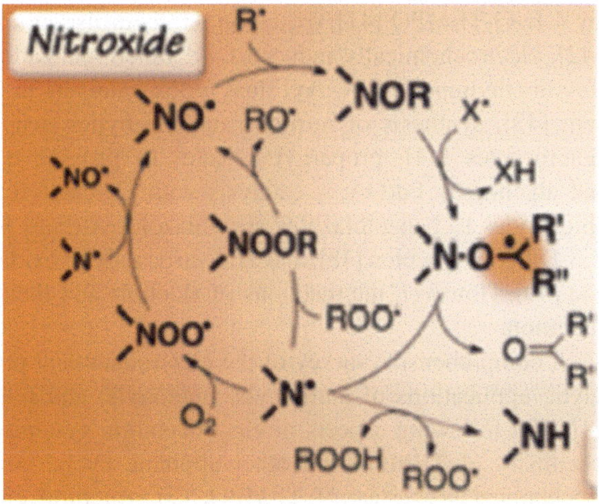

Fig. 3.13 HALS mechanism [31]. Reprinted from [31], Copyright 2012 American Chemical Society [31]

(2) the increase in the polarity of the substituents on the nitroxide moiety destabilizes the reactant state, increasing k_c, and (3) the steric effect is better ascribed to an entropy effect, which is in turn affected by the stabilization/polarity effect.

3.5 Nitroxides in Electrocatalysis

N-Oxyl compounds represent a diverse group of reagents that find widespread use as electro catalysts for the selective oxidation of organic molecules in both laboratory and industrial applications [34–62].

In pioneering research Semmelhack and co-workers, it was found that TEMPO can act as an electrocatalyst for alcohol electrooxidation [34]. The reaction between electrogenerated $TEMPO^+$ with primary alcohols in the presence 2,6-lutidine, as a base, was fast, even at −60 °C, and up to 40 catalytic turnovers were achieved. Secondary alcohols were oxidized at significantly slower rates. On the basis of deuterium kinetic isotope effects and Hammet relationship, four potential routes TEMPO-catalyzed alcohol oxidation were considered: (1) direct hydride abstraction from the alcohol by the oxoammonium to afford hydroxylamine and an oxocarbenium ion that undergoes subsequent deprotonation, (2) electron/proton transfer from alcohol to oxoammonium, generating TEMPO, H^+, and an alkoxyl radical, (3) nucleophilic attack of an alkoxide at the nitrogen atom of $TEMPO^+$ to generate a reactive N–O adduct, and (4) nucleophilic attack of the alkoxide at the oxygen atom of $TEMPO^+$ to generate a reactive O–O adduct.

In the past decades, numerous works on electrocatalysis with nitroxides as catalysts have been carried out. Among them: electrooxidation of menthol [35], [6β-Methyl-3β,5α-dihydroxy-16α,17α-cyclohexanopregnan-20-one [36], carbohydrates and carboxylated cellulose [37], racemic SEC-benzylic alcohols [38] lignin β-O-4 model compound [39], oxidation of primary and secondary amines to give the corresponding nitrile or carbonyl [40], oxidation of tertiary amines catalyzed by 4-BzO-TEMPO [41], oxidation tetrahydroisoquinolines and related compounds [42], electrochemically induced C–H functionalization sing bromide ion/ 2,2,6,6-Tetramethylpiperinyl-N-oxyl dual redox catalysts in a two-phase electrolytic system [43], synthesis of nitriles from aldehydes dehydrogenation of saturated N-heterocycles [44], propargyl acetates to provide α,α-dihaloketone [45], terminal alkenes by $Pd(OAc)_2$ catalysis with Co-catalytic TEMPO [46], oxidation of thioamides to benzothiazoles and thiazolopyridines [47], oxidation of glycerol to 1,3-Dihydroxyaceton [48], and 5) electrocatalytic oxidation of L-tyrosine by a nitroxide [49]. However, the reactions of alcohols and their derivatives attract particular attention.

A comprehensive survey of the electrochemical properties and wide electrocatalytic applications of aminoxyls, imidoxyls, and related reagents were reviewed [4]. The following aspects of electrocatalitic systems were described: (1) electrochemistry of TEMPO and other aliphatic cyclic aminoxyl derivatives; (2) electrochemical properties of TEMPO; (3) structural effects on the electrochemical properties of cyclic aminoxyl radicals; (4) electrochemical reactions mediated by TEMPO and related cyclic aminoxyl radicals; (5) alcohol oxidation mechanism of aminoxyl-mediated process; (6) performance of aminoxyl derivatives, (7) cooperative copper/aminoxyl catalysis; (8) oxidation of synthetic alcohol; (9) amine oxidation; (10) C–H functionalization; (11) palladium co-catalyzed reactions; (12) alkene aminooxygenation; (13) alkyne hydroxybromination; (14) oxidation of sulfur-containing compounds; (15) electrochemistry and electrochemical applications of imidoxyl derivatives including electrochemical properties of N-Hydroxyphthalimide in alcohol oxidation and other NHPI-mediated electrooxidation reactions, and (16) electrochemistry of other N-oxyl radicals.

Numerous quantitative experimental data were tabulated: (1) experimental aminoxyl/oxoammonium redox potentials, (2) redox leveling in the 1 e^{-}/H^{+} hydroxylamine/aminoxyl redox couple, (3) effect of ring structure on the calculated e^{-} potentials for oxidation of cyclic aminoxyl radicals, (4) experimental N-hydroxyimide/imidoxyl redox potentials versus NHE (5) redox potential of the aminoxyl to oxoammonium redox process for various N-oxyl radicals potentials, (6) redox potentials (V vs. NHE) of N-aryl hydroxylamines, hydroxyamic acids, and other acyclic N-hydroxy derivatives as determined by CV or differential pulse voltammetry, (7) redox potential of heterocyclic N-hydroxy derivatives, (8) second-order rate constants of the reaction of phthalimide N-oxyl (PINO) with alcohol substrates, and (9) second-order rate constants of the reaction of PINO with substrates containing activated C–H bonds.

Figure 3.14 demonstrates an example of electrochemical processes involving TEMPO and phthalimide N-oxyl (PINO).

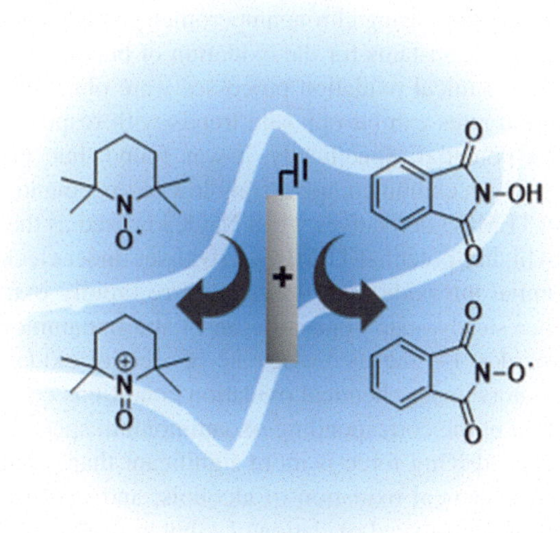

Fig. 3.14 Example of nitroxide redox reactions [4]. Reprinted from [4], Copyright 2012 American Chemical Society

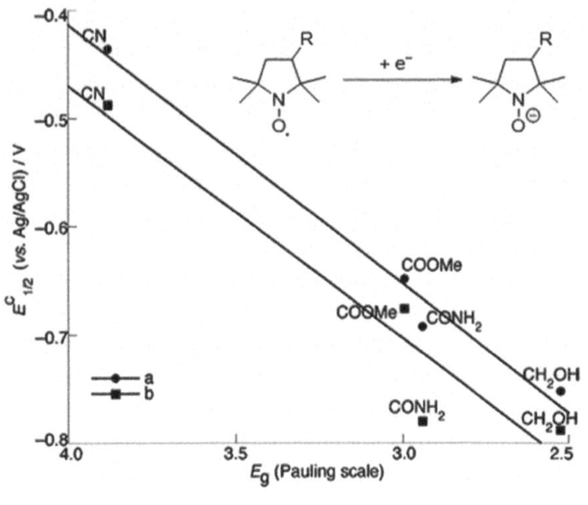

Fig. 3.15 Correlation between the Pauling group electronegativity (E_g) of substituents versus redox potentials ($E_{1/2}$) for the five-membered ring NUTIG hydroxylamine/aminoxyl radicals in **a** MeOH (0.1 M Bu_4NClO_4) and **b** phosphate buffer (0.1 M, pH 7.4) [4]. Reprinted from [4], Copyright 2012 American Chemical Society

A linear relationship between the Pauling group electronegativity of the substituent and the hydroxylamine/aminoxyl redox potential of the corresponding PROXYL derivative is shown in Fig. 3.15.

The voltammetric technique was used to determine the catalytic activity for TEMPO-mediated electrooxidation of a series of primary and secondary alcohols, for example, benzyl alcohol [50]. A mechanism for electrocatalytic oxidation of benzyl alcohol by TEMPO was proposed. Alcohol electrooxidation in CH_3CN with TEMPO as electron–proton transfer mediators and N-methylimidazole as an added base was carried out [51]. A series of TEMPO analogs and bi- and polycyclic nitroxides with a range of redox potentials for the oxidation of alcohols under basic conditions was examined using chronoamperometry and CV techniques [52]. In addition, catalytic rate constants for the oxidation of butanol in both chemical (by NaOCl) and electrochemical oxidation processes were obtained. The experiments showed that reaction rates exhibit opposite trends with respect to the nitroxide/oxoammonium redox potential. Specifically, it was found that: (1) for the less hindered nitroxides, which exhibit comparatively low redox potentials, the reaction run much faster than TEMPO derivatives when NaOCl is used as the oxidant; (2) the rates are faster with higher-potential TEMPO derivatives under electrochemical conditions; (3) low-potential nitroxide mediators are more rapidly oxidized by NaOCl and exhibited higher steady-state concentration of the oxoammonium relative to high-potential nitroxides; (4) the increased rate of alcohol oxidation by low-potential aminoxyl mediators under chemical oxidation conditions correlates with the more rapid generation of the corresponding oxoammonium species under these conditions; (5) the reaction driving force is more significant than steric effects in nitroxide-mediated electrochemical oxidation of alcohols; and (6) linear free energy correlations for aminoxyl-catalyzed oxidation of 1-butanol takes place (Fig. 3.16).

Immobilization of nitroxyl mediators on electrode surfaces significantly enhanced rates of alcohols oxidation and facilitates isolation of products as compare with

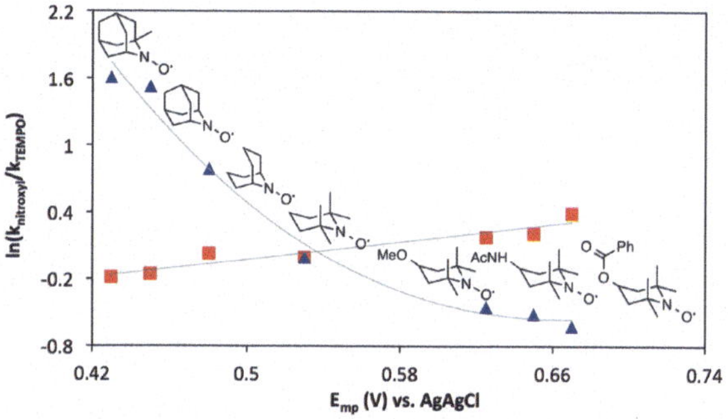

Fig. 3.16 Linear free energy correlations for aminoxyl-catalyzed oxidation of 1-butanol with NaOCl as a chemical oxidant (blue solid triangles) and under electrochemical conditions (red solid squares) [52]. Reprinted from [52], Copyright 2015 American Chemical Society

methods that use dissolved nitroxide mediators [53]. For example, pyrene-tethered TEMPO derivative was noncovalently immobilized at electrodes coated with multi-walled carbon nanotubes [54]. In preparative electrolysis experiments with a series of alcohol substrates and the immobilized catalyst, turnover numbers and frequencies approached 2000 and 4000 h^{-1}, respectively. The reaction conversation was almost 100% and less 5% for the immobilized and free nitroxides, respectively. The oxidation of a sterically hindered hydroxymethylpyrimidine precursor to the blockbuster drug, Rosuvastatin, was also demonstrated.

In paper [55], it was shown that N-Hydroxyphthalimide (NHPI)

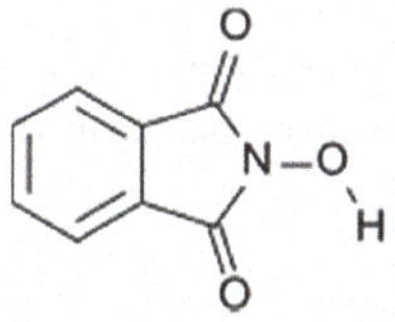

which generates phthalimide *N*-oxyl (PINO) radical is an effective mediator for the oxidation of alcohols by electrolysis. NHPI undergoes one-electron oxidation to form PINO. In CH_3CN, the e^-/H^+ NHPI/PINO couple exhibits a redox potential of 1.44 V versus SCE which is quasi-reversible. It was also found that electrogenerated PINO catalytically oxidates benzyl alcohols [56]. The NHPI/PINO couple was applied to electrocatalytic oxygenation of various compounds including lignin models [57], native lignin [58], benzylic and allylic bonds [59], and allylic C–H bonds oxidation [60]. A mechanism of NHPI-mediated electrooxidation of alcohols was suggested.

The first use of *N*-hydroxyphthalimide as an electrochemical mediator for C–H oxidation to non-oxygenated products provided the basis for direct (in situ) or sequential benzylation of diverse nucleophiles using methylarenes as the alkylating agent [61]. The proposed hydrogen-atom transfer mechanism for C–H iodination allows C–H oxidation to proceed with minimal dependence on the substrate electronic properties and at electrode potentials 0.5–1.2 V lower than that of direct electrochemical C–H oxidation.

The use of NHPI as an electrochemical mediator for iodination of methylarenes was reported [62]. It was found that electrochemical generation of PINO is followed

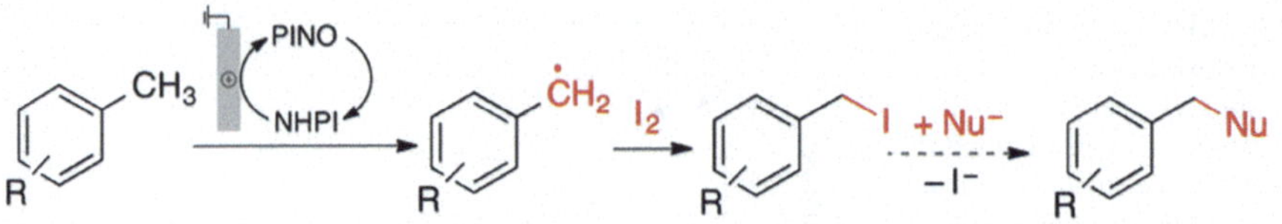

Fig. 3.17 **a** Electrochemical NHPI/PINO-mediated iodination/functionalization of methyl arenes, **b** iodination, and **c** in situ methylarene iodination/alkylation of pyridine [62]. Reprinted from [62], Copyright 2018 American Chemical Society

by H-atom abstraction, and the resulting radical is trapped rapidly by I_2 (Fig. 3.17). Values of experimental N-hydroxyimide/imidoxyl redox potentials versus NHEa of the aminoxyl to oxoammonium redox process for various N-oxyl radicals, N-aryl hydroxylamines, hydroxyamic acids, and other acyclic N-hydroxy derivatives, as determined by CV or differential pulse voltammetry, were listened in separate tables. In addition, the following schemes were described: (1) NHPI-mediated electrochemical oxidation of alcohol substrates, (2) NHPI-mediated electrooxidation of lignin models, (3) electrooxidation and photochemical cleavage of native lignin, benzylic and allylic oxygenation mediated by NHPI, (4) electrochemical NHPI/PINO-mediated oxygenation of benzylic and allylic bonds, (5) comparison of the benzylic oxygenation of heteroaromatic species by NHPI-mediated aerobic and electrochemical method, (6) scope of the oxygenation of allylic C–H bonds mediated by Cl4–NHPI257, (7) NHPI-mediated oxidation of aldehyde acetals, and (8) electrochemical NHPI/PINO-mediated iodination/functionalization of methyl arenes. The electrochemical mediation of iodination reaction was illustrated in Fig. 3.17.

3.6 Nitroxides as Radical Scavenger

Oxidative/nitrosative stress contributes to the etiology of various disorders and associates with the destruction of key macromolecules and inactivation of antioxidant enzymes by reactive oxygen species (ROS) or reactive nitrogen species (RNS). The excess of reactive radicals generated in biological organisms can lead to the damage of vital cellular components such as lipids, proteins, and nucleic acids. These radicals can produce secondary reactive species and then induce a series of cellular response or severely endanger cell health and viability, ultimately leading to irreparable cell damage. Nitroxides have long been ascribed antioxidant activity to underly their chemopreventive and antiaging properties (Chap. 10). The reaction of nitroxides with ascorbate and glutathione (GSH) in solution and biological systems have great importance and been thoroughly studied [63–69]. For most nitroxides, the reaction follows the second-order kinetics law with the only product of nitroxide reduction being hydroxylamine. Nevertheless, when the rate constant of this reaction is low, the reversibility of the first steps may contribute to the kinetics. It is obvious that the biological activity of nitroxides is linked to their electron transfer rates and redox potentials.

Reactivity of nitroxides to quench superoxide radicals as superoxide dismutase (SOD) mimics (Scheme 3.2) was first established by Samuni, Krishna, Goldstein, and co-workers [63].

In the case of oxazolidines (e.g., 2-ethyl-2,5,5-trimethyl-3-oxazolidin-1-yloxyl, OXANO) Scheme 3.2 and its corresponding hydroxylamine, they react with superoxide [64]. The sequence of the SOD mimics reactions for TEMPO is indicated in Scheme 3.3. 2-ethyl-2,5,5-trimethyl-3-oxazolidin-1-yloxyl, OXANO). Within the framework of this mechanism, a nitroxide undergoes repeated reductions and oxidations allowing in catalytic fashion the dismutation of superoxide to oxygen and hydrogen peroxide. Piperidine derivatives such as TEMPO undergo oxidation and reduction reactions to yield the corresponding oxoammonium cation, which then oxidizes superoxide to molecular oxygen. However, this mechanism is not likely physiologically relevant in all cases in cell with cells higher level of SOD enzymes [65]. Employing pulse radiolysis, Goldstein and Samuni demonstrated that nitroxides react readily with peroxyl radicals [66]. Nitroxides can undergo acid-promoted reactions with peroxyl radicals and then be recycled from the resultant oxoammonium ion by reduction with a substrate-derived radical eventually acting as a catalytic radical-trapping antioxidants.

The kinetics and mechanisms of the reactions of TEMPO, as well as an *N*-arylnitroxide and an *N,N*-diarylnitroxide, with alkylperoxyl radicals, the propagating species in lipid peroxidation, were characterized [67]. Inhibited autoxidations of THF in aqueous buffers revealed that nitroxides reduce peroxyl radicals by electron transfer with rate constants ($k \approx 10^6$ to $> 10^7$ M^{-1} s^{-1}) that correlate with the standard potentials of the nitroxides ($E° \approx 0.75$–0.95 V vs. NHE) and that this

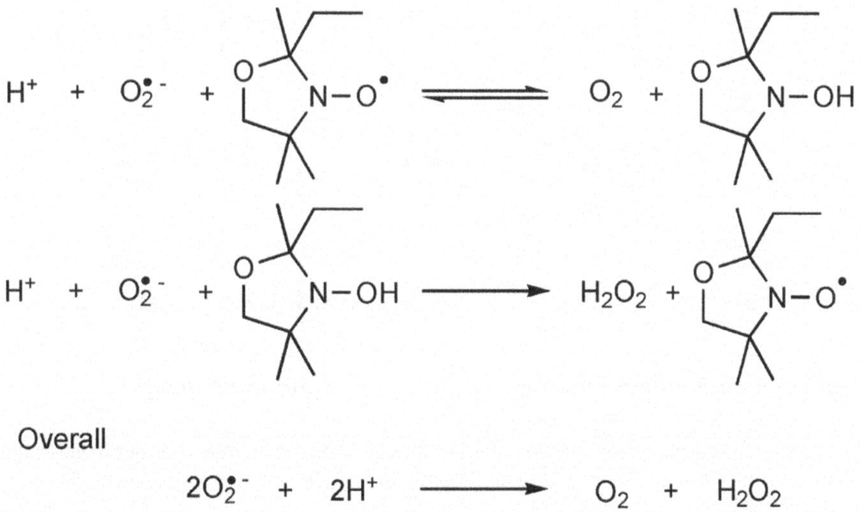

Scheme 3.2 SOD-mimicking behavior of OXANO [2]. Reprinted from [2], Copyright 2007 American Chemical Society

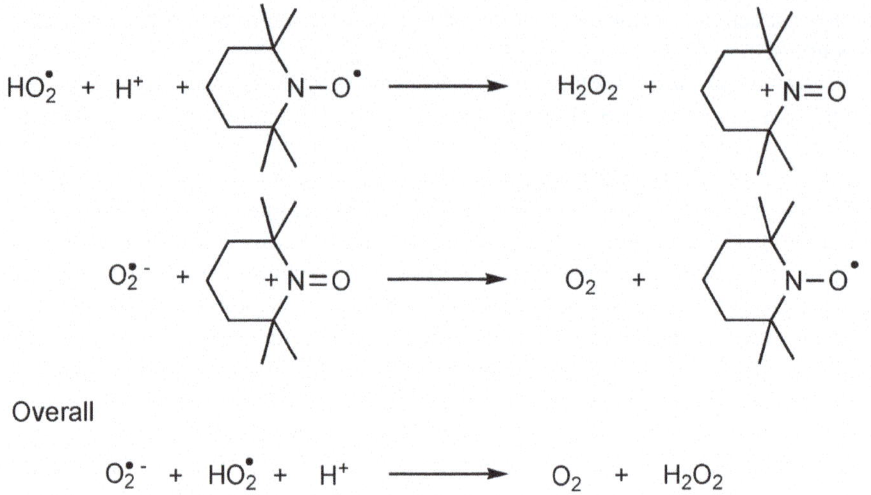

Scheme 3.3 SOD-mimicking behavior of TEMPO [2]. Reprinted from [2], Copyright 2007 American Chemical Society

activity is catalytic regarding nitroxide. Regeneration of the nitroxide was suggested to occur by a two-step process involving hydride transfer from the substrate to the nitroxide-derived oxoammonium ion followed by H-atom transfer from the resultant hydroxylamine to a peroxyl radical. In the presence of a hydride donor (NADPH for example) (Fig. 3.18), TEMPO+ is converted to TEMPOH which can trap another peroxyl radical via H-atom transfer, thereby regenerating TEMPO. Figure 3.19 indicates calculated transition state structure for the reaction of TEMPO$^+$ with THF.

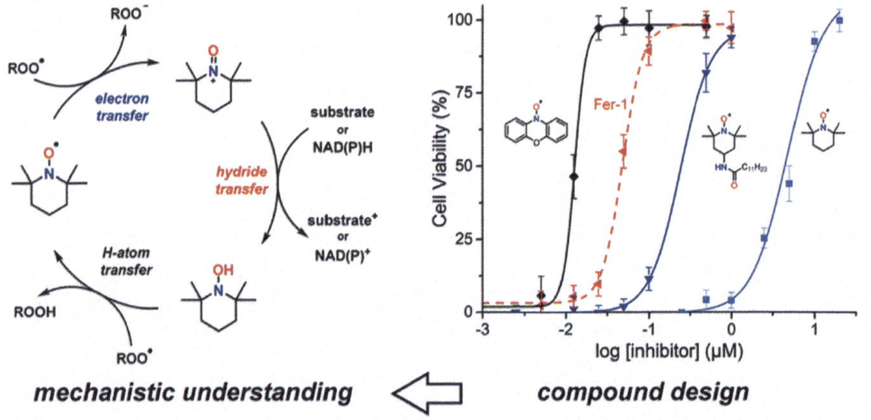

Fig. 3.18 Mechanisms of the reactions of TEMPO, *N*-arylnitroxide *N,N*-diarylnitroxide, with alkylperoxyl radicals and its effect on cell viability [67]. Reprinted from [67], Copyright 2018 American Chemical Society

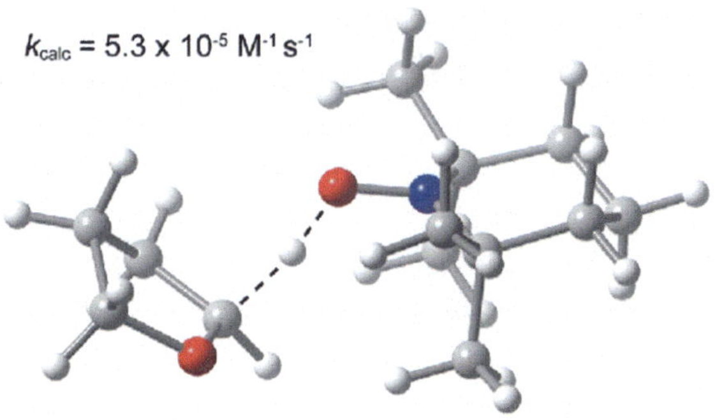

Fig. 3.19 CBS-QB3—calculated transition state structure for the reaction of TEMPO+ with THF [67]. Reprinted from [67], Copyright 2018 American Chemical Society

It was determined that TEMPO can be regenerated from its oxoammonium ion by reaction with alkyl radicals, and this reaction was proposed to be a key step in TEMPO-catalyzed synthetic transformations. Because this process occurs with $k \sim 1\text{–}3 \times 10^{10}$ M^{-1} s^{-1}, it enables to compete with O_2 for alkyl radicals [68]. The addition of weak acids facilitates this reaction, whereas the addition of strong acids slows it by enabling back electron transfer. Another mechanism invoked to account for the biological activity of nitroxides involves the scavenging of alkyl radicals [69]. This reaction is too slow ($k \approx 1 - 3 \times 10^8$ M^{-1} s^{-1}) to compete with the formation of peroxyl radicals from the combination of alkyl radicals and O_2 ($k \approx 3 \times 10^9$ M^{-1} s^{-1}). Nevertheless, a competition can be more unlikely because O_2 is generally present in much higher concentration than nitroxide (mM vs. μM to nM).

Autoxidations of unsaturated hydrocarbons are efficiently inhibited by nitroxides and/or compounds which react readily to produce nitroxides under autoxidative conditions. For example, diarylamines and hindered aliphatic amines comprise two of the three key types of radical-trapping antioxidants (RTAs) to preserve hydrocarbon-based materials (Fig. 3.1a) [64]. Diarylamines can trap peroxyl radicals at ambient temperatures by a catalytic mechanism involving the formation of diarylnitroxide intermediates (the "Korcek Cycle") (Fig. 3. 20) [70]. At elevated temperatures, nitroxides trap radicals in still catalytic manner (the "Denisov Cycle") (Fig. 3.20) [71]. Hindered aliphatic reaction nitroxides do not react with alkylperoxyl radicals, while aryl nitroxides react by addition to the aryl rings [64].

Thermodynamically favored proposed cycle for nitroxide regeneration and competitive amine formation: via H-atom abstraction from an alkoxyamine and subsequent aminyl radical formation is summarized in Fig. 3.21 [31]. Alternative literature mechanisms of nitroxide regeneration were also discussed.

Radical scavenging activity of aromatic and aliphatic mono- and bis-nitroxides was investigated in acetonitrile in the presence of alkyl and peroxyl radicals generated from thermal decomposition of 2,20-azobis(2,4-di-methylvaleronitrile) using EPR and UV–Vis spectroscopy [72]. Antioxidant activity of the nitroxides was evaluated by monitoring conjugated dienes formation during methyl linoleate micelles peroxidation and by measuring carbonyl content in oxidized bovine serum albumin. Suggested scheme showing the scavenging of alkyl (R•) and peroxyl radicals

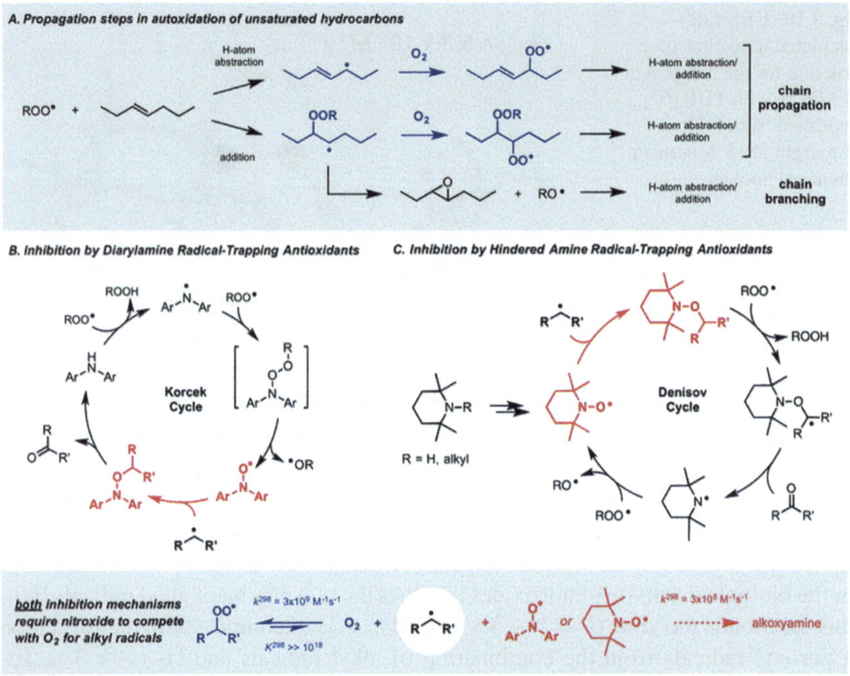

Fig. 3.20 **a** Key propagation steps in the autoxidation of unsaturated hydrocarbons. **b** and **c** The Korcek and Denisov mechanisms believed responsible for the catalytic inhibition of hydrocarbon autoxidation by diarylamine and hindered amine antioxidants [64]. Reprinted from [64], Copyright 2018 RSC

(ROO•) by aromatic and aliphatic nitroxides was proposed. According to the scheme, (1) indolinonic aromatic nitroxides react with peroxyl radical, (2) aliphatic nitroxides react, (3) bis-nitroxides are roughly twice more efficient at inhibiting lipid peroxidation compared to their corresponding mono-derivatives, and (4) aromatic nitroxides are more effective antioxidants than aliphatic ones.

3.7 Nitroxide Reaction with Typical Antioxidants

Within recent two decades, nitroxides are widely used for an analysis antioxidants in biological systems [73–88]. First application of nitroxide for quantitative analysis ascorbic acid in biological liquids was performed in Likhtenshtein group using dual fluorophore–nitroxide probes [77].

It was revealed [73] that tetracarboxylate pyrroline nitroxides are reduced by ascorbate/glutathione (GSH) with second-order rate constants that five orders of magnitude greater than those for gem-diethyl pyrroline nitroxide.

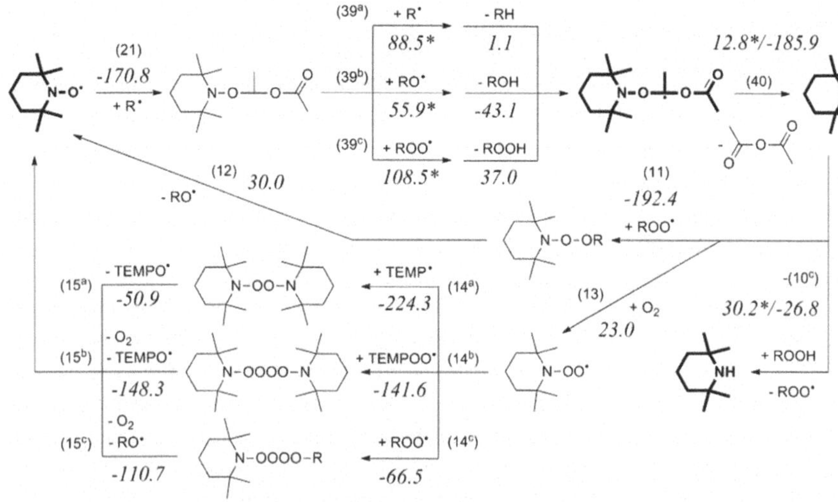

Fig. 3.21 Proposed cycle for nitroxide regeneration and competitive amine formation via H-atom abstraction from an alkoxyamine and subsequent aminyl radical formation. Numbers in italics are calculated as Gibbs free energies (kJ mol^{-1}, gas phase, 25 °C) of the reactions and activation (denoted by an asterisk) [31]. Reprinted from [31], Copyright 2018 American Chemical Society

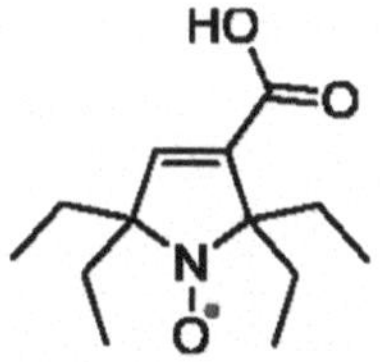

For these nitroxides, the electrochemical reduction potential was found to be less negative (by about 0.8 V), compared with the corresponding gem-diethyl nitroxides, while the oxidation potentials become more positive (by about 0.7 V). Electrochemical potentials as well as rates of reduction with ascorbate/GSH correlate via simple regressions with field/inductive parameters such as Swain/Lupton F-parameters (and/or Charton σ_I-parameters). Rates of reduction with ascorbate/GSH similarly correlate with four pyrroline nitroxides. Exception is the slowest reducing gem-diethyl nitroxide. These results suggest that the electron-withdrawing groups adjacent to the nitroxide moiety have a strong accelerating impact on the reduction rates.

The effects of substitution on nitroxides stability using five different classes of radicals, specifically, piperidine-, imidazolidine-, pyrrolidine-, isoindoline-based nitroxides, and as well as the Finland trityl radical were evaluated [74]. The rate of nitroxide reduction in the presence of ascorbate, cellular extracts, and after injection into oocytes cytoplasmic extracts from *X. laevis* oocytes was measured by continuous-wave EPR spectroscopy. Main results of the study are as follows: (1) Introduction of bulky ethyl groups next to a nitroxide group leads to significant stabilization against reduction by both ascorbic acid and the reductants present in living cells, (2) these radicals are more stable in cells than trityl radical, (3) superior stability against reduction is due to combination of sterical shielding, ring size, and charge, (4) charged and neutral radicals showed different relative stabilities in cell extracts than in cells, and (5) the trityl radical exhibited considerable stability toward reduction, especially in cells. Thermodynamic parameters of nitroxide reduction determined by cyclic voltammetry were tabulated.

With a goal to improve stability of nitroxides to reduction by ascorbic acid, a series of 2,6-substituted nitroxyl radicals was prepared [75]. It was found that tetraethyl-substituted imidazolidine nitroxide is the most stable nitroxide in respect to reduction in ascorbate solutions ($k_1 = 0.02\ M^{-1}\ s^{-1}$). The role of substituents on the stability of new synthesized several nitroxides with different substituents which vary the steric and electronic environment around the N–O moiety the radicals was investigated [76]. Results indicated that: (1) the nitroxide reactivity toward ascorbate correlates with the redox potential of the derivatives, (2) the electronic factors largely determine the radicals' stability, and (3) ab initio calculations confirmed a correlation between the reduction rate and the computed singly occupied molecular orbital–lowest unoccupied molecular orbital energy gap.

To increase the reactivity and selectivity toward the detection of ascorbic acid, the dual fluorophore–nitroxide probe 15-((9-(Ethylimino)-10-methyl-9*H*benzo[*a*]phenoxazin-5-yl)amino)-3,11-dioxa-7-azadispiro hexadecan-7-yloxyl, (Nile-DiPy)

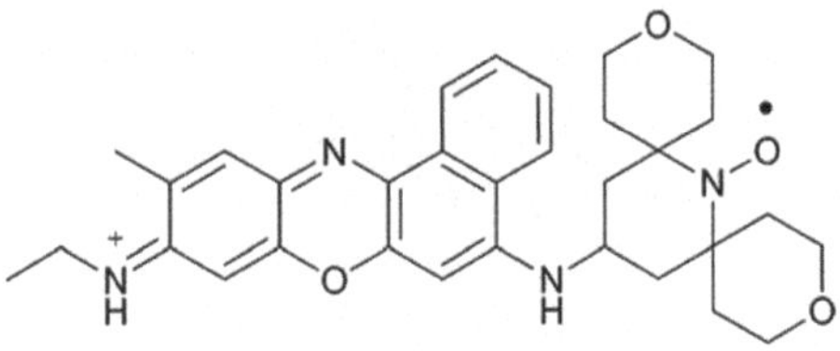

Nile-DiPy

was synthesized and characterized [78]. This fluorophore–nitroxide probe rapidly reacted with ascorbic acid and showed in parallel fluorescence enhancement in PBS at pH 7.4, containing 5% (v/v) DMSO, while other biological reductants, including uric acid, glutathione, NADH, catechin, 2,2,5,7,8-pentamethyl-6-chromanol did not react with the nitroxide in these conditions. In the presence of ascorbic acid, the fluorescence intensity of Nile-DiPy increased in a dose-dependent manner concentrations of ascorbic acid (0.13–8.0 mM). The second-order rate constant for the reaction of Nile-DiPy and of Nile-TEMPO with ascorbic acid was calculated as 246 $M^{-1}\ s^{-1}$ and (17.4 $M^{-1}\ s^{-1}$), respectively. The kinetic isotope effect (KIE) for the detection of ascorbic acid was determined to be 9.77, indicating that Nile-DiPy reacts with ascorbic acid via hydrogen atom transfer. The limit of detection of this fluorometric method was estimated to be 9.72 nM. The usefulness of Naph-DiPy nitroxide for the measurement of ascorbic acid in the plasma of osteogenic disorder Shionogi rats of healthy and streptozotocin-induced diabetic animals was demonstrated.

Mechanistic studies of the reduction of TEMPO and other nitroxides in deaerated solutions of ascorbate were performed [79]. In the kinetics studies, the peak intensity of the low-field component of the triplet EPR spectrum was monitored using an EMX X-band spectrometer. Quantitative kinetics analysis of the nitroxide reduction by ascorbate and hydroxylamine oxidation monitored using an EMX X-band spectrometer was carried out taken into account the following stages:

$$\mathrm{NR} + \mathrm{AscH}^- \underset{k_{-1}}{\overset{k_1}{\rightleftharpoons}} \mathrm{HA} + \mathrm{Asc}^{\cdot -}$$

$$\mathrm{HA} + \mathrm{DHA} \underset{k_{-2}}{\overset{k_2}{\rightleftharpoons}} \mathrm{NR} + \mathrm{Asc}^{\cdot -} + \mathrm{H}^+$$

$$\mathrm{Asc}^{\cdot -} + \mathrm{Asc}^{\cdot -} + \mathrm{H}^+ \underset{k_{-3}}{\overset{k_3}{\rightleftharpoons}} \mathrm{AscH}^- + \mathrm{DHA}$$

$$\mathrm{DHA} \xrightarrow{k_4} \mathrm{DGA}$$

$$\mathrm{NR} + \mathrm{DGA} \xrightarrow{k_5} \mathrm{HA} + \mathrm{OxDGA}$$

where *NR* and *HA* denote nitroxyl radical and its hydroxylamine, respectively; AscH^-, $\mathrm{Asc}^{\cdot -}$, DHA, DGA, *and* OxDGA denote ascorbate anion, ascorbate radical, dehydroascorbic acid, diketogulonic acid, and product of oxidation of diketogulonic acid by NR, respectively. Finally, it was summarized that: (1) The bimolecular rate constants of ascorbate-induced reduction are significantly higher for six-membered ring NR of piperidine types ($k_1 = 3.5\ \mathrm{M}^{-1}\ \mathrm{s}^{-1}$ for TEMPO and $k_1 = 7\ \mathrm{M}^{-1}\ \mathrm{s}^{-1}$ for TEMPOL) than for the five-membered ring NR of pyrrolidine ($k_1 = 0.07$–$0.3\ \mathrm{M}^{-1}\ \mathrm{s}^{-1}$), and imidazolidine ($k_1 = 0.85\ \mathrm{M}^{-1}\mathrm{s}^{-1}$) types; (2) a presence of the double bond at the position 3 in the five-membered ring NR of pyrroline ($k_1 = 0.64$–$1.6\ \mathrm{M}^{-1}\ \mathrm{s}^{-1}$) and imidazoline ($k_1 = 5.6\ \mathrm{M}^{-1}\ \mathrm{s}^{-1}$) types increases their reduction rates by ascorbate, (3) carboxyproxyl NR **1** is resistant against reduction by ascorbate ($k_1 = 0.1\ \mathrm{M}^{-1}\ \mathrm{s}^{-1}$), and (4) tetraethyl-substituted imidazolidine NR 4 is the most stable NR in respect to reduction in ascorbate solutions (e.g., $k_1 = 0.02\ \mathrm{M}^{-1}\mathrm{s}^{-1}$). The equilibrium constants for one-electron reduction of the tetraethyl-substituted NR by ascorbate were found to be in the range from 2.65×10^{-6} to 10^{-5} which is significantly lower than corresponding values for the tetramethyl-substituted NR (more or about 10^{-4}). The reduction of ascorbate radicals by GSH with the rate constant $10\ \mathrm{M}^{-1}\ \mathrm{s}^{-1}$ was examined.

The importance of effect of sterical shielding on the redox properties has been also revealed for various types of nitroxides, including those of isoindoline,4 imidazoline, imidazolidine, piperidine, pyrroline, and pyrrolidine series [80]. Direct measurements of the rate constants of nitroxide of reduction with ascorbate and equilibrium constant K_{eq} in the mixture of the reference isotopically labeled (^{15}N) hydroxylamine ((3-carboxy-1-hydroxy-2,2,5,5-tetramethylpyrrolidine-1-oxyl-1-^{15}N) and the nitroxides of the imidazoline and midazolidine series were carried out. For example, the 2,2,5,5-tetraethyl-substituted pyrrolidine nitroxides demonstrate the highest stability inside cells, exceeding that of trityl radicals. The nitroxide electrochemical

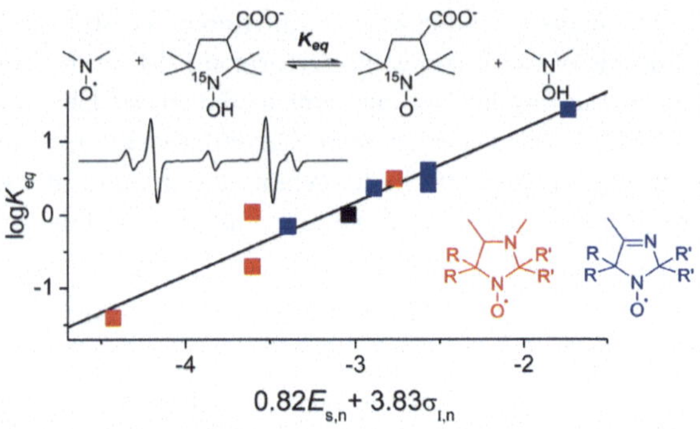

Fig. 3.22 Correlation of $\log(K_{eq})$ versus $f(\text{Es}, \sigma_I)$ [80]. Reprinted from [80], Copyright 2015 American Chemical Society

reduction was also studied. Results showed that increase in the number of bulky alkyl substituents leads to a decrease in the rate of reduction with ascorbate. Oxidant properties of the nitroxides determined by steric and 'electronic effects of the substituents were quantitatively characterized by means of multiple regression using the Fujita steric constant Es and the inductive Hammett constant σ_I (Fig. 3.22).

In order to improve the stability of the radicals through steric and to strength electrostatics shielding, two new pyrrolidine nitroxide radicals, *cis/trans*-2,5-bis(carboxymethyl)-2,5-diethylpyrrolidine 1-oxyl and 2-(carboxymethyl)-2,5,5-triethylpyrrolidine 1-oxyl, were prepared [81]. An ascorbic acid reduction assay proved that the newly synthesized radicals exhibit higher reductive stability than 3-carboxy-PROXYL and 4-carboxy-TEMPO.

Synthesis and antioxidant activity and reactive oxygen species (ROS) scavenging of nitroxides, that is, *N*-[2-(*p*-vinylbenzylidene)benzothiazole] 4-amino-2,2,6,6-tetramethyl-1-piperidinyloxy (3a) and *N*-[2-(*p*-vinylbenzylidene) benzoxazole] 4-amino-2,2,6,6-tetramethyl-1-piperidinyloxy (3b), *N*-[2-(*p*-vinylbenzyl)benzoxazole]4-amino-2,2, 6,6-tetramethyl-1-piperidinyloxy (4a), and *N*-[2-(*p*-vinyl, benzyl)benzoxazole]4-amino-2,2,6,6-tetramethyl-1-piperidinyloxy (4b), were performed [82]. The activities were assayed by the reduction of nitroblue tetrazolium (NBT). The superoxide radical was generated in the *N*-methylphenazonium (PMS) and nicotinamide adenine dinucleotide (reduced form, NADH). The hydroxyl radicals were generated in the reaction of H_2O_2–$FeSO_4$ and colored substance produced by reacting •OH with salicylic acid.. IC50 values of the studied compounds which characterized their effect on ROS scavenging activity were tabulated. The results indicated that the novel compounds exhibit improved antioxidant activity and the superoxide radical scavenging activity which is related to redox property of the compounds.

3.8 Nitroxide Redox Behavior in Biological Systems

The scavenging activities of dual fluorophore–nitronyl compound for superoxide and hydroxyl radicals are described in Section X. Methamphetamine (METH)-induced neurotoxicity is associated with mitochondrial dysfunction and enhanced oxidative stress. The aims of study [83], conducted in the mouse brain repetitively treated with METH, were to examine the redox status using the redox-sensitive imaging probe 3-methoxycarbonyl-2,2,5,5-tetramethylpiperidine-1-oxyl (MCP), and non-invasively visualize the brain redox status with electron paramagnetic resonance imaging. Measurements of the rate of reduction of MCP in a mouse head measured from a series of temporal EPR images were used to construct a two-dimensional map of rate constant. The obtained redox map illustrated the change in redox balance in the METH-treated mouse brain as result of oxidative damage. Enhancement of the reduction reaction of MCP resulted from enzymatic reduction in the mitochondrial respiratory chain and blood-brain barrier (BBB) dysfunction after treatment with METH for 7 days was also revealed.

In vivo physiological ligand citrate can bind iron(II) ions to form the iron(II) citrate complex. EPR and fluorescence spectroscopies were employed to study that TEMPO inhibited •OH production from the Fenton-like reaction of iron(II) citrate with H_2O_2 by up to 90% [84]. The •OH production from the Fenton-like reaction was monitored using the fluorescence probe APF. Reaction of spin-trap DMPO with the generated •hydroxyl was studied employing EPR spectrometer. The spectrophotometrical experiments indicated that this inhibition was due to oxidation of the iron(II) citrate by TEMPO with a stoichiometry of Tempo:Iron(III) citrate 1.1:1.0.

The TEMPO-conjugated gold nanoparticles (Au NPs, Au-PEG-TEMPO NPs) (Fig. 3.23) were used for the culture of hMSCs to investigate their effect on ROS scavenging, proliferation, and osteogenic and adipogenic differentiation of hMSCs [85]. Au NPs with an average size of 40 nm were conjugated with 2,2,6,6-tetramethylpiperidine *N*-oxyl (TEMPO) to endow them with ROS scavenging capacity while holding the beneficial biological effect of Au NPs. The work results indicated that the TEMPO-conjugated Au NPs had high scavenging capacity for overproduced ROS and maintained the promotive effect of Au NPs on osteogenic differentiation of hMSCs without the inhibitory effect of free TEMPO.

A nitroxide radical-containing polymer (NRP), which is composed of poly(4-methylstyrene) possessing nitroxide radicals as a side chain via amine linkage, was coated onto cigarette filters, and its ROS scavenging activity from streaming cigarette smoke was evaluated [86]. The intensity of electron spin resonance signals of the NRP in the filter decreased after exposure to cigarette smoke. It was also demonstrated that the extract of cigarette smoke passed through the NRP-coated filter has a lower cellular toxicity than smoke passed through poly[4 (cyclohexylamino)methylstyrene]- and poly(acrylic acid)-coated filters. Schematic illustrations of the NRP and smoking device were presented. NRP was suggested as a promising material for ROS scavenging from cigarette smoke.

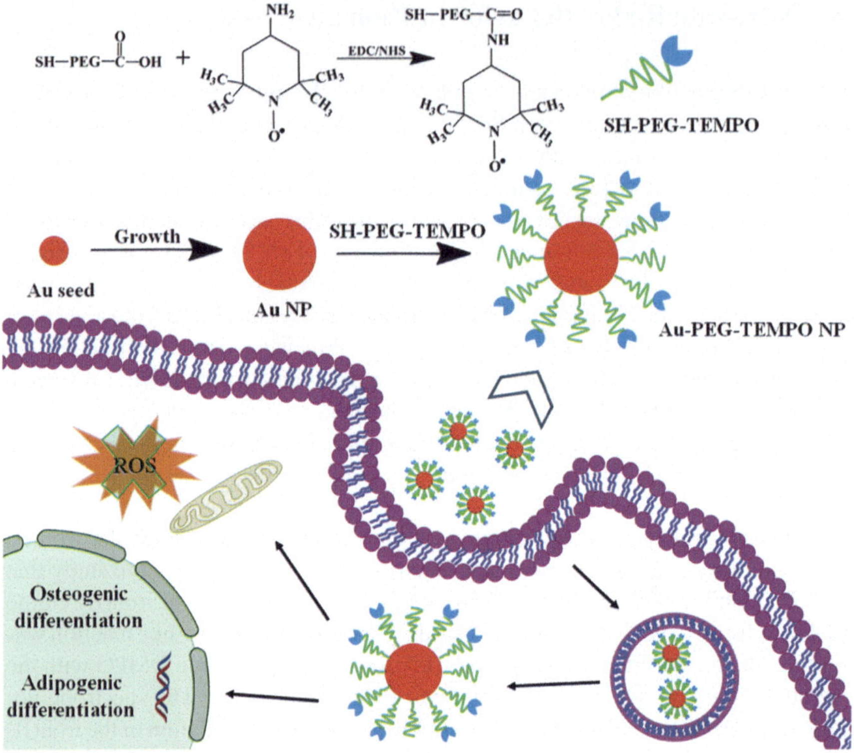

Fig. 3.23 TEMPO-conjugated Au NPs (Au-PEG-TEMPO NPs) [85]. Reprinted from [85], Copyright 2018 American Chemical Society

Pyrrolidine nitroxides, 3,4-bis-(acetoxymethoxycarbonyl)-proxyl (DCP-AM2) and 3-(2-(bis(2-(acetoxymethoxy)-2-oxoethyl)amino)acetamido)-proxyl (DCAP-AM2) which can undergo hydrolysis by cellular esterases to hydrophilic carboxylate derivatives were studied in human aortic endothelial cells (HAEC) using EPR method [87]. EPR measurements in the isolated mitochondrial fraction indicated that mitochondria is the main place where DCP was finally accumulated. TEMPO derivatives showed much faster decay of EPR signal in the cellular fraction, compared to pyrrolidine nitroxides. It was found that (1) supplementation of endothelial cells with 50 nM of DCP-AM2 completely normalized the mitochondrial superoxide level, (2) administration of DCP-AM2 to mice (1.4 mg/kg/day) resulted in substantial nitroxide accumulation in the tissues and significantly reduced hypertension, (3) hydroxylamine derivatives of dicarboxyproxyl nitroxide DCP-AM-H can be used for the detection of superoxide in vivo in angiotensin II model of hypertension, and (4) infusion of DCP-AM-H in mice leads to accumulation of persistent EPR signal of nitroxide in the blood. Thus, the data demonstrate that acetoxymethoxycarbonyl group containing nitroxides accumulate in mitochondria and indicated site-specific antioxidant activity.

The therapeutic and research studies of nitroxide compounds, biologically relevant effects of nitroxides, including their ability to degrade **superoxide** and peroxide, inhibit Fenton reactions, and undergo radical–radical recombination, were reviewed [88].

References

1. M.B. Neiman, É.G. Rozantzev, Y.G. Mamedova, Free radical reactions involving no unpaired electrons. Nature **196**, 472 (1962)
2. J.L. Hodgson, M. Namazian, S.E. Bottle, M.L. Coote, One-electron oxidation and reduction potentials of nitroxide antioxidants: a theoretical study. J. Phys. Chem. A **111**, 13595–13605 (2007)
3. V.A. Golubev, E.G. Rozantsev, M.B Neiman, Some reactions of free iminoxyl radicals with the participation of the unpaired electron. Bull. Acad. Sci. USSR Div. Chem. Sci. **14**, 1898–1904 (1965)
4. J.E. Nutting, M. Rafiee, S.S. Stahl, Tetramethylpiperidine *N*-Oxyl (TEMPO), Phthalimide *N*-Oxyl (PINO), and related *N*-Oxyl species: electrochemical properties and their use in electrocatalytic reactions. Chem. Rev. **118**, 4834–4885 (2018)
5. T. Yamasaki, Y. Matsuoka, F. Mito, M. Yamato, K. Yamada, Redox potential of nitroxides is an index to evaluate superoxide dismutase mimic activity. Asian J. Org. Chem. **2**(5), 388–391. https://doi.org/10.1002/ajoc.201300011
6. Y.G. Budnikova, T.V. Gryaznova, M.K. Kadirov, E.V. Tret'yakov, K.V. Kholin, V.I. Ovcharenko, R.Z. Sagdeev, O.G. Sinyashin, Electrochemistry of nitronyl and imino nitroxides. Russ. J. Phys. Chem. A **83**(11), 1976–1980 (2009)
7. L. Marx, B. Schöllhorn, Intramolecular charge effects in the electrochemical oxidation of aminoxyl radicals. New J. Chem. **30**, 430–434 (2006)
8. J.P. Blinco, J.L. Hodgson, B.J. Morrow, J.R. Walker, G.D. Will, M.L. Coote, S.E. Bottle, Experimental and theoretical studies of the redox potentials of cyclic nitroxide. J. Org. Chem. **73**(17), 6763–6771 (2008)
9. V.D. Sen', I.V. Tikhonovb, L.I. Borodinb, E.M. Plissb, V.A. Golubev, M.A. Syroeshkinc, A. Rusakov, Kinetics and thermodynamics of reversible disproportionation–comproportionation in redox triad oxoammonium cations–nitroxyl radicals–hydroxylamines. J. Phys. Org. Chem. **28**, 17–24 (2015)
10. S. Manda, I. Nakanishi, K. Ohkubo, H. Yakumaru, K.-I. Matsumoto, T. Ozawa, N. Ikota, S. Fukuzumi, K. Anzai, Nitroxyl radicals: electrochemical redox behavior and structure-activity relationships. Org. Biomol. Chem. **5**, 3951–3955 (2007)
11. I.V. Tikhonov, V.D. Sen, L.I. Borodin, E.M. Pliss, V.A. Golubev, A.I. Rusakov, Kinetics and thermodynamics of reversible disproportionation–comproportionation in redox triad oxoammonium cations–nitroxyl radicals–hydroxylamines. J. Phys. Org. Chem. **27**, 114–120 (2014)
12. J.M. Bobbitt, N.A. Eddy, X. Cady, J. Jin, J.A. Gascon, S. Gelpí-Dominguez, J. Zakrzewski, M.D. Morton, Preparation of some homologous TEMPO nitroxides and oxoammonium salts; notes on the NMR spectroscopy of nitroxide free radicals; observed radical nature of oxoammonium salt solutions containing trace amounts of corresponding nitroxides in an equilibrium relationship. J. Org. Chem. **82**(18), 9279–9290 (2017)
13. A. Bobko, V.V. Khramtsov, Redox properties of the nitronyl nitroxide antioxidants studied via their reactions with nitroxyl and ferrocyanide. Free Radical Res. **49**(8), 919–926 (2015)
14. J.A. Bogart, H.B. Lee, M.A. Boreen, M. Jun, E.J. Schelter, Fine-tuning the oxidative ability of persistent radicals: electrochemical and computational studies of substituted 2-pyridylhydroxylamines. J. Org. Chem. **78**, 6344–6349 (2013)

15. J.B. Gerken, Y.Q. Pang, M.B. Lauber, S.S. Stahl, Structural effects on the pH-dependent redox properties of organic nitroxyls: Pourbaix diagrams for TEMPO, ABNO and three TEMPO analogs. J. Org. Chem. **83**(14), 7323–7330 (2018)
16. S. Caron, R.W. Dugger, S.G. Ruggeri, J.A. Ragan, D.H.B. Ripin, Large scale oxidations in the pharmaceutical industry. Chem. Rev. **106**, 2943–2989 (2006)
17. R. Ciriminna, M. Pagliaro, Industrial oxidations with organocatalyst TEMPO and its derivatives. Org. Process Res. Dev. **14**(1), 245–251 (2010)
18. J.M. Bobbitt, C. Brückner, N. Merbouh, Oxoammonium- and nitroxide-catalyzed oxidations of alcohols, in *Organic Reactions* (Wiley: Hoboken, NJ, 2010), pp. 103–424
19. S. Wertz, A. Studer, Nitroxide-catalyzed transition-metal-free aerobic oxidation processes. Green Chem. **15**, 3116–3134 (2013)
20. J.M. Bobbitt, A.L. Bartelson, W.F. Bailey, T.A. Hamlin, C.B. Kelly, Oxoammonium salt oxidations of alcohols in the presence of pyridine bases. J. Org. Chem. **79**(3), 1055–1067 (2014)
21. K.M. Lambert, Z.D. Stempel, S.M. Kiendzior, A.L. Bartelson, W.F. Bailey, Enhancement of the oxidizing power of an oxoammonium salt by electronic modification of a distal group. J. Org. Chem. **82**(21), 11440–11446 (2017)
22. M. Shibuya, S. Nagasawa, Y. Osada, Y. Iwabuchi, Mechanistic insight into aerobic alcohol oxidation using NO_x–nitroxide catalysis based on catalyst structure-activity relationships. J. Org. Chem. **79**(21), 10256–10268 (2014)
23. K.M. Lambert, J.M. Bobbitt, S.A. Eldirany, K.B. Wiberg, W.F. Bailey, Facile oxidation of primary amines to nitriles using an oxoammonium salt. Org. Lett. **16**(24), 6484–6487 (2014)
24. J.B. Gerken, S.S. Stahl, High-potential electrocatalytic O_2 reduction with nitroxyl/NO_x mediators: implications for fuel cells and aerobic oxidation catalysis. ACS Cent. Sci. **1**(5), 234–243 (2015)
25. E.G. Bagryanskaya, S.R.A. Marque, Scavenging of organic C-centered radicals by nitroxides. Chem. Rev. **114**, 5011–5056 (2014)
26. E.G. Bagryanskaya, S.R.A. Marque, Y.P. Tsentalovich, Alkoxyamine re-formation reaction. Effects of the nitroxide fragment: a multiparameter analysis. J. Org. Chem. **77**, 4996–5005 (2012)
27. A.L.J. Beckwith, V.W. Bowry, K.U. Ingold, Kinetics of nitroxide radical trapping. 1. Solvent effects. Am. Chem. Soc. **114**, 4983 (1992)
28. V.W. Bowry, K.U. Ingold, Kinetics of nitroxide radical trapping. 2. Structural effects. Am. Chem. Soc. **114**, 4992 (1992)
29. R.J. Kaptein, Simple rules for chemically induced dynamic nuclear polarization. J. Chem. Soc. D 732–733 (1971)
30. L.J. Berliner, E. Bagyanskaya, in *Multifrequency Electron Paramagnetic Resonance*, ed. by S.K. Misra (Wiley-VCH, Weinheim, Germany, 2011), p. 947
31. G. Gryn'ova, K.U. Ingold, M.L. Coote, New insights into the mechanism of amine/nitroxide cycling during the hindered amine light stabilizer inhibited oxidative degradation of polymers. J. Am. Chem. Soc. **134**(31), 12979 (2012)
32. C. Reichardt, T. Welton, *Solvent and Solvent Effect in Organic Chemistry*, 4th edn. (Wiley-VCH, Weinheim, Germany, 2011)
33. D. Bertin, P.-E. Dufils, I. Durand, D. Gigmes, B. Giovanetti, Y. Guillaneuf, S.R.A. Marque, T. Phan, P. Tordo, Effect of the penultimate unit on the C-ON bond homolysis in SG1-based alkoxyamines. Macromol. Chem. Phys. **209**, 220 (2008)
34. M.F. Semmelhack, C.S. Chou, D.A. Cortes, Nitroxy-mediated electrooxidation of alcohols to aldehydes and ketones. J. Am. Chem. Soc. **105**, 4492–4494 (1983)
35. Y. Demizu, H. Shiigi, T. Oda, Y. Matsumura, O. Onomura, Efficient oxidation of alcohols electrochemically mediated by Azabicyclo-*N*-oxyls. Tetrahedron Lett. **49**, 48–52 (2008)
36. Y.N. Ogibin, I.S. Levina, A.V. Kamernitsky, G.I. Nikishin, A highly efficient, indirect electrooxidation of 6β-Methyl-3β, 5α-dihydroxy-16α,17α-cyclohexanopregnan-20-one to the corresponding 5α-Hydroxy-3,20-dione using a mediatory couple of sodium bromide and substituted 2,2,6,6-Tetramethylpiperidine-N-oxyl (TEMPO). Mendeleev Commun. **5**, 184–185 (1995)

37. P. Parpot, K. Servat, A.P. Bettencourt, H. Huser, TEMPO mediated oxidation of carbohydrates using electrochemical methods. Cellulose **17**, 815–824 (2010)
38. Y. Kashiwagi, F. Kurashima, C. Kikuchi, J.-I. Anzai, T. Osa, J.M. Bobbitt, Enantioselective electrocatalytic oxidation of racemic sec-alcohols using a chiral 1-azaspiro [5.5] undecane-N-oxyl radical. Tetrahedron Lett. **40**, 6469–6472 (1999)
39. Y. Sannami, H. Kamitakahara, T. Takano, TEMPO-mediated electro-oxidation reactions of non-Phenolic β-O-4-type lignin model compounds. Holzforschung **71**, 109–117 (2017)
40. P.-Y. Blanchard, O. Alévêque, T. Breton, E. Levillain, TEMPO mixed SAMs: electrocatalytic efficiency versus surface coverage. Langmuir **28**, 13741–13745 (2012)
41. Y. Kashiwagi, J.-I. Anzai, Selective electrocatalytic oxidation of N-Alkyl-N-ethylanilines to N-alkylformanilides using nitroxyl radical. Chem. Pharm. Bull. **49**, 324–326 (2001)
42. C. Li, C.-C. Zeng, L.-M. Hu, F.-L. Yang, S.J. Yoo, D. Little, Electrochemically induced C-H functionalization using bromide ion/2,2,6,6-tetramethylpiperidinyl-N-oxyl dual redox catalysts in a two-phase electrolytic system. Electrochim. Acta **114**, 560–566 (2013)
43. Q. Chen, C. Fang, Z. Shen, M. Li, Electrochem. Commun. **64**, 51–55 (2016)
44. Y. Wu, H. Yi, A. Lei, Electrochemical acceptorless dehydrogenation of N-heterocycles utilizing TEMPO as an organo-electrocatalyst. ACS Catal. **8**, 1192–1196 (2018)
45. T. Breton, D. Liaigre, E.M. Belgsir, Allylic oxidation: easy synthesis of alkenones from activated alkenes with TEMPO. Tetrahedron Lett. **46**, 2487–2490 (2005)
46. K. Mitsudo, T. Kaide, E. Nakamoto, K. Yoshida, H. Tanaka, Electrochemical generation of cationic Pd catalysts and application to Pd/TEMPO double-mediatory electrooxidative Wacker-type reactions. J. Am. Chem. Soc. **129**, 2246 (2007)
47. X.-Y. Qian, S.-Q. Li, J. Song, H.-C. Xu, TEMPO-catalyzed electrochemical C–H thiolation: synthesis of benzothiazoles and thiazolopyridines from thioamides. ACS Catal. **7**, 2730–2734 (2017)
48. R. Ciriminna, G. Palmisano, C. DellaPina, M. Rossi, M. Pagliaro, One-pot electrocatalytic oxidation of glycerol to DHA. Tetrahedron Lett. **47**, 6993–6995 (2006)
49. X.L. Wen, Y.H. Jia, L. Yang, Z.L. Liu, Electrocatalytic oxidation of L-tyrosine by a nitroxide. Talanta **53**(5), 1031–1036 (2001)
50. M. Rafiee, B. Karimi, S. Alizadeh, Mechanistic study of the electrocatalytic oxidation of alcohols by TEMPO and NHPI. ChemElectroChem **1**, 455–462 (2014)
51. A. Badalyan, S.S. Stahl, Cooperative electrocatalytic alcohol oxidation with electron-proton-transfer mediators. Nature **535**, 406–410 (2016)
52. M. Rafiee, K.C. Miles, S.S. Stahl, Electrocatalytic alcohol oxidation with TEMPO and bicyclic nitroxyl derivatives: driving force trumps steric effects. J. Am. Chem. Soc. **137**, 14751–14757 (2015)
53. R. Ciriminna, G. Palmisano, M. Pagliaro, Electrodes functionalized with the 2,2,6,6-Tetramethylpiperdinyloxy radical for the waste-free oxidation of alcohols. ChemCatChem **7**, 552–558 (2015)
54. A. Das, S.S. Stahl, Noncovalent immobilization of molecular electrocatalysts for chemical synthesis: efficient electrochemical alcohol oxidation with a Pyrene–TEMPO conjugate. Angew. Chem. Int. Ed. **56**, 8892–8897 (2017)
55. M. Masui, T. Ueshima, S.N. Ozaki, Electrolysis. J. Chem. Soc. Chem. Commun. 479–480 (1983)
56. S. Kishioka, S.A. Yamada, Kinetic study of the catalytic oxidation of benzyl alcohols by phthalimide-*N*-oxyl radical electrogenerated in acetonitrile using rotating disk electrode voltammetry. J. Electroanal. Chem. **578**, 71–77 (2005)
57. T. Shiraishi, T. Takano, H. Kamitakahara, F. Nakatsubo, Studies on the electro-oxidation of lignin and lignin model compounds. Part 2: *N*-Hydroxyphthalimide (NHPI)-mediated indirect electrooxidation of non-phenolic lignin model compounds. Holzforschung **66**, 311–315 (2012)
58. I. Bosque, G. Magallanes, M. Rigoulet, M.D. Karkas, C.R.J. Stephenson, Redox catalysis facilitates lignin depolymerization. ACS Cent. Sci. **3**, 621–628 (2017)
59. F. Recupero, F. Punta, Free radical functionalization of organic compounds catalyzed by *N*-hydroxyphthalimide. Chem. Rev. **107**, 3800–3842 (2007)

60. E.J. Horn, B.R. Rosen, Y. Chen, J. Tang, K. Chen, M.D. Eastgate, P.S. Baran, Scalable and sustainable electrochemical allylic C–H oxidation. Nature **533**, 77–81 (2016)
61. W.F. Bailey, J.M. Bobbitt, K.B. Wiberg, Mechanism of the oxidation of alcohols by oxoammonium cations. J. Org. Chem. **72**(12), 4504–4509 (2007)
62. M. Rafiee, F. Wang, D.P. Hruszkewycz, S.S. Stahl, *N*-hydroxyphthalimide-mediated electrochemical iodination of methylarenes and comparison to electron-transfer-initiated C–H functionalization. J. Am. Chem. Soc. **14**, 22–25 (2018)
63. A. Samuni, C.M. Krishna, P.E. Riesz, E. Finkelstein, A. Russo, J. Biol. Chem. **263**(34), 17921–1724 (1988)
64. K.A. Harrison, E.A. Haidasza, M. Griessera, D.A. Pratt, Inhibition of hydrocarbon autoxidation by nitroxide-catalyzed cross-dismutation of hydroperoxyl and alkylperoxyl radicals. Chem. Sci. **9**, 6068–6079 (2018)
65. T. Fukai, M. Ushio-Fukai, Superoxide dismutases: role in redox signaling, vascular function, and diseases. Antioxid. Redox Signal. **15**(6), 1583–1606 (2011)
66. S. Goldstein, A. Samuni, Kinetics and mechanism of peroxyl radical reactions with nitroxides. J. Phys. Chem. A **111**(6), 1066–1072 (2007)
67. M. Griesser, R. Shah, A.T. Van Kessel, O. Zilka, E.A. Haidasz, D.A. Pratt, The catalytic reaction of nitroxides with peroxyl radicals and its relevance to their cytoprotective properties. J. Am. Chem. Soc. **140**(10), 3798–3808 (2018)
68. E.A. Haidasz, D. Meng, R. Amorati, A. Baschieri, K.U. Ingold, L. Valgimigli, D.A. Pratt, Acid is key to the radical-trapping antioxidant activity of nitroxides. J. Am. Chem. Soc. **138**(16), 5290–5298 (2016)
69. Y. Ishida, Y. Okamoto, Y. Matsuoka, A. Tada, J. Janprasit, M. Yamato, N.P. Morales, K.-I. Yamada, Detection and inhibition of lipid-derived radicals in low-density lipoprotein. Free Radical Biol. Med. **113**, 487–493 (2017)
70. R.K. Jensen, S. Korcek, M. Zinbo, J.L. Gerlock, Regeneration of amine in catalytic inhibition of oxidation. J. Org. Chem. **60**, 5396 (1995)
71. E.T. Denisov, Polymer Deg. Stab. **25**, 209 (1989)
72. E. Damiani, R. Castagna, P. Astolfi, L. GRECI Aromatic and aliphatic mono- and bis-nitroxides: a study on their radical scavenging abilities. Free Radical Res. **39**(3), 325–336 (2005)
73. S. Huang, H. Zhang, J.T. Paletta, S. Rajca, A. Rajca, Reduction kinetics and electrochemistry of tetracarboxylate nitroxides. Free Radical Res. **52**(3), 327–333 (2018)
74. A.P. Jagtap, I. Krstic, N.C. Kunjir, R. Hänsel, T.F. Prisner, S.T. Sigurdsson, Sterically shielded spin labels for in-cell EPR spectroscopy: analysis of stability in reducing environment. Free Radic. Res. **49**(1), 78–85 (2015)
75. Y. Kinoshita, K. Yamada, T. Yamasaki, H. Sadasue, K. Sakai, H. Utsumi, Development of novel nitroxyl radicals for controlling reactivity with ascorbic acid. Free Radical Res. **43**, 565–571 (2009)
76. T. Yamasaki, F. Mito, Y. Ito, S. Pandian, Y. Kinoshita, K. Nakano, R. Murugesan, K. Sakai, H. Utsumi, K.J., Yamada, Structure-reactivity relationship of piperidine nitroxide: electrochemical, ESR and computational studies. J. Org. Chem. **76**, 435–440 (2011)
77. E. Lozinsky, V.V. Martin, T.A. Berezina, A.I. Shames, A.L. Weis, G.I. Likhtenshtein, Dual fluorophore–nitroxide probes for analysis of vitamin C in biological liquids. J. Biochem. Biophys. Methods **38**(1), 29–42 (1999)
78. Y. Matsuoka, M. Yamato, T. Yamasaki, F. Mito, K. Yamada, Rapid and convenient detection of ascorbic acid using a fluorescent nitroxide switch. Free Radical Biol. Med. **53**, 2112–2118 (2012)
79. A.A. Bobko, I.A. Kirilyuk, I.A. Grigor'ev, J.L. Zweier, V.V. Khramtsov, Reversible reduction of nitroxides to hydroxylamines: the roles for ascorbate and glutathione. Free Radical Biol. Med. **42**(3), 404–412 (2007)
80. I.A. Kirilyuk, A.A. Bobko, S.V. Semenov, D.A. Komarov, I.G. Irtegova, I.A. Grigor'ev, E. Bagryanskaya, Effect of sterical shielding on the redox properties of imidazoline and imidazolidine nitroxides. J. Org. Chem. **80**, 9118–9125 (2015)

81. L. Lamp, U. Morgenstern, K. Merzweiler, P. Imming, R.W. Seidel, Synthesis and characterization of sterically and electrostatically shielded pyrrolidine nitroxide radicals. J. Mol. Struct. **1182**, 87–94 (2019)
82. H. Zhao, J. Wu, X. Meng, S. Zuo, W. Wang, H. Yuan, Novel piperidine nitroxide derivatives: synthesis, electrochemical and antioxidative evaluation. J. Heterocycl. Chem. **45**(2), 371–376 (2008)
83. M.C. Emoto, M. Yamato, H. Sato-Akaba, K. Yamada, Y. Matsuoka, H.G. Fujii, Brain imaging in methamphetamine-treated mice using a nitroxide contrast agent for EPR imaging of the redox status and a gadolinium contrast agent for MRI observation of blood-brain barrier function. Free Radical Res. **49**(8), 1038–1104 (2015)
84. F. Shi, P. Zhang, Y. Mao, C. Wang, M. Zheng, Z. Zhao, The nitroxide Tempo inhibits hydroxyl radical production from the Fenton-like reaction of iron(II)-citrate with hydrogen peroxide. Biochem. Biophys. Res. Commun. **483**, 159–164 (2017)
85. J. Li, J. Zhang, Y. Chen, N. Kawazoe, G. Chen, Tempo-conjugated gold nanoparticles for reactive oxygen species scavenging and regulation of stem cell differentiation. ACS Appl. Mater. Interfaces **9**, 35683–35692 (2017)
86. T. Yoshitomi, K. Kuramochi, L. Binh Vong, Y. Nagasaki, Development of nitroxide radicals–containing polymer for scavenging reactive oxygen species from cigarette smoke. Sci. Technol. Adv. Mater. **15**, 035002 (2014)
87. S.I. Dikalov, A.E. Dikalova, D.A. Morozov, I.A. Kirilyuk, Cellular accumulation and antioxidant activity of acetoxymethoxycarbonyl pyrrolidine nitroxides. Free Radical Res. **52**(3), 339–350 (2018)
88. B.P. Soule, F. Hyodo, K. Matsumoto, N.L. Simone, J.A. Cook, M.C. Krishna, J.B. Mitchell, The chemistry and biology of nitroxide compounds. Free Radical Biol. Med. **42**(11), 1632–1650 (2007)

Chapter 4
ESR and NMR as Tools for Nitroxides Studies

4.1 Introduction

The electron spin resonance (ESR) phenomena involve the resonance absorption or dispersion of a microwave frequency (0.3–250 GHz) of electromagnetic field (ν) by a system of particles with the intrinsic spin moment of an unpaired electron in a constant magnetic field of strength H_0 [1–5]. The absorption leads to magnetization in the excited state of the system. Accordingly, the electron magnetic resonance condition is

$$\boldsymbol{h}\,\nu = g_e \beta_B\, \boldsymbol{H}_0, \tag{4.1}$$

where g_e is a g-factor, characterizing the value of the intrinsic electron spin moment (free electron g-value is 2.002319), and β_B is the Bohr magneton (9.27400968 × 10^{-24} J T^{-1}). The values of g-factors and magnetic field strength H_0 dictate the position of resonance frequencies in ESR spectra. The hyperfine interaction between the electron and the nuclear spins consists of the isotropic Fermi contact interaction and the anisotropic dipole–dipole interaction. The hyperfine interaction is manifested by characteristic splittings of ESR spectra (Fig. 4.1).

The main ESR features are [1–5]: (1) integral intensity and amplitude of the ESR spectrum, (2) the position of the spectral features depending on the value of the g-factor according to the resonance condition, (3) the ESR spectral line shape which can be either homogenous (single spin packet) or heterogeneous (overlap of packets), (4) spin–lattice (T_1) and spin–spin (T_2) relaxation times, (5) the ESR spectra hyperfine splitting (hfs) attributed to the contact and dipolar interactions of electron spin with nuclear spin I. (6) the ESR spectral fine splitting (fs) caused by electron spin–electron spin exchange and dipolar interactions, and (7) the response of the ESR signal to progressive saturation; and (8) The degree of electron spin polarization.

Equation 4.1 predicts a dependence of position of EPR spectra on g-factor value and improvement of the resolution of the spectra for species of different g-factor with increasing frequency (Fig. 4.1). One of the most important advances has been

G. I. Likhtenshtein, *Nitroxides*, Springer Series in Materials Science 292,
https://doi.org/10.1007/978-3-030-34822-9_4

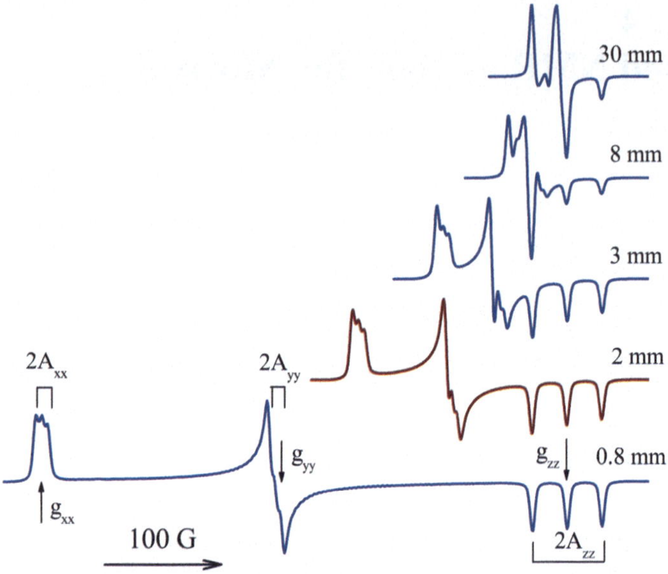

Fig. 4.1 EPR spectra of nitroxide radical calculated for different spin precession frequencies (microwave linewidths) [14]. http://hf-epr.awardspace.us/index.htm

the extension of ESR to high magnetic fields and high frequencies. High-frequency–high-field EPR (HF–HF EPR) was pioneered by Lebedev group [6]. Since then, theoretical and experimental bases of the method were essentially developed and found numerous applications [7–10]. The utilization of quasi-optical methods, especially above 150 GHz markedly expanded the HF–HF EPR efficiency [8].

Spin relaxations times are important parameters in many aspects of electron spin resonance theory and application. Spin–lattice relaxation time (T_1) characterizes the recovery of the induced magnetization to the ground state on account of a transfer of magnetic energy to energy of media (lattice). Another type of relaxation, spin–spin time (T_2), is related to time of return of the spin system to equilibrium in the excited magnetic state as a result of spin–spin interaction with environment. The second approach to the electron spin relaxation is based on the properties of the spin electron Larmor precession, which is the precession of the magnetic moments of electrons about external magnetic field. In the frame of Bloch model [11] after the microwave resonance absorption, the dynamic of spin magnetization $M = (M_x, M_y, M_z)$ is a function of longitudinal relaxation times T_1 and transverse relaxation time or spin phase memory time (T_m, T_2) which are synonyms of spin–lattice and spin phase memory relaxation time, respectively.

The two fundamental phenomena, spin exchange and dipole–dipole coupling, constitute the basis of the spin electron–spin electron and spin electron–spin nuclear interactions. Electrons and nuclei can be thought as tiny magnets and interact with each other through space. The exchange process occurs via direct or indirect overlap of the orbitals of the interacting species.

An arsenal of experimental ESR methods including conventional, electrically, and optically detected ESR (CW ESR, ED ESR, and OD ESR, respectively) and advanced pulse techniques is widely used in chemistry and biology [1–10]. The CW technique is employed for collecting information of interest through the analysis of ESR spectra, saturation curves, electron–electron double resonance (ELDOR) or double electron–electron resonance (DEER), electron-nuclear double-resonance (ENDOR) multifrequency ESR (MF ESR), two-dimensional ESR (2D ESR), two-dimensional electron–electron double resonance (2D-ELDOR), ENDOR with circularly polarized radiofrequency fields (CP-ENDOR), electron–nuclear–nuclear resonance (double ENDOR), proton–electron double-resonance imaging (PEDRI), time-domain ESR, and electron–nuclear–nuclear triple resonance (TRIPLE), reaction yield detection magnetic resonance (RYDMR), and magnetically affected reaction yield (MARY).

The more important developments of pulse EPR include [1–5]: Fourier transform ESR (FT ESR), two-dimensional Fourier transform ESR (2D FT ESR), electron spin echo (ESE), electron spin echo envelope modulation (ESEEM), hyperfine sublevel correlation spectroscopy (HYSCORE), double nuclear coherence transfer hyperfine sublevel correlation electron spin echo (DONUT-HYSCORE), dynamic nuclear polarization (DNP), electron spin echo-detected magnetization transfer (ESE MT), two-dimensional electron spin echo correlation spectroscopy (2D SECSY), pulse electron-nuclear double resonance (pulse ENDOR), pulse electron–electron double resonance (PELDOR), electron spin transient nutation (ESTN), two-dimensional electron spin transient nutation (2D ESTN), phase-inverted echo-amplitude detected nutation (PEANUT), saturation recovery ESR (SR ESR), pulse multifrequency ESR (PMF ESR), magnetic isotope effect (MEF), quantum beats effect (QBE), double quantum coherence pulsed ESR (DQC ESR), ESR spectra hole burning, A "211" electron spin echo method, relaxation enhancement (RE), the pulsed triple electron resonance (TRIER), pulse ESR-based electron spin nutation (ESN) spectroscopy, two-dimentional (2D-ESN) spectroscopy, relaxation-induced dipolar modulation enhancement (RIDME), and hyperfine-correlated electron-nuclear double-resonance spectroscopy (HYEND).

4.2 Spin Electron–Spin Electron Interactions. Distance Determination

The method of double spin labeling (DSL), pioneered by Likhtenshtein in 1968 [12, 13], is based on specific modification of chosen groups in the object of interest by two or several spin labels, nitroxides, or complexes of paramagnetic metal, followed by the analysis of the effects of the spin–spin interactions on the label ESR spectra or spin relaxation times. Nowadays, various modifications of the double labeling method and their modifications are widely employed for solving numerous problems of structure and molecular dynamics of biological molecules [13–55, 57–68].

The first applications of dipole–dipole spin–spin interactions to the investigation of protein surface topography were based on the changes in the line shape of nitroxide radical ESR spectra arising from interactions with a second nitroxide radical or a paramagnetic ion [12–17]. This approach allows one to estimate distances between the paramagnetic centers up to 2.5 nm. Later, the higher sensitivity of power saturation curves of a radical to interactions between the radical and paramagnetic ions up to 10 nm was demonstrated [18–22]. The effects of the spin–spin-exchange interaction on the ESR spectra and spin relaxation were first used in the Likhteshtein group for establishing the structure of systems under investigation, such as nitrogenase and non-heme protein [23].

Methods of determination of spin electron–spin electron distance based on analysis the CW EPR spectra were described in comprehensive reviews [24, 25]. A detailed description of the continuous wave microwave saturation recovery technique which allows to measure spin relaxation times, T_1 and T_2 and corresponding theory was presented in [26–28].

Factors affecting dipolar interaction between two spins in a static magnetic field (H_0) are [18, 30–36]: (i) magnetic moments, (ii) interspin distance, (iii) the angle between the interspin vector and the external field, (iv) the spin–spin (transverse, phase memory, T_2 T_m) and spin–lattice (T_1) relaxation times, the ESR frequencies (ω) of both interacting paramagnetics. When the product of relaxation time T_1 for the second center and the interaction energy ($\Delta\omega$) expressed in frequency units $\Delta\omega^2 T_1^2 \gg 1$, the effect of the second spin may be regarded as an interaction with a permanent dipole moment (slow relaxing spin). When $\Delta\omega^2 T_1^2 \ll 1$, the interaction may be considered as a weak perturbation because of the fast relaxation (fast relaxing spins).

The Hamiltonian that describes the interaction between two permanent spins in a point-dipole approximation is given by

$$\hat{H}_{SS} = \frac{g_e^2 \alpha^2}{8} \sum_{i \neq j} \left[\frac{\hat{s}(i)\hat{s}(j)}{r_{ij}^3} - 3\frac{(\hat{s}(i)\mathbf{r}_{ij})(\hat{s}(j)\mathbf{r}_{ij})}{r_{ij}^5} \right] \tag{4.2}$$

here, $\mathbf{r}_{ij} = \mathbf{r}_i - \mathbf{r}_j$ for electrons i and j at positions $\mathbf{r}_i$ and $\mathbf{r}_j$, and with spins $\hat{s}(i)$ and $\hat{s}(j)$, g_e is the free electron g-value. The dipolar coupling between two spin leads to splitting of the NMR and ESR lines with formation of the Pake doublet (Fig. 4.2) [37].

In high magnetic fields, the dipole–dipole interaction is expressed with the secular dipole–dipole coupling constant d_{12} including the angle Θ between the axis connecting the two spins and the external magnetic field vector $\mathbf{B}_0$ as follows:

$$d_{12} = b_{12}\frac{1}{2}(3\cos^2\Theta_{12} - 1) \tag{4.3}$$

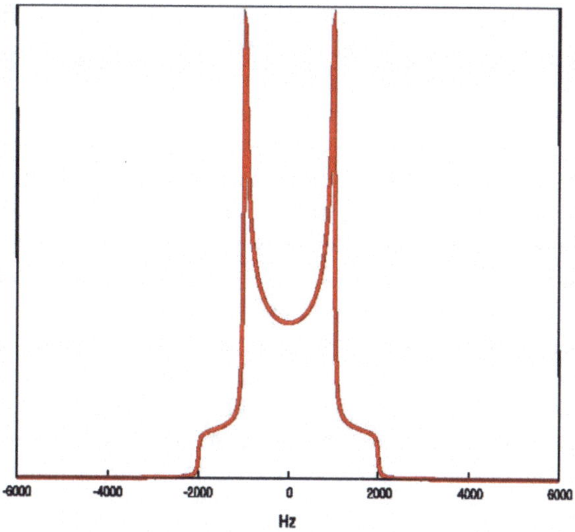

Fig. 4.2 Pake doublet https://en.wikipedia.org/wiki/Pake_doublet

where

$$b_{12} = -\frac{\mu_0}{4\pi}\frac{\gamma_1\gamma_2\hbar}{r_{12}^3} \tag{4.4}$$

and μ_0 is the vacuum magnetic permeability.

Within the Larmor precession model, in the laboratory frame, the static magnetic field B is assumed to be parallel to the z-axis and an applied microwave field B_1 parallel to the x-axis. In the rotating frame, the B_1 component rotating with frequency w_0 can be on resonance with the precessing magnetization vector M_0 if $\omega_L = \omega_0$. The magnetization vector will rotate around the x-axis in the zy-plane for the duration of the applied B_1 field. The tip angle α by which M_0 is rotated is given by

$$\alpha = -\gamma|B_1|t_p \tag{4.5}$$

Thus, the desirable angle can be regulated.

For the two-pulse primary echo experiment and the three-pulse stimulated echo-detected (ED)experiments, the EPR line shape is obtained when the magnetic field B is scanned across the resonance line [38]. The two-pulse ED spectra (pulse sequence: π/2–τ–π–τ–echo) experiment is used to detect processes, on the timescale of the phase memory time (T_m), in the nanosecond regime. Echo-detected spectra from the three-pulse stimulated spin echo (pulse sequence: π/2–τ–π/2–T–π/2–τ–echo) experiment are used to detect slower processes, due to its dependence on the longer T_1 process.

Since the pioneer work of Milov, Salikhov, and Tsvetkov based on instantaneous diffusion that arises from dipolar spin–spin interactions between different spins [39], the use of pulse electron paramagnetic resonance (Echo-detected ELDOR, pulse

ELDOR, PELDOR, double electron–electron resonance DEER) for long distance spin–spin interaction has been essentially developed and wildly applied for the distance determination up to 10 nm [40–55, 57–68]. In a system of two spins (observed and pumped), the effects of diffusion coupled with the action of the applied pulse on the precession frequency of the observed spin were detected [39, 40]. In three-pulse PELDOR, the ESR spectrum is excited by two electron spin echo (ESE) pulses and an additional pumping pulse. For a system of two spins A and B, coupled by the dipole–dipole interaction the experiment consists of two π/2–π-pulse electron ESE subsequence with a mixed interpulse delay τ at the observer frequency w_a for the spin A and a pump π pulse at frequency w_b for the spin B. For these spins A and B, coupled by the dipole–dipole interaction at a distance R and given the angle between the direction of the external magnetic field and the vector which connects the paramagnetic centers (θ), the ratio of the primary echo signal amplitude $V_p(2\tau)$ with the pumping pulse to the ESE signal amplitude $V_0(2\tau)$ without the pumping pulse is

$$V(T) = \frac{V_p(2\tau)}{V_0(2\tau)} = 1 - p_B[1 - \cos(DT)]. \tag{4.6}$$

where p_B is the probability of a spin B flip under the action of the pumping pulse; D is the splitting (in rad/s) of the resonance spin A line due to interaction with spin B and has the analytical form

$$D = \frac{\gamma^2 \hbar}{R^3}\left(1 - 3\cos^2\Theta\right) + J, \tag{4.7}$$

which includes the dipolar and J exchange interaction.

A "211" electron spin echo method developed by Raitsimring and Tsvetkov is complementary to the pulse ELDOR [41]. The difference between the PELDOR and "2 + 1" methods consists in the choice of the carrier frequency of the second microwave pulse. The "2 + 1" pulse train allows to measure the dipole–dipole interactions between paramagnetic centers which are substantially weaker than those that can be measured by the ordinary two-pulse train and between spins with overlapping EPR spectra.

In the four-pulse PELDOR experiment invented by the Jeschke group, the detection sequence $\pi/2–\tau_1–\pi–\tau_1–\text{echo}_1–\tau_2–\pi$ is applied at a microwave frequency ν_A (Fig. 4.3) [42]. The four-pulse techniques allow one to avoid a number of artifacts.

In a five-pulse DEER sequence, an extra pulse at the pump frequency is added and compared with standard four-pulse DEER (Fig. 4.4) [43]. The position of the extra pulse is fixed relative to the three pulses of the detection sequence. This procedure significantly reduces the effect of nuclear spin diffusion on the electron spin phase relaxation, thereby enabling longer dipolar evolution times that are required to measure longer distances.

Using spin-labeled *T*4 lysozyme at a concentration less than 50 μM, as an example, it was shown that the evolution time increases by a factor of 1.8 in protonated solution and 1.4 in deuterated solution to 8 and 12 μs, respectively (Fig. 4.5).

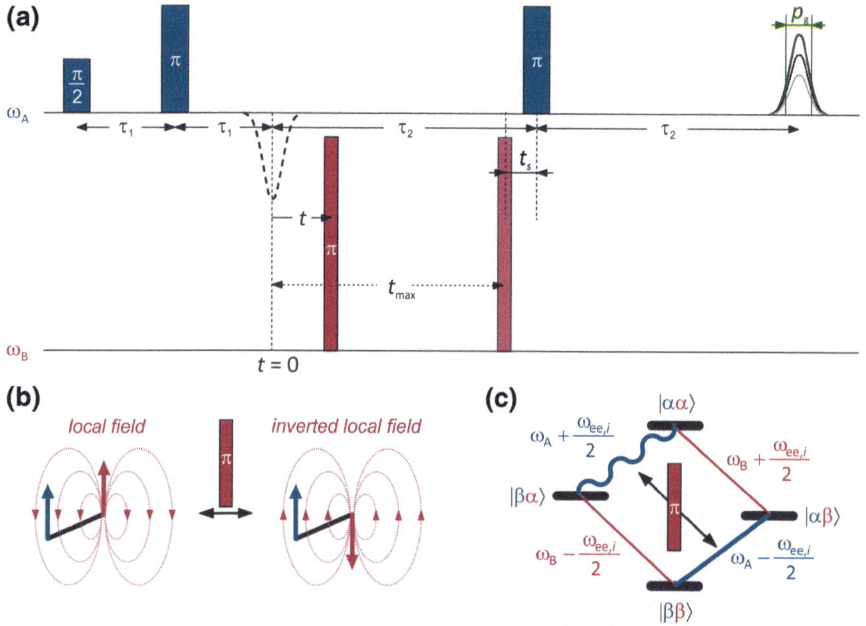

Fig. 4.3 Four-pulse DEER experiment. **a** Pulse sequence. Time t is varied from $t < 0$ to t_{max}, and variation of the integral echo intensity in the window of length p_g is recorded. **b** Local field picture. The π pump pulse at frequency ω_B inverts the state of spin B (gray), thus inverting the local field imposed by spin B at the site of spin A (black). **c** Energy level diagram. Inversion of the local field at spin A exchanges coherence between the two transitions of spin A that differ in frequency by $\omega_{ee,i}$ [42]. Private communication from Prof. Jeschke

A Refocused Out-Of-Phase DEER (ROOPh-DEER) seven-pulse sequence detection scheme which acquires only the modulated fraction of the dipolar DEER signal was recently proposed for a case when Zeeman splitting is small [45]. The application of seven-pulse ROOPh-DEER sequence to a model biradical yielded the interspin distance identical to the one obtained with the conventional four-pulse DEER, however, without the unmodulated background present as a dominant fraction in the latter signal. The four-pulse DEER-RELOAD scheme applied to two membrane protein complexes labeled with nitroxide may improve SNR by a factor of >3 as it has been demonstrated for one of these two membrane proteins.

Compared to four-pulse DEER, five-pulse DEER suffers from additional artifacts that stem from pulse imperfection and excitation band overlap. To eliminate experimentally the artifacts in five-pulse DEER due to partial excitation and excitation band overlap at frequencies, a data post-processing method was introduced [46]. The method removes the partial excitation artifact without relying on previous knowledge of its amplitude and without sensitivity loss on acquisition of two traces with shifted positions of the artifact and computation of the artifact shape from the difference of the two traces. Experimental conditions that suppress additional artifact

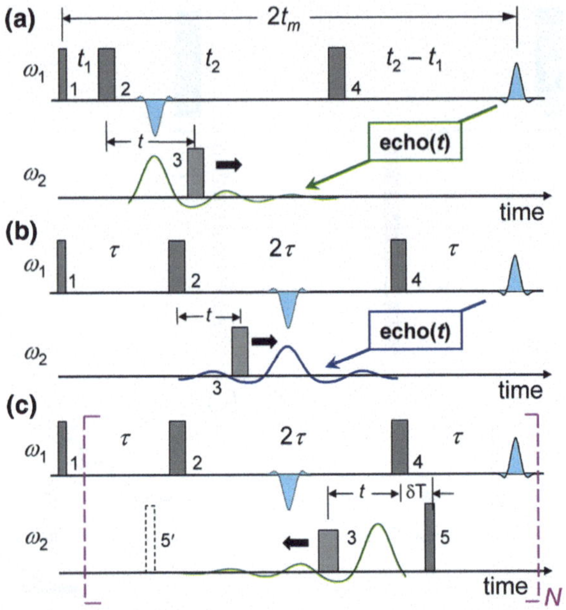

Fig. 4.4 **a** Standard four-pulse DEER sequence with the respective dipolar modulation pattern plotted in green. **b** The four-pulse sequence modified for t_{2}-$2t_{1}$-2τ to minimize nuclear spin diffusion, thus increasing the signal, but this shifts the dipolar modulation (in blue) to the middle of the second interval, thereby losing half of the dipolar modulation because the halves are identical. **c** Placing the second pump pulse, 5, after the pulse 4 shifts dipolar modulation toward pulse 4, thereby recovering the full time span, 2τ. The dipolar modulation (green) is reversed in time compared with panel a. Pulse 5 could also be placed at position 5′ before pulse 2, reversing the modulated time trace. Note that time period, t_m, available for recording the dipolar signal is **a** t_2 and (**b**, **c**) 2τ [43]. Reprint from [43], Copyright 2013, American Chemical Society

contributions stemming from overlap between the excitation bands of the microwave pulses that introduce additional dipolar evolution pathways were analyzed in detail.

An ESR version of multiple quantum coherence (ESR MQC) was predicted by Tang and Noris [47] and observed in experiments on spin-correlated pairs formed by laser-induced dissociation by Dzuba et al. [48]. A novel six-pulse pulsed ESR technique for the detection of double quantum coherence (DQC), which yields high-quality dipolar spectra for distance measurements, was developed by Freed with coworkers [49, 50]. The suggested six-pulse DQC sequence is shown in Fig. 4.6. The detection of double quantum coherence (DQC) yields high-quality dipolar spectra for distance measurements allow one to extract distance distributions. The main virtues of DQC ESR are (1) detection of weak dipolar interactions between paramagnetic molecules in the ESR signal and reliable measurement of distances and distance distributions between them up to 80 Å, (2) determination of the angular geometry of the biradical, (3) determination of the asymmetric *g*-factor and hyperfine tensors, (4) characterization of many body spin systems, and (5) investigation of conformers.

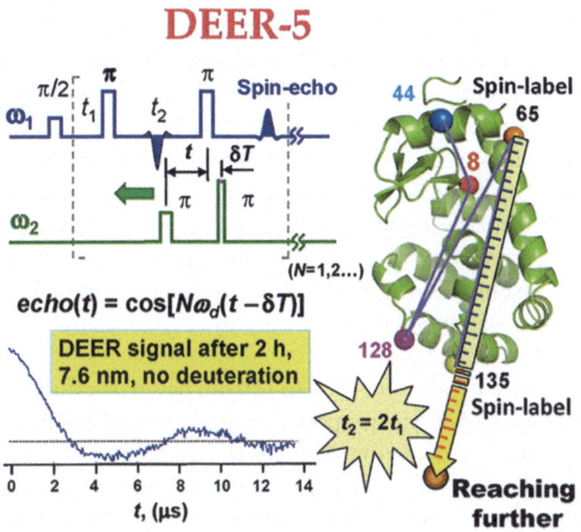

Fig. 4.5 Schematic illustration five-pulse DEER application to spin-labeled *T*4 lysozyme [43]. Reprint from [43], Copyright 2013, American Chemical Society

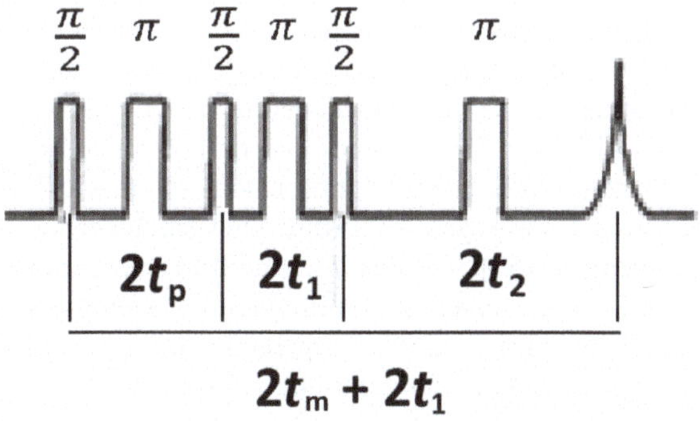

Fig. 4.6 Six-pulse DQC sequence **a** is shown with the corresponding coherence pathways [49]

Two-dimensional electron–electron double-resonance (2D-ELDOR) technique was invented, developed, and applied by the Freed group [51]. The following time periods in a three-pulse 2D-ELDOR experiment are defined as follows: (1) the preparation period, (2) the mixing period, *and* (3) the final π/2 pulse. 2D-ELDOR has proven to be a technique that is sensitive to the fast and slow dynamics processes.

The hole burning method has been invented by Schweiger et al. [52] and developed by Dzuba and Kawamori [53]. For inhomogeneously broadened electron paramagnetic resonance spectra of solids and viscous liquids, spectral diffusion can be studied by selective hole burning of homogeneous linewidth (0.5–0.05 G) using a free-induction decay (FID). The pulse sequence for the hole-burning experiment is π– τ–π/2–τ–π/2. Sources of spectral diffusion which can be investigated by these methods are as follows: (1) chemical exchange which can be induced by chemical reactions, by transitions between different molecular conformations or isomers, and by rotation about some particular molecular bond, (2) molecular tumbling in a viscous liquid, (3) spin–lattice relaxation, and (4) spin diffusion, a process of mutual flip-flops in the spin system leading to a spatial diffusion of the Zeeman energy. For selective hole-burning relaxation-induced dipolar modulation enhancement (RIDME) experiments with stable nitroxide biradical in a frozen solution, it allowed to eliminate the major unwanted contributions to the FID modulation originating from spectral diffusion and from excitation of forbidden electron-nuclear spin transitions by mw pulses [54]. The proposed method can be employed to measure interspin distances in the range of 13 Å $\leq r \leq$ 25 Å.

A method of distance measurement up to 150 nm based on effect of fast relaxation spin on slow relaxation spin was proposed by Kulikov and Likhtenshtein in 1974 [18, 19, 56] in parallel of invention of the pulse ELDOR [39]. According to the Solomon and Bloembergen theory [55], the electron relaxation enhancement depends on magnetic moments μ, and interspin distance (r), electron spin (S), the spin-phase (T_2) and spin–lattice (T_1) relaxation times, the ESR frequencies (ω) of both interacting paramagnetics, the angle (θ) between the interspin vector and the external field (H_0), and the interaction energy expressed in frequency units ($\Delta\omega$).

For estimation of distances between paramagnetic centers using the Solomon-Bloembergen theory, it is necessary to know T_{1f} and T_{2f} for the radical. A method of estimation of the relaxation times by analysis of CW power saturation curves in solids from saturation curves recorded under conditions of rapid passage was applied to nitroxide radicals [18–21]. These values can be determined directly by pulse methods. In the case of nitroxide TEMPOL in ethanolic solution at 77 K, the obtained relaxation rates were reasonably agreed with direct measurements by pulse saturation and spin echo measurements [57]. In solids, T_{1f} usually decreases monotonically with increasing temperature, so one can find experimentally the temperature at which the dipolar contribution is maximum. At this temperature, $1/T_{1s} = \omega$ and the Solomon-Bloembergen theory predicted dipolar contribution given by (4.8)

$$\Delta\left(\frac{1}{T_{1s}}\right) = \frac{\mu_f^2 \gamma^2}{15 r^6 \Delta\omega} \tag{4.8}$$

The theoretical maximum distances (r_{max}) to a second paramagnetic center with spin S_f that could be determined were estimated as r_{max} = 100 Å for S_f = 1/2 and r_{max} = 150 Å for S_f = 5/2 [18, 21]. The effect of a rapidly relaxing spin on T_1 of a slowly relaxing spin also was considered in [44, 58]. Application of this method was illustrated by the estimation of distances between the heme group of human hemoglobin and the nitroxide fragment of three spin labels attached covalently to the α-93 SH-group [21]. The determined values of r (in Å with an accuracy 5–6%) between heme and the nitroxide fragment are in reasonable agreement with each other and with the x-ray diffraction structure of hemoglobin [59].

In work [60], the relaxation enhancement (RE) value Δk was defined as the difference of the inverse relaxation times of a slowly relaxing species in the presence T_{1f} and absence of a fast relaxing species $T_{1f,0}$. Specifically, the final intermolecular RE time traces employing the nitroxide echo-detected (ED) EPR were rescaled and fitted by multiexponential fit curves to extract T_1 times that were used as a measure of the average relaxation enhancement Δk, which is related to the spin–spin distance (r) (4.9).

$$\Delta k = \frac{C T_{1f}}{r^6} \tag{4.9}$$

The approach was tested on a water-soluble $T4$ lysozyme labeled by nitroxide and lanthanide complex and showed the spin–spin distance value of about 3.2 nm.

Pulse-ESR-based electron spin nutation (ESN) spectroscopy was suggested as a tool to determine relatively short distances between weakly exchange-coupled electron spins [61, 62]. Two-dimentional (2D-ESN) spectroscopy was used to determine spin dipolar interactions for spin distances within 2.0 nm in a system of weakly exchange-coupled molecular spins in non-oriented media. Thus, ESN spectroscopy has been proved to be a useful complement to other modern pulse-ESR methods.

A method of singular value decomposition (SVD) to determine distance distributions in pulsed dipolar electron spin resonance was invented [63]. Even very small amounts of noise (e.g., signal-to-noise ratio SNR $\approx$ 850) are sufficient to corrupt measurement of spin–spin distance distribution. SVD can be employed directly on the denoised data, using pulse dipolar electron spin resonance experiments as an example. This method is useful in measuring distances and their distributions, $P(r)$, between spin labels on proteins. SVD can be employed directly on the denoised data using pulse dipolar electron spin resonance experiments. A new wavelet transform-based method of denoising experimental signals to pulse-dipolar electron spin resonance spectroscopy (PDS) was developed [64]. The method allowed to reduce the signal averaging times of the time-domain signals by as much as two orders of magnitude, while retaining the fidelity of the underlying signals, and to achieve excellent signal recovery when the initial noisy signal has an SNR $\gtrsim$ 3. The authors expected that the method can be applicable to other time-domain spectroscopies.

The "chirp-induced dipolar modulation enhancement" (CIDME) six-pulse sequence was introduced in [65]. CIDME eliminates limitation of frequency-swept pump pulses in DEER experiments associated with a need for the pump pulses to be short in comparison with dipolar evolution periods by means of longitudinal storage during the application of one single or two consecutive pump pulses. In pulse EPR correlation techniques, the monochromatic excitation often cannot uniformly excite the entire spectrum. Restrictions in pulse EPR correlation techniques were alleviated for nitroxide spin labels at Q-band microwave frequencies around 35 GHz T by substitution of monochromatic pulses by frequency-swept chirp pulses tailored for uniform excitation [66]. The dipole–dipole interaction can be separated by a constant-time Zeeman-compensated solid echo sequence called SIFTER. EPR-correlated dipolar spectra can be obtained when the four pulses are replaced by chirp pulses. These two-dimensional spectra encode additional information on the geometrical arrangement of the two spin labels. In a new approach to DEER for distance determination involving nitroxide spin labels at dilute concentrations (nDEER), non-selective chirp pulses that refocus all relevant spin pairs are combined with DEER [67]. A key advantage of nDEER is the high fidelity of the chirp refocusing pulses, which is important for nDEER schemes that incorporate dynamical decoupling to access longer distances.

The pulsed triple electron resonance (TRIER) experiment, which correlates dipolar frequencies of molecules with three electron spins, was introduced [68]. These correlation patterns in combination with DEER allow to interpret distance distributions of biological systems that exist in more than one conformation. This method is able to obtain two-dimensional distance correlation maps of the previously investigated model compounds as well as of spin-labeled proteins by application of two-dimensional approximate Pake transformation to TRIER data. This enabled to get distance correlation plots from two triple-labeled protein samples that were in good agreement with DEER data and simulations.

4.2.1 *Spin Electron–Spin Nuclear Interactions*

The electron-nuclear spin interaction appears from the coupling of the nuclear magnetic moment to the magnetic field generated by the electron magnetic moment or equivalently from the coupling of the electron magnetic moment to the magnetic field generated by the nuclear magnetic moment [69, 70]. This interaction causes superfine splitting in EPR spectra and effects on nucleus NMR properties. The nuclear magnetic moment $\mathrm{m_I}$ originating from the nuclear spin quantum number I is $m_I = 2I + 1$.

The total Hamiltonian of the system of an electron and many nuclear spins is given by

$$\hat{H} = \hat{H}_{\mathrm{e}} + \hat{H}_{\mathrm{N}} + \hat{H}_{\mathrm{eN}} + \hat{H}_{\mathrm{NN}}, \tag{4.10}$$

and composed of the single spin Zeeman energies, Hamiltonian for interaction in the applied magnetic field along the z-axis, the hyperfine interaction, and the intrinsic nuclear–nuclear interaction, respectively. The electron-nuclear spin dipole–dipole interactions for an electron in an s-symmetry orbital and in a non-s-symmetry orbital are expressed as follows:

$$\begin{aligned} H_{\mathrm{HF}} &= \tfrac{16\pi}{3}\gamma_{\mathrm{I}}\mu_{\mathrm{B}}\mu_{\mathrm{N}}\delta(\mathbf{r})[\mathbf{S}\cdot\mathbf{I}], \quad l = 0 \\ H_{\mathrm{HF}} &= \tfrac{2\gamma_{\mathrm{I}}\mu_{\mathrm{B}}\mu_{\mathrm{N}}}{r^3}\left[(\mathbf{L}-\mathbf{S})\cdot\mathbf{I} + 3\tfrac{(\mathbf{S}\cdot\mathbf{r})(\mathbf{I}\cdot\mathbf{r})}{r^2}\right], \quad l \neq 0. \end{aligned} \tag{4.11}$$

where μ_{I}, μ_{B}, and μ_{N} are the gyromagnetic factor of the nuclear spin, the Bohr magneton, and the nuclear magneton, respectively. S and I are the spin operators for the electron and the nucleus, and L is the angular momentum operator for the electron.

The contact Fermi electron–nucleus interaction, which is caused by the probability to find the electron spin at the same point in space as the nuclear spin, is given by

$$\hat{H}_{\mathrm{eN}} = \sum_{n} a_{\mathrm{n}} \hat{\mathbf{S}}_{\mathrm{e}} \cdot \hat{\mathbf{J}}_{\mathrm{n}}, \tag{4.12}$$

where $\hat{S}_e$ and $\hat{J}_n$ are the spin and nucleus operators, respectively, and

$$a_n = \frac{\mu_0}{4\pi} \gamma_{\mathrm{e}} \gamma_{\mathrm{n}} \frac{8\pi}{3} |\Psi(\mathbf{R}_{\mathrm{n}})|^2, \tag{4.13}$$

where μ_0 is the vacuum magnetic permeability, $\mathbf{R}_n$ denotes the coordinates of the *n*th nucleus, γ_n and γ_e are the nuclear gyromagnetic ratio, and the electron gyromagnetic ratio, respectively; $|\psi(R_n)|^2$ is the spin density of s-electron on the nucleus.

The main sources of information on the electron–nucleus interactions are CW EPR spectra, electron-nuclear double resonance (ENDOR), two-dimensional electron–electron double resonance (2D-ELDOR), electron spin echo (ESE), electron spin echo envelope modulation (ESEEM), ENDOR with circularly polarized radiofrequency fields (CP-ENDOR), electron–nuclear–nuclear resonance (double ENDOR), proton–electron double-resonance imaging (PEDRI), time-domain ESR, and electron–nuclear–nuclear triple resonance (TRIPLE), hyperfine sublevel correlation spectroscopy (HYSCORE), double nuclear coherence transfer hyperfine sublevel correlation spectroscopy (DONUT-HYSCORE), magnetic isotope effect (MEF), ESR nutation spectroscopy (ESN), and two-dimensional ESN (2D-ESN).

In a continuous wave electron-nuclear double resonance (CW-ENDOR) introduced by Feher [71], an EPR transition is partly saturated by microwave radiation of amplitude B_1, and, in parallel, a driving radio frequency (rf) field of amplitude B_2 induces nuclear transitions. While the magnetic field is swept through the EPR spectrum, the rf frequency follows the Zeeman frequency of the nucleus. Transitions occur at frequencies υ_1 and υ_2 obeying the NMR selection rules $\Delta M_{\mathrm{I}} = \pm 1$ and ESR selection rule $\Delta M_{\mathrm{S}} = 0$. The ENDOR spectrum can reveal both the hyperfine coupling constant (a) and the nuclear Larmor frequencies (υ_n), using EPR which is markedly more sensitive than that NMR [71–74]. Thus, the hyperfine enhancement effect manifests itself both in CW and pulse ENDOR.

With the aim to identify by CW EPR techniques electron-nuclear dipolar and contact interaction, and electron spin–electron spin dipolar and exchange interactions, a method based on the phenomenon of continuous wave electron–electron double resonance (CW ELDOR) was independently introduced in 1968 by Hyde and Freed [72–75] and Bendersky and Blumenfeld [76] groups. In the CW ELDOR technique, one part of the EPR spectrum of a paramagnetic sample is irradiated with an intense saturating microwave pump field, and the effect of this intense field on other parts of the spectrum is assessed by a second weak microwave probe field. The EPR signals detected by the weak microwave probe field are reduced in intensity due to saturation transfer effects when the two frequencies are separated either by an integral number of hyperfine energy differences or by a *g*-factor difference of two or more paramagnetic species.

Electron **s**pin **e**cho **e**nvelope modulation (ESEEM) arises from a coupling between the EPR-active electron being probed and nuclei [77]. ESEEM contains information on the type of nuclei, on the distances between nuclei, on the spin density distribution (hyperfine interaction), and on the electric field gradient at the nuclei (nuclear quadrupole interaction). Pulse-ENDOR techniques make use of a combination of microwave and radiofrequency pulses. ESEEM is suited for measuring weak hyperfine couplings, e.g., of the order of the free nuclear Larmor frequency, while continuous wave ENDOR is better suited for strong hyperfine couplings [77–80]. The two most common forms of ESEEM experiments use either a two-pulse (π/2–τ–π–τ–echo) or three-pulse (π/2–τ–π/2–T–π/2–τ + T–echo) sequences. In ESEEM experiments, the modulation of the echo intensity is monitored as a function of the delay between the microwave pulses. Electron spin echoes, as created by the two-pulse ESEEM sequence, are a superposition of Hahn echoes [78] arising from allowed and forbidden nuclear transitions. This leads to a non-oscillatory or unmodulated part. Coherence transfer echoes, whose amplitude oscillates at the sum and difference frequencies of the (weakly coupled) local nuclei, form the modulated part. T Anisotropy of *g*-, hyperfine anisotropy, field inhomogeneities *g*-, and *A*-strains caused an extreme inhomogeneous broadening in the solid state can be revealed by these methods.

In the frame of a three-pulse sequence (π/2–τ–π/2–T–π/2–τ–echo), a set of stimulated ESEEM patterns is recorded as a function of time T [80, 81]. The four-pulse method ESEEM, which exploits additional filtering of the relevant signal, is advantageous for measuring combination lines compared to two-pulse ESEEM. To improve the ESEEM modulation depth, five-pulse ESEEM, with the sequence ($\pi/2$–τ_1–π–τ_1–$\pi/2$–T–$\pi/2$–τ_2–π–τ_2–echo) has been proposed. In Davies ENDOR of an $S = 1/2$, $I = 1/2$ system a microwave (mw) inversion recovery pulse sequence (π–T–π/2–τ–π–τ–echo) was used. Davies ENDOR is useful for systems with large hfs while applications of Mims ENDOR are limited to relatively small hfs. constants ($B_1 > a$) 0.13 [81].

A time-domain pulsed method in which polarization transfer ENDOR occurs can be performed with selective as well as non-selective mw pulses. In these experiments, the mixing period consists of two rf π/2 pulses separated by a variable time interval *T*' [82]. Hyperfine-correlated electron-nuclear double-resonance spectroscopy (HYEND), where the nuclear transition frequencies are correlated with the corresponding hyperfine frequencies, can be used for considerable simplification of the interpretation of the spectra.

The HYSCORE is a two-dimensional ESEEM method first published by Hofer et al. in 1986 [83]. The following pulse sequence of a HYSCORE experiment is used: p/2–t–p/2–t_1–p–t_2–p/2–t–echo. The experiment can be separated into three steps:

generation of nuclear coherence, the evolution time with inversion pulse, and the detection sequence. After 2D Fourier transformation, the HYSCORE spectrum is a 2D plot related to nuclear frequencies of different spin manifolds of the paramagnetic system for a certain orientation [83]. Due to the high spectral resolution and reasonable number of pulses, HYSCORE experiments can be performed on samples with a low concentration of paramagnetic species. DONUT-HYSCORE (double nuclear coherence transfer hyperfine sublevel correlation) is a two-dimensional experiment, which was designed to obtain correlations between nuclear frequencies belonging to the same electron spin manifold [84]. The sequence employed is $\pi/2$–τ_1–$\pi/2$–t_1–π–τ_2–π–t_2–$\pi/2$–τ_1–echo. The echo is in turn measured as a function of t_1 and t_2; whereas, τ_1 and τ_2 are held constant. The DONUT-HYSCORE experiment was designed to improve the resolution of the ESEEM frequencies and to resolve the assignment of ESEEM frequencies for the case of an electron spin, $S = {}^1/_2$, interacting with a number of nuclear spins with $I \geq 1$ (e.g., ^{14}N nuclei) with non-negligible nuclear quadrupole interactions.

Two-dimensional electron–electron double-resonance (2D-ELDOR) technique was invented, developed, and applied by the Freed group [85]. The FID, obtained during the evolution period, t_1 is collected as a function of t_2. The experiments are repeated for a series of mixing time, T_m, at several temperatures. The signals are then doubly Fourier transformed and can be presented as magnitude spectra. 2D-ELDOR is sensitive to the dynamical processes affecting spin labels in complex fluid environments. In ordered fluids, such as membrane vesicles, the 2D-ELDOR experiment is affected by the molecular tumbling in the locally ordered environment.

Two-dimentional Fourier transfer ESR approach includes correlation spectroscopy (COSY) and spin echo correlation spectroscopy (SECSY) [86]. In the standard COSY experiment, the first $\pi/2$ pulse creates transverse magnetization (± 1 coherences). The signal is measured after a second $\pi/2$ pulse. The SECSY signal is related to the COSY signal by the transformation $t_2 \rightarrow t_2 + t_1$, and inhomogeneities are refocused in t_1. In the SECSY experiment with a hard pulse, the homogeneous linewidth $\left(T_2^{-1}\right)$ across the spectrum can be obtained in a single experiment The linear dependence T_2^{-1} versus τ_R^a was τ_R which is the rotational correlation time and allows one to determine τ_R in the slow motion region (10^{-7}–10^{-8} s).

Effects of electron spin on nuclear spins on NMR spectra and nuclear spins relaxation were widely investigated [87]. The total observed NMR chemical shift (δ_{tot}) includes both a diamagnetic or orbital contribution (δ_{dia}) from paired electrons and a hyperfine contribution (δ_{hf}) from unpaired electrons of the electron-nuclear term of Hamiltonian. Effects of electron spin on nuclear spins on position of NMR spectra are revealed in Fermi contact (δ_{FC})

$$\delta_{FC} = m(S+1)\rho_{\alpha\beta}/T \tag{4.14}$$

depending on the spin state (S) of the system, the spin density at the at the nucleus ($\rho_{\alpha\beta}$), temperature and a collection of fundamental physical constants, and pseudocontact (δpc) term:

$$\delta^{\mathrm{pcs}} = \frac{1}{12\pi r^3}\left[\Delta\chi_{\mathrm{ax}}\left(3\cos^2\theta - 1\right) + \frac{3}{2}\Delta\chi_{\mathrm{th}}\sin^2\theta\cos 2\varphi\right]$$
$$\Delta\chi_{\mathrm{ax}} = \chi_{zz} - \frac{\chi_{\mathrm{xx}} + \chi_{\mathrm{yy}}}{2} \text{ and } \Delta\chi_{\mathrm{th}} = \chi_{\mathrm{xx}} - \chi_{\mathrm{yy}} \tag{4.15}$$

where r is the distance between observed nuclei and electron, $\Delta\chi_{\mathrm{ax}}$ and $\Delta\chi_{\mathrm{rh}}$ are the axial and rhombic anisotropy parameters of the magnetic susceptibility tensor of the system, and θ and φ are the polar coordinates of the nucleus in the frame of the electronic magnetic susceptibility tensor.

The unpaired electron–nucleus dipolar interaction causes R_1 (longitudinal) and R_2 (transverse) nuclear relaxation because of its modulation. For the Solomon and Curie contributions, proton relaxation enhancement (PRS) is described by equations [87]

$$R_2^{\mathrm{PRE}} = \frac{k_{\mathrm{Solomon}} + k_{\mathrm{Curie}}}{r^6} \quad (3)$$
$$k_{\mathrm{Solbmon}} = \left(\frac{\mu_0}{4\pi}\right)^2 \frac{\gamma_{\mathrm{I}}^2 g_e^2 \mu_{\mathrm{B}}^2 S(S+1)}{15}\left[4\tau_{\mathrm{c}} + \frac{\tau_{\mathrm{c}}}{1 + (\omega_{\mathrm{I}} - \omega_{\mathrm{S}})^2\tau_{\mathrm{c}}^2}\right.$$
$$\left. + \frac{3\tau_{\mathrm{c}}}{1 + \omega_{\mathrm{I}}^2\tau_{\mathrm{c}}^2} + \frac{6\tau_{\mathrm{c}}}{1 + (\omega_{\mathrm{I}} + \omega_{\mathrm{S}})^2\tau_{\mathrm{c}}^2} + \frac{6\tau_{\mathrm{c}}}{1 + \omega_{\mathrm{S}}^2\tau_{\mathrm{c}}^2}\right]$$
$$k_{\mathrm{Curie}} = \frac{1}{5}\left(\frac{\mu_0}{4\pi}\right)^2 \frac{\omega_{\mathrm{I}}^2 g_e^4 \mu_{\mathrm{B}}^4 S^2(S+1)^2}{(3kT)^2}\left[4\tau_{\mathrm{Curie}} + \frac{3\tau_{\mathrm{Curie}}}{1 + \omega_{\mathrm{I}}^2\tau_{\mathrm{Curie}}^2}\right] \tag{4.16}$$

As it is seen from (4.16), the interaction depends on the nuclear Larmor frequency, electron Larmor frequency, electron–nucleus distance, electron g-factor, electron Bohr magneton, nuclear magnetogyric ratio, Boltzmann constant, temperature, *and* the electron spin quantum number. In certain conditions, (4.14–4.16) allow to estimate spin density on nucleus and determine electron–nucleus distances. Recently, a new method for extracting probability densities from pseudocontact shift (PCS) data that relies on Tikhonov-regularised 3D reconstruction was described [89].

Dynamic nuclear polarization (DNP) [88, 92–94] results from transferring spin polarization from electrons to nuclei, thereby enhancing the nuclear spin polarization. Electrons at a given magnetic field and temperature in thermal equilibrium can be aligned to a higher degree by chemical reactions (Chemical-induced DNP, CIDNP), optical pumping and spin injection. The polarization transfer between electrons and nuclei can occur spontaneously when electron spin polarization deviates from its thermal equilibrium value through electron-nuclear cross-relaxation and/or spin-state mixing among electrons and nuclei. The general static DNP Hamiltonian can be written as [90, 91]

$$\mathrm{H} = H = H_E + H_N + H_{EN} = \omega_{0E}E_z - \omega_{0N}N_z + H_{EN}^{is} + H_{EN}^{di}$$
$$= \omega_{0Z}E_z - \omega_{0N}N_z + K_{SE}(E_Z N_Z + E_X N_X + E_Y N_Y) + K_{PSE}E_X N_Z \tag{4.17}$$

where H_E and H_N are the Hamiltonians for the electron and nucleus, respectively; H_{EN} is the hyperfine coupling, which is separated into the isotropic hyperfine interaction H_{EN}^{is} and the anisotropic dipolar coupling H_{EN}^{di} between the electron and the nucleus; the coefficients K_{SE} and K_{PSE} denote the secular and pseudosecular hyperfine interactions. ω_{0E} and ω_{0N} are the electron and nuclear Larmor frequencies. The nuclear Overhauser effect (NOE), the solid effect (SE), the cross-effect (CE), and thermal mixing (TM) can be mechanisms for microwave-driven DNP processes. The Overhauser effect accounts for the perturbation of nuclear spin level populations observed in free radicals when electron spin transitions are saturated by the microwave irradiation [96]. The solid effect occurs when an electron–nucleus mutual spin flip transition in an electron–nucleus two spin system is excited by microwave irradiation [97]. The cross-effect requires two unpaired electrons as the source of high polarization [87]. The thermal mixing effect is an energy exchange phenomenon between the electron spin ensemble and the nuclear spin, using the electron spin ensemble as a whole to provide hypernuclear polarization.

An important advantage of the DNP-NMR method is the ability to significant NMR signal enhancement [98, 99]. For example, the technique of dynamic nuclear polarization (DNP) can significantly improve the sensitivity of solid state NMR spectroscopy by transferring the large electron spin polarization to the nuclear spin system, through microwave irradiation at or near the electron paramagnetic resonance frequency. Studies of the temperature, magnetic field, and microwave power dependence of the DNP enhancement can optimize the signal enhancements. In some cases, magnetic or chemical interactions between the radicals and the target molecules can lead to attenuation of the NMR signal through paramagnetic quenching and/or radical decomposition [99]. To avoid this problem, polarizing materials incorporating nitroxide radicals of the solids to minimize interactions between the radicals and the solute were introduced. For example, these materials can hyperpolarize pure pyruvic acid.

A prepared series of 37 dinitroxide biradicals was used as polarizing agents in cross-effect DNP-NMR experiments at 9.4 T and 100 K in 1,1,2,2-tetrachloroethane [100]. It was observed that in this regime the DNP: (1) Performance is strongly correlated with the substituents on the polarizing agents. (2) Electron and nuclear spin relaxation times, with longer relaxation times lead to better enhancements. (3) Deuteration of the radicals generally leads to better DNP enhancement with longer build up time. To avoid suffering many polarizing agents from an unfavorable field and magic angle spinning (MAS) frequency dependence, a series of new hybrid biradicals, α,γ-bisdiphenylene-β-phenylallyl (BDPA), tethered to a broad line nitroxide was prepared and investigated [95]. By tuning the distance between the two electrons and the substituents at the nitroxide moiety, correlations between the electron–electron interactions and the electron spin relaxation times on the one hand and the BDPA DNP enhancement factors on the other hand were established. The best radical in this series has a short methylene linker and bears bulky phenyl spirocyclohexyl ligands.

References

1. W.R. Hagen, *Biomolecular EPR Spectroscopy* (CRC, 2008)
2. G.R. Eaton, S.S. Eaton, D.P. Barr, R.T. Weber, *Quantitative EPR* (Springer, Berlin, 2010)
3. A. Savitsky, K. Mobius, *High-Field EPR Spectroscopy on Proteins and their Model Systems* (Wiley-VCH Verlag GmbH, 2009)
4. J. Jamauci, Fundamentals of electron spin resonance, in *Nitroxides: Application in Chemistry, Biomedicine, and Materials Science*, ed. by G.I. Likhtenshtein, J. Yamauchi, S. Nakatsuji, A. Smirnov, R. Tamura (WILEY-VCH, Weinhem, 2008), pp. 71–120
5. G.I. Likhtenshtein, *Electron Spin in Chemistry and Biology: Fundamentals, Methods, Reactions Mechanisms, Magnetic Phenomena, Structure Investigation* (Springer, Berlin, 2016)
6. O.Ya. Grinberg, A.A. Dubinskii, V.F. Shuvalov, L.G. Oranskii, V.I. Kurochkin, Ya.S. Lebedev, EPR submillimeter spectroscopy of free radicals. Doklady Akademii Nauk SSSR **230**, 884–887 (1976)
7. M. Malferrari, A. Savitsky, W. Lubitz, K. Mobius, G. Venturoli, Protein immobilization capabilities of sucrose and trehalose glasses: the effect of protein/sugar concentration unraveled by high-field EPR. J. Phys. Chem. Lett. **7**(23), 4871–4877 (2016)
8. J.H. Freed, New technologies in electron spin resonance. Ann. Rev. Phys. Chem. **51**, 655–689 (2000)
9. K. Moebius, A.N. Savitsky, *High-Field EPR spectroscopy of Proteins and their Model Systems: Characterization of Transient Paramagnetic States* (Royal Society of Chemistry, 2009
10. A.I. Smirnov, Resent advantages in ESR techniques used in nitroxide application, in G.I. Likhtenshtein, J. Yamauchi, S. Nakatsuji, A. Smirnov, R. Tamura, Nitroxides: *Application in Chemistry, Biomedicine, and Materials Science* (WILEY-VCH, Weinhem, 2008), pp. 121–160
11. F. Bloch, Nuclear induction. Phys. Rev. **70**, 460–473 (1946)
12. G.I. Likhtenshtein, Determination of the topography of proteins groups using specific paramagnetic labels. Mol. Biol. (Moscow) **2**, 234–240 (1968)
13. G.I. Likhtenshtein, *Spin Labeling Method in Molecular Biology* (Wiley Interscience, New York, NY, 1976)
14. J.C. Taylor, J.S. Leigh, M. Cohn, The effect of dipole-dipole interaction between nitroxide radical and a paramagnetic ion on the line shape of the ESR spectra of radical. Proc. Natl. Acad. Sci. U.S.A. **64**, 206–219 (1969)
15. G.I. Likhtenshtein, Study on the proteins microstructure by method of spin-label paramagnetic probe. Mol. Biol. (Moscow) **4**, 782–789 (1970)
16. A.V. Kulikov, G.I. Likhtenshtein, E.G. Rozantzev, V. Suskina, A.V. Shapiro, Nitroxide bi- and polyradicals as standard models for distance estimation between the nitroxide moities. Biofizika **17**, 42–49 (1972)
17. A.I. Kokorin, K.I. Zamaraev, G.L. Grigoryan, V.P. Ivanov, E.G. Rozantsev, Distance estimation between nitroxyl radicals. Biofizika **17**, 34–41 (1972)
18. A.V. Kulikov, G.I. Likhtenstein, Application of saturation curves for evaluating distances in biological objects by the method of double spin-labels. Biofisika **19**, 420–424 (1974)
19. G.I. Likhtenshtein, Depth of immersion of paramagnatic centers, in *Magnetic Resonance in Biology*, ed. by L. Berliner, S. Eaton, G. Eaton (Kluwer Academic Publishers, Dordrecht, 2000), pp. 1–36
20. G.D. Case, J.S. Leigh, Intramitochondrial position of cytochrome haem groups determined by dipolar interaction with paramagnetic cations. Biochem. J. **160**, 769–783 (1976)
21. A.V. Kulikov, G.I. Likhtenshtein, The use of spin-relaxation phenomena in the investigation of the structure of model and biological systems by method of spin labels. Adv. Mol. Relax Proc. **10**, 47–78 (1977)
22. M.K. Bowman, J.R. Norris, Cross relaxation of free radicals in partially ordered solids. J. Phys. Chem. **86**, 3385–3390 (1982)

23. L.A. Syrtsova, L.A. Levchenko, E.N. Frolov, G.I. Likhtenshtein, T.N. Pisarscaya, L.V. Vorob'ev, V.A. Gromoglasova, Structure and function of the nitrogenase components from *Azotobacter vinelandii*. Mol. Biol. (Moscow) **5**, 726–734 (1971)
24. E.J. Hustedt, A.H. Beth, Structural information from CW-EPR spectra of dipolar coupled nitroxide spin label, in *Biological Magnetic Resonance: Distance Measurements in Biological Systems by EPR*, vol. 19, ed. by L.J. Berliner, G.R. Eaton, S.S. Eaton (Kluwer Academic, New York, 2000), pp. 155–184
25. W. Xiao, Y.K. Shin, EPR spectroscopic ruler: the deconvolution method and its application, in *Biological Magnetic Resonance: Distance Measurements in Biological Systems by EPR*, vol. 19, ed. by L.J. Berliner, G.R. Eaton, S.S. Eaton (Kluwer Academic, New York, 2000), pp. 249–276
26. J.H. Freed, S. Wang, Theory of relaxation, applied to saturation and double resonance in ESR spectra.VI. Saturation recovery. J. Phys. Chem. 1155–1157 (1974)
27. M. Houisjen, J.S. Hyde, Saturation recovery measurements of electron spin relaxation time of free radicals in solution. J. Chem. Phys. **60**, 1682–1683 (1974)
28. S.S. Eaton, G.R. Eaton, Relaxation times of organic radicals and transition metal ions, in *Biological Magnetic Resonance: Distance Measurements in Biological Systems by EPR*, vol. 19, ed. by L.J. Berliner, G.R. Eaton, S.S. Eaton SS, (Kluwer Academic, New York, 2000), pp. 29–154
29. G.M. Smith, D.J. Keeble, O. Schiemann (eds.), *Introduction to Modern EPR Spectroscopy* (CRC, 2010)
30. S.S. Eaton, G.R. Eaton, L.J. Berliner (eds.), Biomedical EPR—part B: methodology, instrumentation, and dynamics, in *Biological Magnetic Resonance*, vol. 24 (Springer, Berlin, 2005)
31. L.J. Berliner, S.S. Eaton S, G.R. Eaton (eds.), Distance measurements in biological systems by EPR, in *Biological Magnetic Resonance*, vol. 19 (Springer, Berlin, 2001)
32. G.R. Eaton, S.S. Eaton, K.M. Salikhov, *Foundations of Modern EPR* (World Scientific, 1998)
33. G.I. Likhtenstein, *Biophysical Labeling Methods in Molecular Biology* (Cambridge University Press, New York, Cambridge, 1993), pp. 46–79
34. G. Jeschke, Dipolar spectroscopy—double-resonance methods. eMagRes **5**(3), 1459–1476 (2016)
35. O. Krumkacheva, E. Bagryanskaya, EPR-based distance measurements at ambient temperature. J. Magn. Reson. **280**, 117–126 (2017)
36. A. Feintuch, G. Otting, D. Goldfarb, Gd^{3+} spin labeling for measuring distances in biomacromolecules: why and how? in *Methods in Enzymology*, vol. 563 (2015), pp. 416–457
37. G.E. Pake, Nuclear resonance absorption in hydrated crystals: fine structure of the proton line. J. Chem. Phys. **16**, 327–336 (1948)
38. S.A. Dzuba, E.S. Salnikov, L.V. Kulik, CW EPR, echo-detected EPR, and field-step ELDOR study of molecular motions of nitroxides in o-terphenyl glass: Dynamical transition, dynamical heterogeneity and β-relaxation. Appl. Magn. Resn. **30**, 637–650 (2006)
39. A.D. Milov, K.M. Salikhov, Y.D. Tsvetkov, Phase relaxation of hydrogen atoms stabilized in an amorphous matrix. Sov. Phy. Solid State **15**, 802–806 (1973)
40. A.D. Milov, R.I. Samoilova, Y.D. Tsvetkov, M. De Zotti, C. Toniolo, J. Raap, PELDOR conformational analysis of bis-labeled Alamethicin aggregated in phospholipid vesicles. J. Phys. Chem. B **112**, 13469–13472 (2008)
41. V.V. Kurshev, A.M. Raitsimring, Yu.D. Tsvetkov, Selection of dipolar interaction by the "2+ 1" pulse train ESE. J. Magn. Reson. **81**, 441 (1989)
42. G. Jeschke, DEER distance measurements in proteins. Annu. Rev. Phys. Chem. **63**, 419–446 (2012)
43. P.P. Borbat, E.R. Georgieva, J.H. Freed, Improved sensitivity for long-distance measurements in biomolecules: five-pulse double electron-electron resonance. J. Phys. Chem. Lett. **4**, 170–175 (2013)
44. G.I. Likhtenshtein, Depth of immersion of paramagnatic centers, in *Magnetic Resonance in Biology*, ed. by L. Berliner, S. Eaton, G. Eaton (Kluwer Academic Publishers, Dordrecht, 2000), pp. 309–347

45. S. Milikisiyants, M.A. Voinov, A.I. Smirnov, Refocused Out-Of-Phase (ROOPh) DEER: a pulse scheme for suppressing an unmodulated background in double electron-electron resonance experiments. J. Magn. Reson. **293**, 9–18 (2018)
46. F.D. Breitgoff, Y.O. Polyhach, G. Jeschke, Reliable nanometre-range distance distributions from 5-pulse double electron electron resonance. Phys. Chem. Chem. Phys. **19**(24), 15754–15765 (2017)
47. J. Tang, J.R. Noris, Multiple quantum EPR coherence in a spin correlated pair system. Chem. Phys. Lett. **233**, 192–200 (1995)
48. S.A. Dzuba, M.K. Bosh, A.Y. Hoff, Electron and spin-echo detection of quantum beats and double quantum coherence in spin-correlated pairs of protonated photosynthetic reaction centers. Chem. Phys. Lett. **248**, 427–433 (1996)
49. P.P. Borbat, J.H. Freed, Double-quantum ESR and distance measurements, in *Distance Measurements in Biological Systems by EPR*, Part of the Biological Magnetic Resonance book series, vol. 19 (Springer, Boston, 2002), pp. 383–459
50. P.P. Borbat, J.H. Freed, Multiple-quantum ESR and distance measurements. Chem. Phys. Lett. **313**, 145–154 (1999)
51. Y.-W. Chiang, A.J. Costa-Filho, B. Baird, J.H. Freed, 2D-ELDOR study of heterogeneity and domain structure changes in plasma membrane vesicles upon cross-linking of receptors. J. Phys. Chem. B **115**, 10462–10469 (2011)
52. A. Schweiger, C. Gemperle, R.R. Ernst, Soft pulse electron-spin-echo-envelope modulation spectroscopy (Soft ESEEM). J. Magn. Reson. **86**, 70–81 (1990)
53. S.A. Dzuba, A. Kawamori, Selective hole burning in EPR: spectral diffusion and dipolar broadening. Concepts Magn. Reson. **8**, 49–61 (1996)
54. K.B. Konov, A.A. Knyazev, Y.G. Galyametdinov, N.P. Isaev, L.V. Kulik, Selective hole-burning in RIDME experiment: dead-time free measurement of dipolar modulation. Appl. Magn. Reson. **44**(8), 949–966 (2013)
55. J. Solomon, N. Bloembergen, Nuclear magnetic interaction in HF molecule. J. Chem. Phys **25**, 261–266 (1956)
56. A.V. Kulikov, Determination of distance between the nitroxide label and a paramagnetic center in spin-labeled proteins from the parameters of the saturation curve of the ESR spectrum of the label at 77K. Mol. Biol. (Moscow) **10**, 109–116 (1976)
57. K.M. Salikhov, A.G. Semenov, Yu.D. Tsvetkov, *Electron Spin Echo and Its Application* (Nauka, Novosibirsk, 1976)
58. S. Eaton, G. Eaton, Determination of distance based on T_1 and T_2 effects, in *Magnetic Resonance in Biology*, ed. by L. Berliner, S. Eaton, G. Eaton (Kluwer Academic Publishers, Dordrecht), pp. 348–382
59. M.F. Perutz, Mechanisms of cooperativity and allosteric regulation in proteins. Quart. Rev. Biophys. **22**, 139–236 (1989)
60. S. Razzaghi, E.K. Brooks, E. Bordignon, W.L. Hubbell, M. Yulikov, G. Jeschke, EPR relaxation-enhancement-based distance measurements on orthogonally spin-labeled T4-lysozyme. Chembiochem **14**(14), 1883–1890 (2013)
61. S. Weber, G. Kothe, J.R. Norris, Transient nutation electron spin resonance spectroscopy Transient nutation electron spin resonance spectroscopy on spin-correlated radical pairs: a theoretical analysis on hyperfine-induced nuclear modulations. J. Chem. Phys. **106**, 6248–6261 (1997)
62. K. Ayabe, K. Sato, S. Nishida, T. Ise, S. Nakazawa, K. Sugisaki, Y. Morita, K. Toyota, D. Shiomi, M. Kitagawa, T. Takui, Pulsed electron spin nutations spectroscopy of weakly exchange-coupled biradicals: a general theoretical approach and determination of the spin dipolar interaction. Phys. Chem. Chem. Phys. **14**, 9137–9148 (2012)
63. M. Srivastava, J.H. Freed, Singular value decomposition method to determine distance distributions in pulsed dipolar electron spin resonance: II Estimating uncertainty. J. Phys. Chem. A **123**(1), 359–370 (2019)
64. M. Srivastava, E.R. Georgieva, J.H. Freed, A new wavelet denoising method for experimental time-domain signals: pulsed dipolar electron spin resonance. J. Phys. Chem. A **121**(12), 2452–2465 (2017)

65. A. Doll, M. Qi, A. Godt, G. Jeschke, CIDME: short distances measured with long chirp pulses. J. Magn. Reson. **273**, 73–82 (2016)
66. A. Doll, G. Jeschke, EPR-correlated dipolar spectroscopy by Q-band chirp SIFTER. Phys. Chem. Chem. Phys. **18**(33), 23111–23120 (2016)
67. A. Doll, G. Jeschke, Double electron-electron resonance with multiple non-selective chirp refocusing. Phys. Chem. Chem. Phys. **19**(2), 1039–1053 (2017)
68. M. Sajid, M. Hulsmann, A. Godt, G. Jeschke, Two-dimensional distance correlation maps from pulsed triple electron resonance (TRIER) on proteins with three paramagnetic centers. Appl. Magn. Reson. **49**(11), 1253–1279 (2018)
69. H.M. McConnell, R.E. Robertson, Isotropic nuclear resonance shifts. J. Chem. Phys. **29**, 1361–1366 (1958)
70. M. Karplus, Contact electron-spin coupling of nuclear magnetic moments. J. Chem. Phys. **30**, 11–16 (1959)
71. G. Feher, Observation of nuclear magnetic resonances via the electron spin resonance line. Phys. Rev. **103**(3), 834–835 (1956)
72. J.S. Hyde, ENDOR of free radicals in solution. J. Chem. Phys. **43**, 1806–1812 (1965)
73. H. Kurreck, B. Kirste, W. Lubitz, *Electron Nuclear Double Resonance Spectroscopy of Radicals in Solution* (VCH Publishers, New York, 1988)
74. D.V. Murphy, R.D. Farley, Principles and applications of ENDOR spectroscopy for structure determination in solution and disordered matrices. Chem. Soc. Rev. **35**, 249–268 (2006)
75. J.S. Hyde, J.C.W. Chien, J.H. Freed, Electoron–electron double resonance of free radicals in solution. J. Chem. Phys. **48**, 4211 (1968)
76. V.A. Benderskii, L.A. Blyumenfel'd, P.A. Stunzhas, F.A. Sokolov, Double electron-electron resonance of triplet excitons in ion-radical salts. Nature **220**, 365–367 (1968)
77. W.B. Mims, Amplitudes of superhyperfine frequencies in electron-spin-echo-envelope. Phys. Rev. B **6**, 3543–3545 (1973)
78. E.L. Hahn, Spin echoes. Phys. Rev. **80**, 580–594 (1950)
79. B. Kasumaj, H. Dube, N. Zoelch, F. Diederich, G. Jeschke, Relaxation and modulation interference effects in two-pulse electron spin echo envelope modulation (ESEEM). J. Magn. Reson. **223**, 187–197 (2012)
80. L.V. Kulik, I.A. Grigor'ev, E.S. Salnikov, S.A. Dzuba, Y.D. Zvetkov, Electron spin–echo envelope modulation induced by slow intramolecular motion. J. Phys. Chem. A **107**, 3692–3695 (2003)
81. A. Schweiger, G. Jeschke, *Principles of Pulse Electron Paramagnetic Resonance* (Oxford University Press, Oxford, 2001)
82. M. Mehring, P. Hofer, A. Grupp, Pulsed electron nuclear double and triple resonance schemes. Ber. Bunsenges. Phys. Chem. **91**, 1132–1137 (1987)
83. P. Hofer, A. Grupp, H. Nebenfuhr, M. Mehring, Hypern sublevel correlation (HYSCORE) spectroscopy—a 2D ESR investigation of the squaric acid radical. Chem. Phys. Lett. **132**, 279–282 (1986)
84. D. Goldfarb, V. Kofman, J. Libman, A. Shanzer, R. Rahmatouline, S. Van Doorslaer, A. Schweiger, Double nuclear coherence transfer (DONUT)-HYSCORE: a new tool for the assignment of nuclear frequencies in pulsed EPR experiments. J. Am. Chem. Soc. **120**, 7020–7029 (1998)
85. B. Fresch, D. Frezzato, G.J. Moro, G. Kothe, J.H. Freed, Collective fluctuations in ordered fluids investigated by two-dimensional electron-electron double resonance spectroscopy. J. Phys. Chem. B **110**, 24238–24254 (2006)
86. S. Lee, D. Budil, J.H. Freed, Theory of two-dimensional Fourier transform electron spin resonance for ordered and viscous fluids. J. Chem. Phys. **101**, 5529 (30 pages) (1994)
87. I.A. Solomon, Relaxation processes in a system of two spins. Phys. Rev. **99**, 559–565 (1955)
88. I. Bertini, C. Luchinat, M. Nagulapalli, G. Parigia, E. Ravera, Paramagnetic relaxation enhancement for the characterization of the conformational heterogeneity in two-domain proteins. Phys. Chem. Chem. Phys. **14**, 9149–9156 (2012)

89. E.A. Suturina, D. Haussinger, K. Zimmermann, L. Garbuio, M. Yulikov, G. Jeschke, I. Kuprov, Model-free extraction of spin label position distributions shift data from pseudocontact. Chem. Sci. **8**(4), 2751–2757 (2017)
90. Y.N. Molin (ed.), *Spin Polarization and Magnetic Effects in Radical Reactions* (Elsevier, Amsterdam, 1984)
91. L.J. Berliner, E. Bagryanskaya, Chemically induced electron and nuclear polarization, in *Multifrequency Electron Paramagnetic Resonance*, ed. by S. Misra (Willey-VCH, 2011), pp. 947–992
92. A. Abragam, M. Goldman, Principles of dynamic nuclear polarisation. Rep. Prog. Phys. **41**, 395–467 (1976)
93. C. Griesinger, M. Bennati, H.M. Vieth, C. Luchinat, G. Parigi, P. Höfer, F. Engelke, S.J. Glaser, V. Denysenkov, T.F. Prisner, Dynamic nuclear polarization at high magnetic fields in liquids. Prog. Nucl. Magn. Reson. Spectrosc. **64**, 4–28 (2012)
94. H. Hayashi, *Introduction to dynamic spin chemistry* (World Scientific, Singapore, 2004)
95. D. Wisser, G. Karthikeyan, A. Lund, G. Casano, H. Karoui, M. Yulikov, G. Menzildjian, A.C. Pinon, A. Purea, F. Engelke, S.R. Chaudhari, D. Kubicki, A.J. Rossini, I.B. Moroz, D. Gajan, C. Copéret, G. Jeschke, M. Lelli, L. Emsley, A. Lesage, O. Ouari, BDPA-nitroxide biradicals tailored for efficient dynamic nuclear polarization enhanced solid-state NMR at magnetic fields up to 21.1 T. J. Am. Chem. Soc. **140**(41), 13340–13349 (2018)
96. A.W. Overhauser, Polarization of nuclei in metals. Phys. Rev. **92**, 411–415 (1953)
97. A.A. Smith, B. Corzilius, A.B. Barnes, T. Maly, R.G. Griffin, Solid effect dynamic nuclear polarization and polarization pathways. J. Chem. Phys. **136**(1), 015101 (2012)
98. K.H. Sze, Q. Wu, H.S. Tse, G. Zhu, Dynamic nuclear polarization: new methodology and applications. Top. Curr. Chem. **326**, 215–242 (2012)
99. D.L. Silverio, H.A van Kalkeren, T.C. Ong, M. Baudin, M. Yulikov, L. Veyre, P. Berruyer, Chaudhari, G. Sachin, D. Gajan, D. Baudouin, C. Matthieu, V. Basile, G. Jeschke, G. Bodenhausen, A. Lesage, L. Emsley, S. Jannin, C. Thieuleux, C. Copéret. Tailored polarizing hybrid solids with nitroxide radicals localized in mesostructured silica walls. Helv. Chim. Acta **100**(6) (2017)
100. D. Kubicki, G. Casano, M. Schwarzwalder, S. Abel, C. Sauvee, K. Ganesan, M. Yulikov, A.J. Rossini, G. Jeschke, C. Coperet, A. Lesage, P. Tordo, O. Ouari, L. Emsley, Rational design of dinitroxide biradicals for efficient cross-effect dynamic nuclear polarization. Chem. Sci. **7**(1), 550–558 (2016)

Chapter 5
Nitroxide Biradicals

Alexander I. Kokorin

Abstract This chapter is devoted to the current state in the area of nitroxide biradicals shortly describing their history, effects of the electron spin exchange in nitroxide biradicals, structural investigations including comparing the X-ray structure with DFT calculations, features of the intramolecular dynamics in biradicals, and their possible applications.

5.1 Introduction—Historical Notes

The first information about nitroxide biradicals (NB) has appeared in press [1–4] less than in a year after classical papers by Rozantsev and Neiman [5, 6]. Afterward, the number of publications on this item increased exponentially and a lot of them were cited in many books and reviews, e.g., in [7–17]. The figures presented in these articles revealed the wide diversity of EPR spectra of different biradicals and their variability on chemical composition, temperature and solvent nature, which initiated discussions concerning the mechanisms of the intramolecular spin exchange in such systems with two stable unpaired electrons. In principle, any biradical molecule can be reproduced as R˙-Z- R˙, where R˙ means a paramagnetic radical group (e.g., see Fig. 5.1), and Z denotes organic or metal-organic bridge connecting two radical fragments [15].

Evidently, the "world" of stable nitroxide radicals centers which were involved to constructing NB molecules is much more diversed including such groups as are shown in Fig. 5.2.

These compounds were synthesized and described, and their properties were investigated, e.g., in: (*A*) [18], (*B*) [19], (*C*) [20, 21], (*D*) [21], (*E*) [22], (*F*) [23], (*G*) [24], (*H*) [25]. Also, such interesting works as [26, 27] should be mentioned in which ferromagnetic interactions in a crystalline nitroxide biradical: 1,3,5,7-tetramethyl-2,6-diazaadamantane *N*,*N*′-dioxyl [26] and the structure–properties' relationships

A. I. Kokorin (✉)
N.N. Semenov Federal Research Center for Chemical Physics, Russian Academy of Sciences, Moscow, Russian Federation
e-mail: alex-kokorin@yandex.ru; kokorin@chph.ras.ru

G. I. Likhtenshtein, *Nitroxides*, Springer Series in Materials Science 292,
https://doi.org/10.1007/978-3-030-34822-9_5

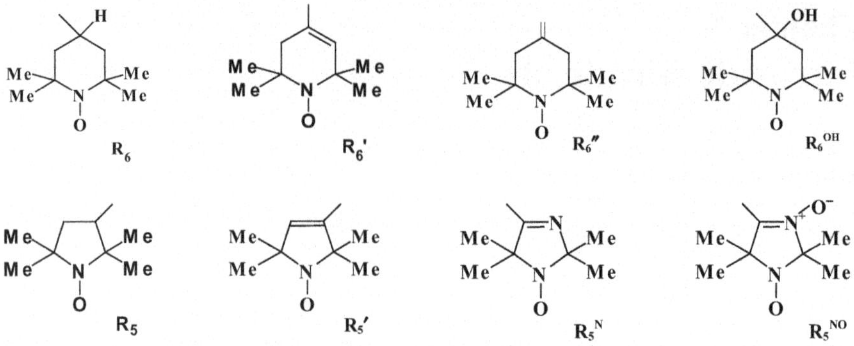

Fig. 5.1 Schematic structures of some nitroxide rings

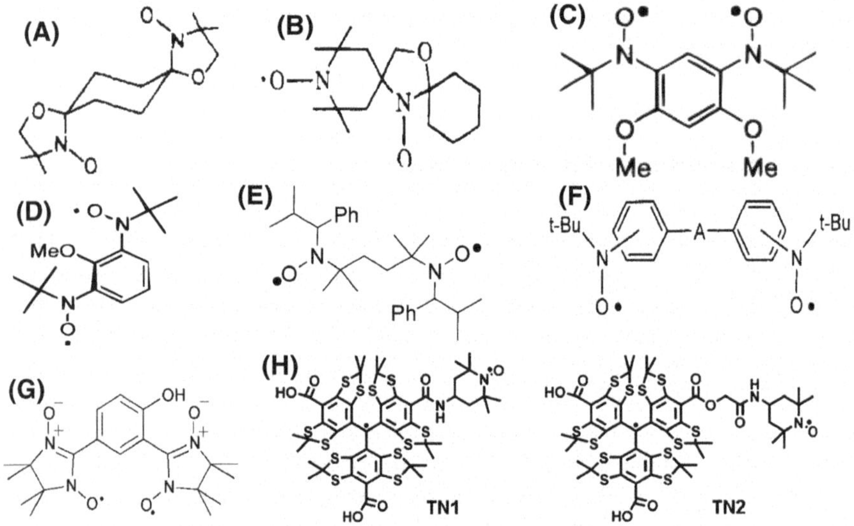

Fig. 5.2 Structures of some novel nitroxide groups

were analyzed in trimethylenemethane-type biradicals which were synthesized and studied by EPR spectral characterization of biradicals [27].

A lot of publications were devoted to a huge class of nitronyl nitroxide biradicals, their structure, and properties, e.g., [13, 17, 24, 28–30] and many others. As an example, magnetic properties of phenol- and phenoxide-substituted nitronyl nitroxide biradicals were revealed to be perspective as building blocks of organic-salt ferrimagnets [28], and some nitronyl nitroxide biradicals and their copper (II) complexes showed interesting structures, and magnetic properties in a solid state [29]. A nice family of stable organic triradicals with quartet ground states, which consisted of different nitroxide radicals, their syntheses, and magnetic properties were described in [31].

During the last decade, a new class of mixed biradicals, first of all, trityl-nitroxide biradicals as unique molecular probes which exhibited enhanced sensitivity and stability for rapid and simultaneous measurement of redox status and oxygenation by EPR spectroscopy [25]. First results showed that these biradical probes provided an opportunity for the design of new spin probes with the sensitivity of nitroxides to measure redox status while the narrow singlet signal of the trityl group enhanced sensitivity and speed of detection for greatly increased resolution of EPR imaging. Such new probes have high potential for the simultaneous EPR measurement of redox status and oxygenation in a wide variety of chemical and biological systems.

The value of exchange interaction J in organic biradicals strongly depends on their composition and can be modulated by changing the linker. In [32], for the first time the effect of chiral configurations of radical parts on the J value was demonstrated in the case of trityl-nitroxide biradicals (TN, Fig. 5.2h). Four diastereoisomers were synthesized and purified by the conjugation of a racemic nitroxide with the racemic trityl radical via l-proline. The absolute configurations of these diastereoisomers were assigned by comparing experimental and calculated electronic circular dichroism (ECD) spectra. EPR spectra showed that the configuration of the nitroxide part instead of the trityl part was dominant in controlling the exchange interaction and the order of the J value at room temperature. Measured $|J|$ values clearly showed this effect, and in some cases, the $|J|$ values varied with temperature and the polarity of the solvent due to their flexible linker, whereas the J values of another biradical were almost insensitive to these factors due to the rigidity of their linkers [32].

Simultaneous evaluation of redox status and oxygenation in biological systems is of great importance for the understanding of biological functions, and an approach of their measuring using EPR spectroscopy using nitroxide radicals has already interesting applications but were still limited by rather low oxygen sensitivity and low EPR resolution due to the moderately broad EPR triplet and spin quenching through biochemical processes. Authors of [33] showed that these problems can be overcome by the use of new trityl-nitroxide biradicals (TNB) contained ^{14}N pyrrolidinyl nitroxide and a trityl and its isotopically labeled by ^{15}N analog. Both TNB exhibited much stronger spin–spin interaction with $|J| > 40$ mT compared with that of the previously synthesized TNB with longer linker chain length. The enhanced stability of new TNB was confirmed, and the effect of different types of cyclodextrins on its stability was also studied. New biradicals are sensitive to redox status, and their corresponding trityl-hydroxylamines formed from the reduction of the biradicals by ascorbate shared the same oxygen sensitivity. The ^{15}N-labeled TNB exhibited higher EPR signal amplitude as compared with that of ^{14}N analog. Cyclic voltammetric studies verified the electrochemical behavior of these TNB [33].

Several trityl-nitroxide biradicals (TNB) with highly asymmetric exchange coupling revealed specific magnetic properties provided new possibilities for the application in physicochemical, biophysical, and biological studies [34]. The effect of the linker length on the spin–spin interaction in TNB was investigated. It was shown that the magnitude of the spin–spin coupling $|J|$ could be easily tuned from ~0.4 mT to over 120 mT using various linkers separated two radical moieties and under varying temperatures. Computer simulation of EPR spectra was carried out to estimate $|J|$

values in TNB. The spin–spin interaction in TNB, their hyperfine splitting (hfs) constants g, and zero-field splitting D were measured at 220 K. Changes of the spin–spin coupling with variation of linker length and temperature provide a way to develop new TNB for applications in relevant fields [34].

TN biradicals were successfully used to study supramolecular host–guest interactions of TNB with methyl-β-cyclodextrin, hydroxypropyl-β-cyclodextrin, and γ-cyclodextrin [35]. EPR spectroscopy showed that in the presence of cyclodextrins, host–guest complexes were formed where the nitroxide and linker parts could interact with the cyclodextrins' cavities. Complexation with cyclodextrins led to the suppression of the intramolecular through-space spin–spin-exchange coupling allowing the determination of the through-bond spin–spin-exchange coupling value which was calculated as 0.16 mT using EPR simulations. Different types of cyclodextrins had various binding affinities with TNB. The effect of the linker in TNB on the host–guest forming was studied. The complexes of TNB with cyclodextrins could noticeably change the spin–spin-exchange coupling due to the size of linkers, and the stability of TNB toward ascorbate ions was significantly enhanced after the attaching to cyclodextrins. This approach seems to be a useful method to modulate the magnitude of the spin–spin interaction and redox sensitivity of TNB.

5.2 Electron Spin Exchange in Nitroxide Biradicals

Theory of the electron spin exchange in stable nitroxide biradicals with explanation of the changes observed in the EPR spectra has been developed and published in several independent papers practically at the same time [3, 4, 36–38], was detailed in papers [39, 40], and then enhanced to both areas of fast and slow exchange coupling by Parmon et al. in [41, 42]. A bit later, all data concerning physical–chemical and EPR spectroscopy features of nitroxide biradicals were collected and critically analyzed in [11, 12].

In liquid solutions with a low viscosity, the spin Hamiltonian $\hat{H}$ comprises the isotropic hyperfine interactions, the Zeeman splitting, and the exchange coupling. In those cases when both radical fragments are identical and each bears only one nucleus with a nonzero nuclear spin I, the following equation is valid [37, 38]:

$$\hat{H} = \gamma_e H_0\left(S_z^{(1)} + S_z^{(2)}\right) + a\left(S_z^{(1)} I_z^{(1)} + S_z^{(2)} I_z^{(2)}\right) + J S^{(1)} S^{(2)} \qquad (5.1)$$

The spin Hamiltonian here is written in frequency units; superscripts 1 and 2 denote different radical fragments; $S^{(k)}$ are electron spin operators; $Sz^{(k)}$ and $Iz^{(m)}$ are projections of the electron and nuclear spins to the z-axis, respectively; γ_e is the electron magnetogyric ratio; H_0 is the external magnetic field; a is the ^{14}N isotropic hfs constant of the radical fragments, and J is the exchange integral. In a low-viscous solvent, the dipole–dipole coupling tensor is averaged to zero owing to the fast rotational motion of the biradical molecules [12, 42]. For any individual conformation, only one value $|J|$ should correctly describe the position and integral intensity of each line in the EPR spectrum. In some cases, J in (5.1) is the mean value of two rather similar exchange integrals corresponding to two conformations, which one cannot distinguish in the experimental setup because of very fast hopping between them, e.g., [43–45].

The manifestation of exchange interaction in EPR spectra is caused by the ratio between the exchange integral J and the isotropic hfs constant a. If $|J| \ll |a|$, the hyperfine components of the spectrum can be split; in the zero approximation, a value of $|J|$ can be calculated by the parameter $|J/a|$; and in this case, an exchange integral value is easily determined. With increasing of $|J/a|$, the splitted lines become asymmetric relating to hfs lines at $J = 0$, and their intensities are no longer equal. All such changes at different $|J/a|$ values are schematically shown in many papers for the case of ^{14}N-^{14}N nitroxide biradicals [X, Y, Z, 12, 46]. At high J values, a five lines' spectrum with the ratio 1:2:3:2:1 is observed, $|J| \gg a$, and precise determination of the J value is not available from the EPR spectrum.

An example of changes experimentally observed in the row of homological nitroxide biradicals containing only acetylene or *para*-phenylene groups in the bridge connecting two identical radical rings is shown in Fig. 5.3. R_6' is shown in Fig. 5.1.

$|J|$ values determined from these spectra from parameter $|J/a|$ presented in the related articles according to recommendations given in [12, 46] and confirmed by EPR spectra and DFT calculations are listed in Fig. 5.3. One can conclude that upon increasing the length of the bridge between two nitroxide rings, the value of the

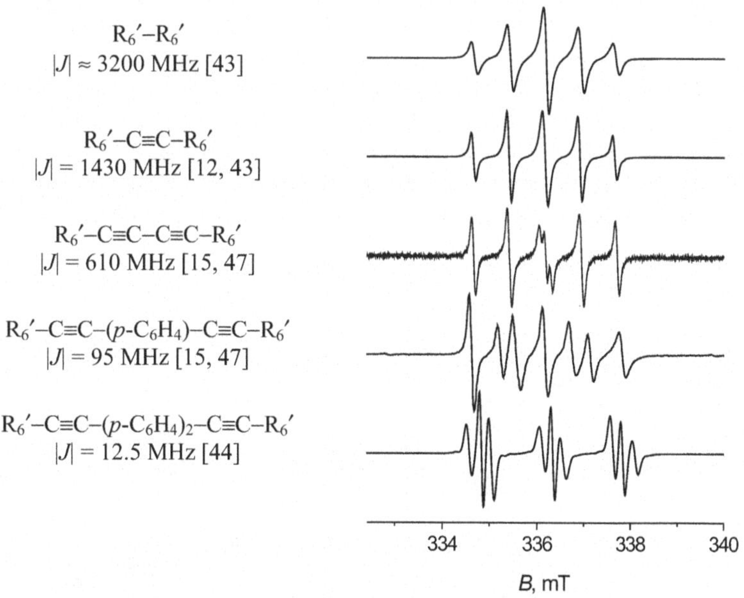

Fig. 5.3 Structures and the corresponding EPR spectra of some conjugated NB with acetylene and *p*-phenylene groups in the bridge dissolved in toluene at 313 K

exchange integral is noticeably decrease. A logical question arises: What is the reason of such changes? This could be caused either by the geometric elongation of the bridge or because of inserting various chemical groups in the bridge, their nature, and mutual orientation. In many theoretical works, old ones [48, 49], and much newer [50, 51], authors tried to distinguish two different mechanisms of the intramolecular spin exchange: "direct," via straight collisions of two paramagnetic centers, and "indirect," by delocalization of the electron spin density through the bridge of atoms and bonds [12]. In flexible long-chain biradicals, the direct mechanism should be realized, and in rigid stick-like molecules, one would assume that the indirect mechanism of spin exchange can occur. In the case of short-chain biradicals without conjugation, choosing of the mechanism is much more complicated as it was discussed, e.g., in [52, 53].

For the set of homological biradicals shown in Fig. 5.3 and such ones as R_6'–$(C{\equiv}C)_3$–R_6' and R_6'–C≡C–*p*–C_6H_4–C≡C–≡C–R_6' [12], the dependence of the exchange integral |*J*| as a function of the distance $r_{NO\text{-}NO}$ between the unpaired electrons has been calculated. It follows from Fig. 5.4 that calculations carried out using the spin-flip TD-DFT method show very nice agreement with the experimental data while the broken symmetry calculations showed the tendency only qualitatively.

EPR spectra of biradicals with several acetylene and *p*-phenylene groups in the spacer linking two nitroxide rings (Fig. 5.3) are weakly temperature dependent. |*J*/*a*| values measured for all of them allowed authors to suggest that the efficiency of spin density delocalization of unpaired electrons should be described with the use of the "coefficient of attenuation" $\gamma_k = |J_i/J_k|$ for the biradicals with ($|J_k/a|$) and without ($|J_i/a|$) a certain group or atom in the bridge [12]. In such linear biradicals, the intramolecular spin exchange can be realized only via the mechanism without straight collisions of the two N–O• groups but rather due to overlapping of spin orbitals of two unpaired electrons by the indirect mechanism. $|J_k/a|$ and $|J_i/a|$ values should both be measured at the same temperature and in the same solvent [11, 12]. This attenuation coefficient γ_k is a characteristic parameter of the functional group or atom, which does not depend on other groups forming the bridge. This fact was

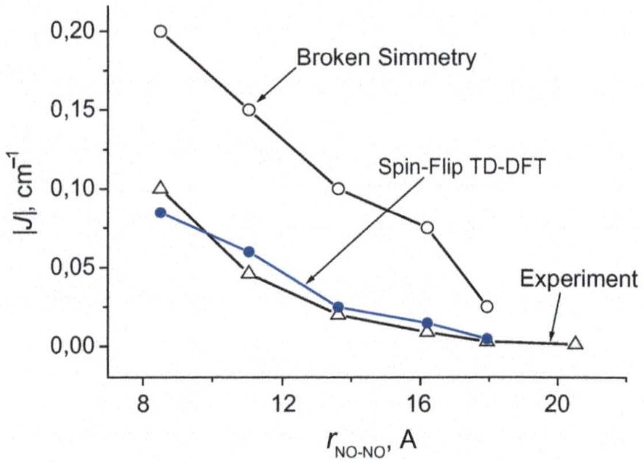

Fig. 5.4 Electron spin-exchange integral value |*J*| as a function of the distance $r_{NO\text{-}NO}$ between the unpaired electrons in biradicals' X-Y: experimental and DFT calculated

discussed in detail in [12] and then confirmed and enlarged with new examples in [15, 44, 45]. Many γ_k coefficients were determined and are listed in [15].

$$|J_{RR}| = |J_{Bir}| \cdot \gamma_k \cdot \gamma_l \cdot \gamma_m \cdot \ldots, \tag{5.2}$$

where γ_k, γ_l, γ_m are the attenuation coefficients for k-, l-, or m-atom or a group in the bridge between two R_6' or other radical fragments in a real biradical with its $|J_{Bir}|$. These γ_k values were experimentally measured for many biradicals with different nitroxide rings, e.g., R_6', R_5, R_5', R_5^N, R_5^{NO} (Fig. 5.1), and various compositions of the linker [15]. Note that knowledge of all γ_k values for all groups in the bridge makes it possible to estimate the spin density ρ_C localized on the carbon atom in the fourth position of the piperidine ring connected with the first atom of the biradical bridge [44, 45]. Thus, the exchange integral value $|J_{RR}|$ for a biradical R_6'–R_6' (the first one in Fig. 5.3) can be easily calculated by (5.2) knowing the γ_k, γ_l, γ_m, and $|J_{Bir}|$ parameters for a certain biradical.

As an example, using this procedure and keeping in mind that $|J/a| = 2.2$ and 14.5 for R_6'–C≡C–*p*–C_6H_4–C≡C–R_6' and R_6'–C≡C–C≡C–R_6' at 298 K in toluene, respectively [54], one can calculate $\gamma_{Ph} = 6.6$ (Ph is the same as *p*–C_6H_4) which is close to the values published earlier [47, 54]. Thus, the electron spin-exchange integral $|J|$ decreases by *ca.* a factor of 6.6 passing through the *p*-C_6H_4 group. It is known from [15] that a value of $\gamma_{C\equiv C} = 2.2 \pm 0.15$; therefore, one can calculate a value of $|J_{RR}| \approx 113$ mT, which is close to $|J_{RR}| \approx 122$ mT estimated in [44] and is in a good correlation with $|J_{RR}| \approx 0.12\ \text{cm}^{-1} \approx 128.5$ mT obtained in [43]. The similarity between these values obtained independently using various pairs of nitroxide biradicals confirms the correctness and universality of the approach envisaged by (5.2).

Similar estimating the spin density ρ_C localized on the carbon atom in the fourth position of the piperidine or in the third position of the five-membered ring connected with the nearest (first) atom of the biradical bridge, or the exchange integral value $|J_{RR}|$ for biradicals $R_{5,6}$–Z–$R_{5,6}$, real or hypothetic, was calculated and is given in Table 5.1.

It follows from the table that $|J_{RR}|$ parameters, hence, the ρ_C values, essentially depend on the structure of the radical ring. Indeed, in the case of R_5'–R_5' biradical this value exceeds that of the R_6'–R_6' analog because the C3 atom in a five-membered ring with a double bond is located much closer to the paramagnetic N–O$^\bullet$ group than the C4 atom in the six-membered ring in R_6'–R_6' [44]. In $R_5^{NO} - Z - R_5^{NO}$ biradicals

Table 5.1 $|J_{RR}|$ values estimated from the data given in Tables 1–3 in [15] for R_6'–R_6' and some other hypothetic nitroxide biradicals

Biradical	R_6'–R_6'	R_6''–R_6''	R_5–R_5	R_5'–R_5'	R_5^N–R_5^N	R_5^{NO}–R_5^{NO}
$|J_{RR}|$, MGz	3400	750[a]	1100	4500	42	3600

The precision of calculations is *ca.* ± 10%
[a]Calculated by data from [40, 55]

[13, 15], the spin density is more delocalized by the imidazolium ring comparing to similar $R_5^N - Z - R_5^N$ structures; therefore, $|J_{RR}|$ values of the latters are *ca.* 80–100 times smaller than in those containing R_5^{NO} rings. In the case of R_6''=N–N=R_6'' biradical [40, 55], the $|J_{RR}|$ value (Table 5.1) is sufficiently less comparing to that of the R_6'–R_6' system due to (i) the geometry of the R_6' ring is a semi-chair and R_6'' ring is a twisted-form [56, 57], and (ii) the presence of a double bond in the R_6' ring helps to realize better conjugation between nitroxide groups and the bridge.

5.3 X-Ray Structures and DFT Calculations

One of the most important and informative characteristics of a nitroxide biradical is its structure obtained usually from the X-ray analysis data [56, 57]. Knowledge of the exact geometry allows understanding and prediction of some features of a biradical molecule and its behavior in liquid solutions. X-ray studies report often about geometries of a few conformers existing even in a solid state, in a crystal: see, e.g., [58, 59].

A very valuable advantage for evaluating structural, physical–chemical and dynamic properties of biradicals in a liquid phase can be obtained by the use of quantum chemical and especially the density functional theory (DFT) computations [60–63]. DFT calculations became very popular during the last 10–15 years due to increasing the calculation possibilities of new computer clusters and further development of specialized program packages. DFT calculations permit not only to predict the geometry of the molecule but also to determine correctly the spin Hamiltonian parameters such as a g-tensor, hyperfine splitting (hfs) A-tensor, the dipole–dipole coupling constant D, the intramolecular electron spin-exchange integral J and to characterize the dynamic behavior of the intramolecular motions in a biradical as well as transitions between different conformations [43–45, 64]. Figure 5.5 illustrates a comparison between the X-ray and calculated DFT structural parameters for several biradicals.

Numerical comparison geometries (bond lengths and angle) for these and more nitroxides demonstrated perfect agreement of the X-ray and DFT results which were represented in the tables (see references above). At several examples, we will describe the abilities of the DFT method. Relating to the structure of the R_6'–C≡C–R_6' biradical obtained by DFT calculations is shown in Fig. 5.5a [43]. The effect of toluene as a solvent calculated in the scope of COSMO model did not reveal any significant changes in bond lengths, angles, or the $r_{\text{NO–NO}}$ distance in the biradical comparing to the X-ray results. The zero-field splitting parameter E was not observed as well as in the case of R_6'–R_6' biradical and was not taken into consideration at EPR spectra simulations. DFT calculation predicted E to be about three orders smaller than D parameter. The g-tensor axes were oriented along the N–O bond and were assumed to be collinear in both radical rings. Angles which represent the internal rotation around the main molecular axis connecting two N–O groups were included in fitting

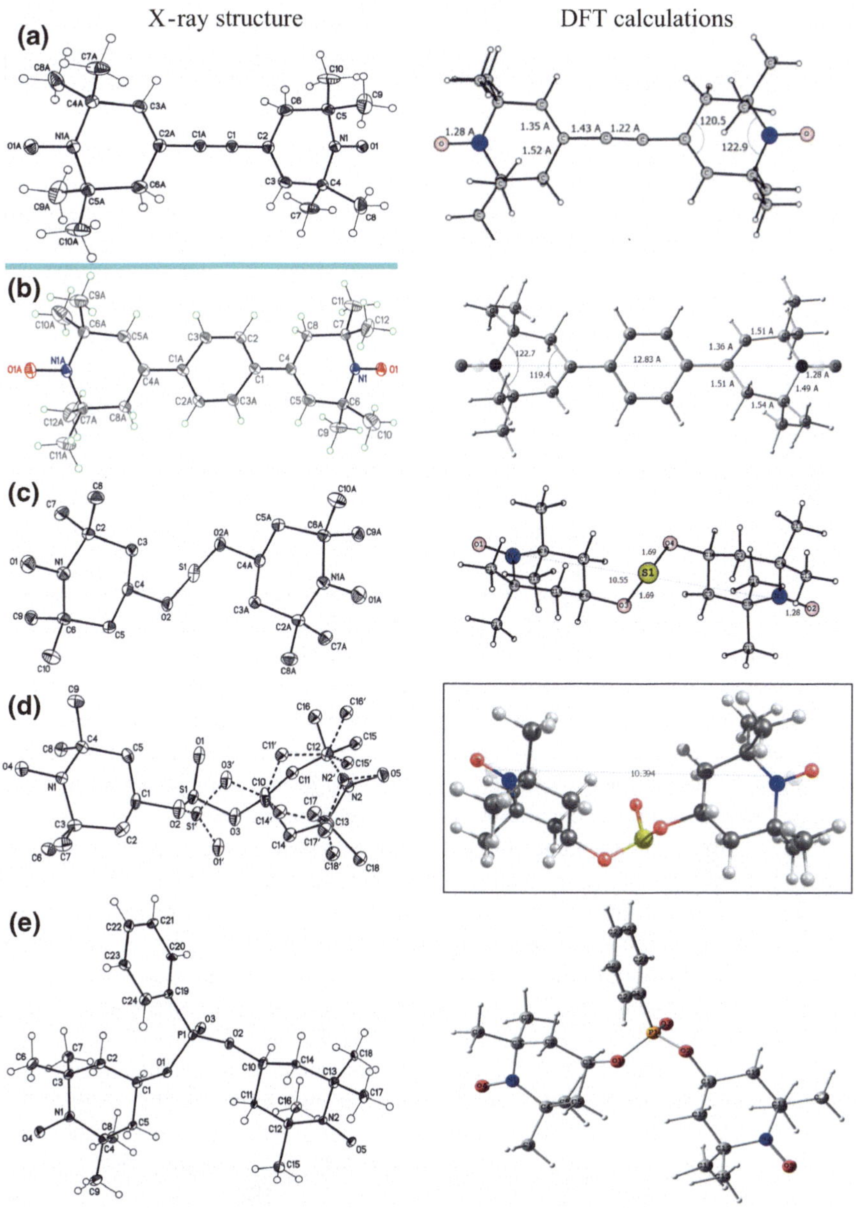

Fig. 5.5 Comparison of the X-ray experimental (*left*) and DFT calculated (*right*) structures of several biradicals extracted from: **a** [43], **b** [45], **c** and **d** [58], and **e** [59]

procedure. Parameters D calculated by DFT were in perfect agreement with experimental ones. The COSMO model slightly increased the delocalization of the spin density, and the calculation taken into account the solvent influence allowed obtaining the D parameters in a better agreement with the experiment. r_{NO-NO} distances calculated using the point-dipole approximation from EPR spectra for both biradicals differed both from the X-ray data and DFT-calculated r_{NO-NO} distances and that implied assuming some delocalization of the electron spin density though the major part of the spin density is localized on the N and O atoms of the N–O$^{\bullet}$ bond [43].

The geometry of biradical R_6'–p–C_6H_4–R_6' calculated at the UDFT/B3LYP level with a split valence basis set cc-PVDZ [65] is shown in Fig. 5.5b. Comparing the computational results with the X-ray pattern (Fig. 5.5 and Table 2 in [45]), one can conclude that DFT calculations provide correct information on the biradical molecular structure with a precision comparable with X-ray diffraction data. DFT calculations revealed the presence of several conformers of R_6'–p–C_6H_4–R_6' biradical which were characterized by identical r_{NO-NO} distances and identical zero-field splitting values D [45]. $|J|$ values were likely to be varied for these conformations due to significantly different mutual orientations of the two nitroxide moieties; see below in detail.

DFT calculations were applied to a model mixed-valence system presented a double-exchange phenomenon [62]. Due to the intrinsic multireference character of the spin states, it was shown that the couplings involved in the double-exchange model cannot be extracted from the DFT energies as it is usually done, but it was possible to extract the interactions of a generalized Hubbard Hamiltonian from different DFT single determinant energies, from which then the double-exchange spectrum could be evaluated. The problems generated by the charge and spin polarization were discussed in both symmetric and non-symmetric geometries [62].

The most important advantage of DFT method consists in the possibility to characterize not only structural peculiarities of the biradical molecule but also in distinguishing different conformations in which biradicals can exist, energetic features of transitions between them, calculating the spin Hamiltonian parameters, evaluating the spin density distribution in the system, etc. For example, characteristics such as the g-tensor, hyperfine splitting A-tensor, D, and r_{NO-NO} values for biradicals R_6'–R_6', R_6'–C≡C–R_6', R_6'–C≡C–C≡C–R_6', R_6'–p–C_6H_4–R_6', R_6'–C≡C–p–C_6H_4–C≡C–R_6', $S(OR_6)_2$, $O{=}S(OR_6)_2$, $O{=}P(C_6H_5)(OR_6)_2$, and others were calculated with high precision and demonstrated very good agreement with experimental results [43–45, 58, 59, 64]. At the same time, estimating exchange integral values seems to be rather more complicated problem: If $|J| > 1000$ MHz, its magnitude and sign c\n be determined with high accuracy; at $150 < |J| < 1000$ MHz, discrepancy with the experiment can achieve two- or threefold, and at $|J| \leq 100$ MHz, its evaluation is usually impossible due to very small energy magnitudes.

It should be noted that while calculations of the dipolar components of the hfs tensor can be systematically tuned by going to a much larger basis set of common use, isotropic hfs constants a are usually reproduced better using a specialized basis set described in [66]. Two important results were pointed out in [45]: First, specialized basis sets, such as N07 family or EPR family, enable much better agreement with the experimental data than the conventional basis sets even in the case of very small hfs coupling. Second, for some reasons, PBE0 functional gives better agreement of calculated ^{13}C isotropic hfs constant with the experimental one in the particular case of ^{13}C-labeled R_6'–C≡C–*p*–C_6H_4–C≡C–R_6' biradical [45], while the functional B3LYP, in turn, underestimates it, though both approaches yield results which are within the experimental error limits and agree well with the experiment. It was concluded that combination of PBE0/N07D or PBE0/EPR-II can be recommended as a cheap and precise way to compute small hfs couplings in nitroxide biradicals. A more detailed comparison of the experimental results obtained in different solvents with calculated ones taking into account various polarities were obtained and discussed in [67].

Concerning the number of conformations in which biradical molecules can exist in solutions: Fig. 5.6 demonstrates basing on DFT results that biradical R_5'–C≡C–(*p*–$C_6H_4)_2$–C≡C–R_5' (**B$_5$**) realizes cis and trans conformations R_6'–C≡C–(*p*–$C_6H_4)_2$–C≡C–R_6' (**B$_6$**) biradical has 8 energy equivalent conformers with the angle between SOMOs varying from 0°, this conformer corresponds to the maximal $|J|$ value, to ~60° at conformers with much lower $|J|$ values [44]. It makes a simple two-conformational model [12, 15, 36–39] inapplicable in a case of the latter biradicals **B$_5$** and **B$_6$**. Biradical **B$_5$** exhibited fast inversions of R_6' ring and internal rotations around the axis passing through the main molecular axis of the bridge even at low temperatures, averaging the $|J/a|$ values. Low barriers' activation energy E_a allows "mixing," i.e., averaging the conformer's spectral parameters at high temperatures effectively as have been demonstrated in [43, 47].

Another important for biradicals' result of DFT method is calculating the spin density distribution in nitroxide radicals and polyradicals. Figure 5.7 presents several examples of spin density distribution in R_6'–R_6', R_6'–C≡C–R_6', and O=P(C_6H_5)($OR_6)_2$ biradicals in triplet and singlet states.

One can see that spin densities of biradicals shown in Fig. 5.8 are strongly localized at the N–O$^•$ groups and the spin density value at the nitrogen atom is equal to 0.43 in both cases [47].

Very important from theoretical, spectroscopic, and application points of view results were reported in [53] where authors have synthesized new stable radical 2,5-di(*tert*-butyl)-3-ethoxycarbonyl-4-hydroxy-1-pyrroloxyl (**R**), and biradical 2,2′,5,5′-tetra(*tert*-butyl)-4,4′-bis(ethoxycarbonyl)-3,3′-bipyrrolyl-1,1′-dioxyl (**BR**), in which the unpaired electrons were highly delocalized by the pyrroloxyl rings. Both were investigated using X-ray crystallography, EPR spectroscopy, and quantum chemical calculations. Besides, using this radical **R** and several linkers of various lengths, saturated and unsaturated, a group of new biradical products was also synthesized and studied using EPR and DFT methods.

To study the electronic structure, the localized SOMOs were calculated for three configurations: two coplanar at 0° and 180° and a perpendicular one at −90°. In all configurations, the SOMOs were π-orbitals delocalized over the N–O group and the parent ring system [53]. It was revealed that in the coplanar configurations even the localized SOMOs had significant contributions on the neighboring atoms: The population of the SOMOs on the N–O fragments was estimated as 91% and up to

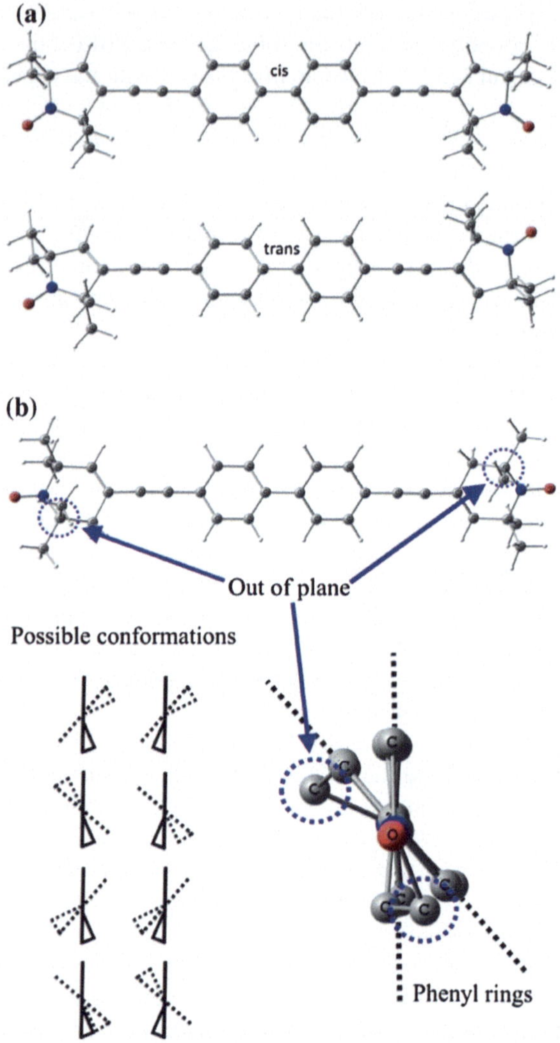

Fig. 5.6 Calculated geometries of biradicals R_5'–C≡C–(*p*–C_6H_4)$_2$–C≡C–R_5' (**a**) in cis and trans conformations (**a**) and R_6'-C≡C–(*p*–C_6H_4)$_2$–C≡C–R_6' (**b**)

9% delocalization, and in the perpendicular configuration, the SOMOs showed small tails into the σ-system of ~2%.

For three biradicals differed only by their bridges: (i) a saturated one with a –$C_{10}H_{20}$ linker, (ii) the unsaturated (conjugated) –$C_{10}H_{10}$ linker, and (iii) for the partially saturated linker –$C_{10}H_{12}$ localized SOMOs were calculated [53]. Note that all structures with the allyl linkers are planar. DFT results obtained were astonishing: The ratio between spin density in the nitroxide ring to that by the linker was equal to (i) 97:3%, (ii) 74:26%, and (iii) 78:22%, correspondingly. These results are very important for solving an old problem of spin-exchange coupling: "through bonds" or "through space" as well as for correct determining the dipole–dipole coupling

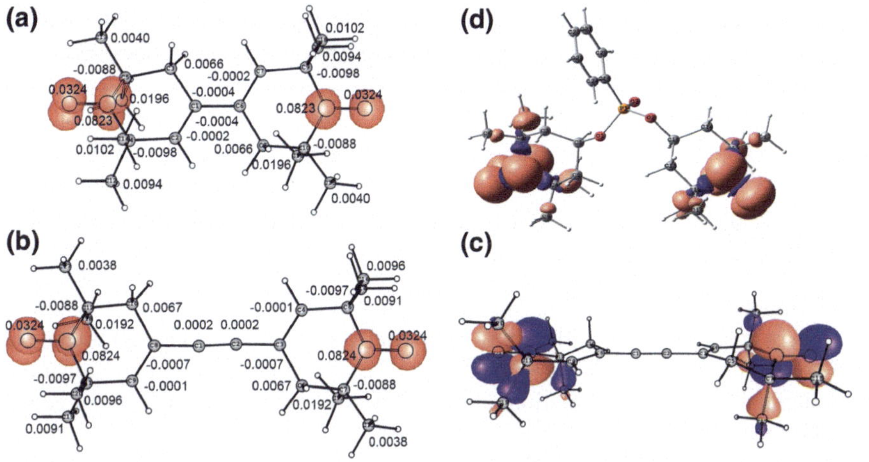

Fig. 5.7 Spin density distribution in biradicals R_6'–R_6' (**a**) and R_6'–C≡C–R_6' in triplet (**b**) [43] and in singlet (**c**) state including atomic spin densities at the C, N, O nuclei (a.u.). **d** shows spin densities distribution in biradical $O{=}P(C_6H_5)(OR_6)_2$ in the open-shell triplet state [59]. Contour value of 0.01 a.u. was used in all cases

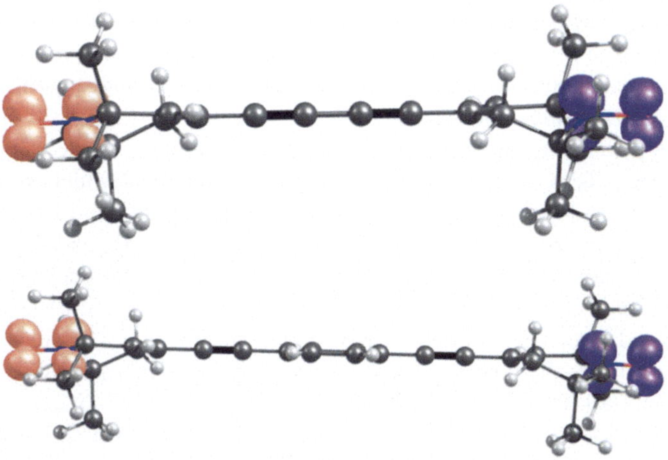

Fig. 5.8 Spin densities' distribution in biradicals R_6'–C≡C–C≡C–R_6' and R_6'–C≡C–*p*–C_6H_4–C≡C–R_6' in open-shell singlet states [47]

parameter D and the distance r_{NO-NO} between unpaired electrons from EPR spectra. Concerning the influence of unsaturated (or conjugated) groups in a linker, we have already discussed above.

The analysis presented in [53] suggests that the point-dipole approximation for dinitroxide **BR** is not completely correct due to the close proximity of the two electronic systems and the significant electron delocalization, which leads to large quantum mechanical corrections to the point-dipole approximation, and the authors could estimate at what distance and separation of two spin systems does the native point-dipole approximation become accurate.

It was understood when the delocalization of the spin density strongly affects the D value and led to errors in the interspin distances obtained using the point-dipole model. Spins at closer distances much stronger depend on the dipole–dipole interaction operator than larger distances, and even small delocalization affects to errors in derived distances and become dramatic for aromatic nitroxides or if there are conjugated groups between radical centers. In this case, the D value deviates from the expected behavior becoming significantly sensitive to the relative orientation of the two radical fragments. If the spin label or the bridge separated two paramagnetic centers is rather unsaturated, it is better to avoid the point-dipole approximation, and quantum chemical calculations could be of great help in the analysis of actual experiments.

A nitroxide biradical R_6'–13C≡C–p–C_6H_4–C≡^{13}C–R_6' has been investigated by X-band EPR and electron-nuclear double-resonance (ENDOR) spectroscopy [64]. Spin density distribution and hfs constant value on ^{13}C atoms in the bridge were calculated using B3LYP and PBE0 functionals and several different basis sets including N07 family. These results were compared with the experimental value of the hfs constant on ^{13}C atoms measured from ENDOR spectra. A few recommendations concerning the calculation of electron spin density distribution and the isotropic hfs constant values in nitroxide biradicals were given, and the results presented in that paper confirmed the fact that intramolecular electron spin exchange in this biradical is realized by the indirect mechanism [64].

An isotope-labeling approach has been continued with a specially synthesized nitroxide R_6'–C≡^{13}C–(p–C_6H_4)$_2$–^{13}C≡C–R_6' biradical which was investigated using X- and W-band EPR, echo-detected EPR and W-band ENDOR spectroscopy in comparison with two radicals: R_6'–C≡^{13}C–p–C_6H_4 and R_6'–C≡^{13}CH [69]. DFT calculations were performed with ORCA 3.0.3 program package [70]. The biradical and radical geometries were optimized on UKS/B3LYP/cc-pVDZ level and showed a good agreement with previously reported results [44]. Hfc constants were calculated using density functional theory with PBE0 functional and N07D full electron basis set [66]. Fine Lebedev 770 angular grid and 10^{-10} Eh SCF convergence tolerance were used. Solvent effects were simulated with the COSMO model [71].

The hyperfine splitting constants on ^{14}N, ^{1}H, and ^{13}C atoms were experimentally determined and compared with those obtained for previously investigated R_6'–^{13}C≡C–p–C_6H_4–C≡^{13}C–R_6' biradical [64]. It was concluded that the current quantum chemical approaches do not allow determining precise values of the hfs constants on the β-^{13}C atoms in the bridge connecting two paramagnetic nitroxide rings in the biradical, though it gave good results in calculating hfs constants on α-^{13}C atoms.

We can conclude that in the case of quantum chemical calculations of rather short-spacer biradicals such as R_6'–C≡C–R_6' or O=S(OR_6)$_2$ [43, 58, 67], the parameters $|D|$, $r_{NO\text{-}NO}$, and A_{zz} correlate well with experimental ones, while the hfs constants A_{xx}, A_{yy}, and a_{iso} differ significantly. On the other hand, the results presented in [43–45, 64, 68, 69] regarding calculations of the geometry, dipolar coupling, and intramolecular dynamics confirm the fact that the electron spin exchange in all polyacetylene-poly-*para*-phenylene biradicals is realized via the indirect mechanism.

5.4 Intramolecular Dynamics in Biradicals

Evidently, nitroxide biradicals can be absolutely rigid only in solid crystals or in a frozen state at temperatures around or below 77 K. In liquid solutions at low biradical concentrations, when *intermolecular* collisions and interactions become negligible [72], one can distinguish three different types of *intramolecular* motions and transitions [12, 43, 67]: (i) in short-linked flexible biradicals, (ii) in long-chain flexible ones, and (iii) in so-called linear (conjugated) rather chemically rigid structures.

The typical temperature changes in EPR spectra of several biradicals R_6'–$(CH_2)_4$–R_6' (**BA**); R_6–$(CH_2)_4$–R_6 (**BB**), $S[(CH_2)_4$-$COOR_6]_2$ (**BC**), and $S(OR_6)_2$ (**BD**) are shown in Fig. 5.9, which presents two essentially different types of spectral changes for biradicals **BA** and **BB** of similar composition, the structure of the bridge and slightly different nitroxide rings. In the case of **BB**, the multilinear spectrum is observed providing the exact measurement of the $|J_1|$ value in one conformation while for **BA** it is only possible to estimate that $|J_1| \gg a$. Temperature behavior of EPR spectra for both biradicals shows that they exist in solution in two conformations with fast transitions between them [12]. For both **BA** and **BB**, $J = 0$ [73].

Theory for correct quantitative description or both biradicals **BA** and **BB** has been given in [12, 41, 42] connected experimental parameters ln|J/a| or $\ln(\Delta B_2 - \Delta B_3)$, where ΔB_2 is shown in Fig. 5.9C, and ΔB_3 is the line width of the corresponding central line, with temperature, T^{-1} [74–77]. EPR spectra in Fig. 5.9C look very similar to those presented in Fig. 5.9A, but there is a principle difference between them: Since all changes in EPR spectra of **BA** are completely described by two-conformational mode, a quantitative description of **BC** could be done the more complex system of fast and slow transitions among three conformations [11, 74–77]. Correction of the three-conformational model has been confirmed in many papers of different research groups. Bridges linked two nitroxide rings were varied widely, e.g., $-(CH_2)_n-$, $-OOC-(CH_2)_n-COO-$, $-NHCO-(CH_2)_n-CONH-$, $-(OCH_2CH_2)_n-$, $-COO(OCH_2CH_2)_n-OOC-$, $S[(CH_2)_m-COO-]_2$, $-[O-(SiCH_2)_2]_n-O-$. [7, 12, 13, 78]. The most attractive in such long-chain biradicals is that they are nicely modeling structural and dynamic properties of oligomers in liquid solutions including room temperature ionic liquids [79–81].

Figure 5.9D illustrates changes in the EPR spectra of a short and very flexible $S(OR_6)_2$ biradical at different temperatures which were nicely confirmed by theoretical calculations [58]. Such changes are characteristic for fast transitions between two conformers with values of the differences in enthalpies, ΔH, and entropies, ΔS, of these conformations equal to 13.3 ± 1.1 kJ/mol and 37.2 ± 4 J/mol K respectively, calculated by (5.3) and (5.4) [58] :

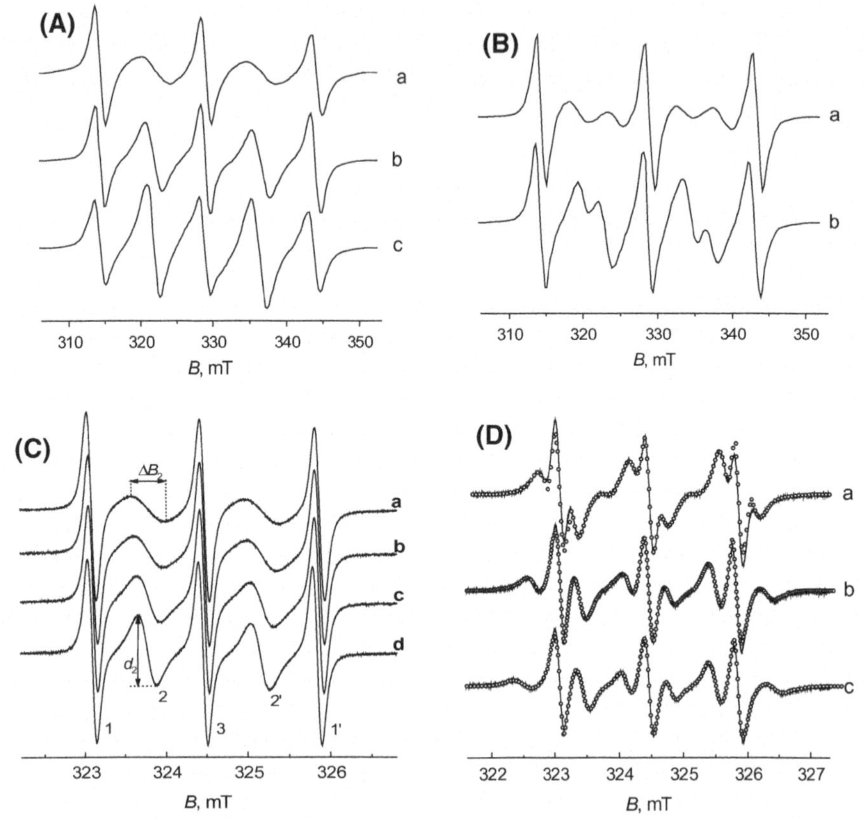

Fig. 5.9 Experimental (lines) and calculated (open circles) EPR spectra of biradicals in toluene solutions: **A** R_6'–$(CH_2)_4$–R_6' at 250 (**a**), 291 (**b**), and 345 K (**c**); **B** biradical R_6–$(CH_2)_4$–R_6 at 287 (**a**) and 348 K (**b**) [73]. **C** biradical $S[(CH_2)_4$–$COOR_6]_2$ at 295 (**a**), 304 (**b**), 324 (**c**), and 344 K (**d**) [74]. **D** biradical $S(OR_6)_2$ at 293 (**a**), 323 (**b**), and 343 K (**c**) [58]

$$\ln J^*/a| = \Delta S/R - \Delta H/RT, \quad (5.3)$$

$$J^* = (J_1\tau_1 + J_2\tau_2)/(\tau_1 + \tau_2), \quad (5.4)$$

where J^* is the experimentally measured from EPR spectra effective (averaged in time) value of the exchange integral; $|J_1| < |J_2|$; J_1, J_2, and τ_1, τ_2 are exchange integrals and the characteristic lifetimes of these conformations, respectively. In the case of $S(OR_6)_2$, $J_1 = 0$, hence, (5.4) becomes even simpler.

Another type of intramolecular rotational mobility has been observed for biradicals of the polyacetylene line (Fig. 5.3). The slight decrease of $|J/a|$ values with the increase of temperature for these biradicals cannot be explained by the population of distinct conformations with lower $|J|$, but rather a higher accessibility of larger

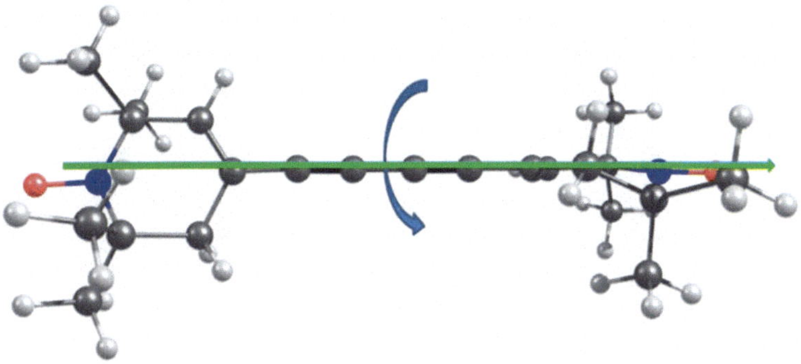

Fig. 5.10 Rotational conformers of biradical R_6'–C≡C–C≡C–R_6'

deviations from equilibrium geometries leading to lowering of the exchange coupling. Such probable rotational movements are shown in Fig. 5.10. Obviously, this type of rotations is also possible for R_6'–C≡C–*p*–C_6H_4–C≡C–R_6' and R_6'–C≡C–(*p*–C_6H_4)$_2$–C≡C–R_6' biradicals. In the case of polyacetylene biradicals, the idea that $|J/a|$ changes on temperature are caused by temperature changes of the hfs constant on ^{14}N a is also not correct. Indeed, it was shown in [15] that a values decrease with temperature in the case of all studied piperidine-type radicals and biradicals except O=R_6 radical [15], but a increases with temperature for five-membered pyrroline- and pyrrolidine-type nitroxides [15, 82]. Thus, $|J/a|$ value dependences *vs* temperature are similar to those of a. Values of the enthalpy ΔH and activation energy E_a, as results from the DFT calculations, in the range of 0–8 kJ/mol [44, 47], are responsible for practically free rotation or librations of the nitroxide rings around the bridge axis; i.e., the EPR spectra and measured values of $|J/a|$ do not characterize individual conformations of such linear biradicals but an averaged pattern with very fast transitions between several rotamers, and the measured value of $|J/a|$ is averaged by all these conformations.

It has been reported above that nitroxide biradicals of the R_6'–(C≡C)$_n$–(*p*–C_6H_4)$_m$–(C≡C)$_n$–R_6' series, where $n = 0, 1, \ldots$ and $m = 1, 2, \ldots$ undergo fast internal rotations with low barriers. In the case of R_6'–*p*–C_6H_4–R_6', two types of barriers of approximately 5 and 10 kJ/mol were revealed as one can see in Fig. 5.11 [45]. One can identify up to 10 possible conformers of R_6'–*p*–C_6H_4–R_6' biradical schematically shown in Fig. 5.11, by changing the bend direction between the two planes in the nitroxide ring (see Fig. 5.11); the respective barrier was estimated to ~15 kJ/mol in [44].

These conformations can be divided into three groups by their energy: The most low-lying conformers are nos. 6, 8, 9, and 10 in Fig. 5.11. The energy of conformers nos. 2, 4, 5, and 7 with respect to the lowest energy conformers is higher by *ca.* 1.5 kJ/mol and that of conformers nos. 1 and 3 is higher by *ca.* 3.0 kJ/mol [45]. The full potential energy surface (PES) projection on the C5–C4–C1–C3A dihedral angle coordinate (Fig. 5.12B) is equivalent to a rotation about one of the Ph–R_6' bonds. It may include three sets of conformations: No. 1–2–3–4, No. 4–8–2–6, and No.

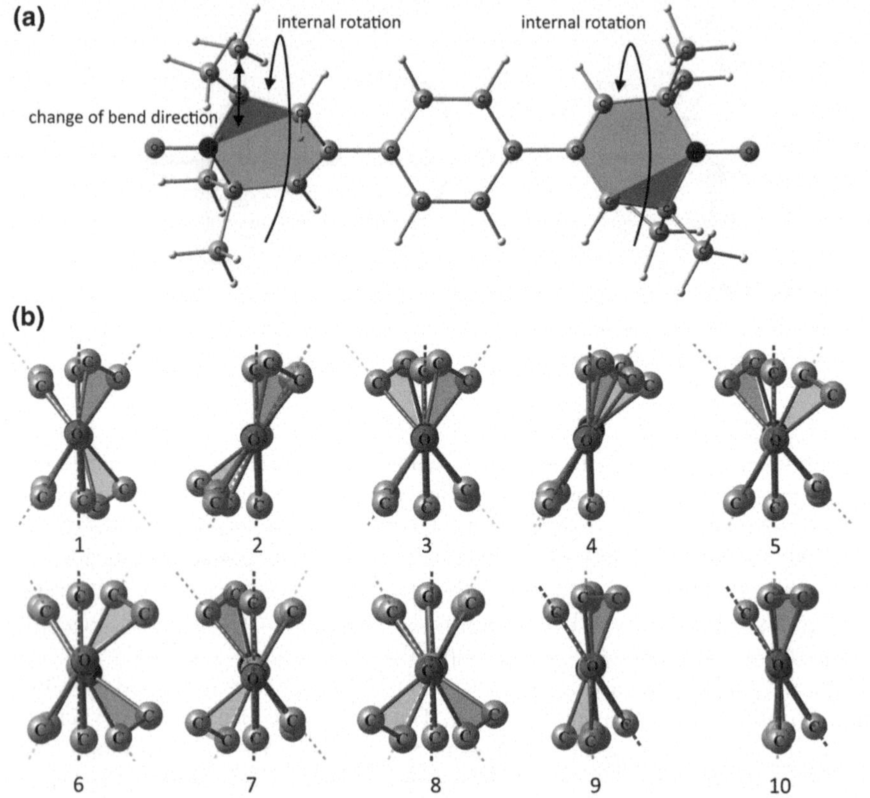

Fig. 5.11 **a** Conformation "1" of biradical R_6'-*p*-C_6H_4–R_6' with color designation of 1-oxyl-2,2,6,6-tetramethyl-1,2,5,6-tetrahydropyridine ring planes: N1–C2–C3–C4–C5 (by IUPAC nomenclature, colored in light gray) plane and N1–C5–C6 (colored in dark gray) plane. **b** Possible conformers of the biradical [45]

5–10–7–9 (see Fig. 5.11). Transitions between trajectories No. 1–2–3–4 and No. 4–8–2–6 require additional rotation about the second Ph–R_6' bond. Transitions among trajectories No. 1–2–3–4 and No. 5–10–7–9 need changing the ring bend direction of one of the –R_6' groups [45].

All the conformers of the discussed biradical were characterized by identical $r_{NO\text{-}NO}$ distances and identical zero-field splitting values *D*. The |*J*| value is likely to be varied for these conformations due to significantly different mutual orientations of the two nitroxide moieties. Its value may range from *ca.* zero in conformers 6 and 8 up to 0.00124 cm^{-1} in conformers 9 and 10 (Fig. 5.11).

One can see from Fig. 5.12 that the internal rotation over the angle θ in R_6'–R_6' should be hindered due to steric factors; the calculated barriers are 17.6 kJ/mol for the perpendicular orientation of two nitroxide rings and 19.2 kJ/mol for a ~ 180° rotation from the equilibrium [45]. The energy difference between the equilibrium geometry and the local minimum at $\theta = -54°$ is equal to ~9 kJ/mol. Including the acetylene

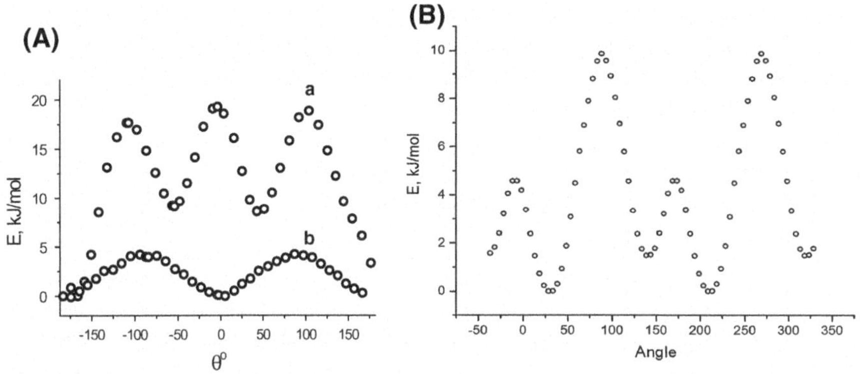

Fig. 5.12 Relaxed PES scan over θ in biradicals **A**: R_6'–R_6' (**a**) and R_6'–C≡C–R_6' (**b**) in triplet state, and **B** for biradical R_6'–*p*–C_6H_4–R_6' over C5–C4–C1–C3A (for atoms numerated in Fig. 5.5b) dihedral angle equivalent to the rotation about one of the Ph–R_6' bonds

spacer into the bridge reduces the rotation barrier in R_6'–C≡C–R_6' significantly: The steric factors are almost vanished, and it becomes difficult to locate the transition states (TSs) at $\theta = 0°$. TS at ~90° is still existing, but the barrier is equal to only 4.0 kJ/mol. Both minima at R_6'–C≡C–R_6' PES are equal within hundredth of kJ/mol [45]. In the case of R_6'–C≡C–C≡C–R_6', the barrier is less than 1.0 kJ/mol, i.e., becomes negligible [47]. This result reveals an almost free internal rotation in two latter biradicals at room temperature, and the exchange integral value $|J|$ measured by EPR spectroscopy can be averaged by the internal rotation.

Very subtle effects can be understood by combined usage of X-, Q-, and W-band EPR (operating frequencies are 9.74 GHz, 34.18 GHz, and 94.21 GHz, respectively), supplied with the standard commercially available flexline cavities, X-ray diffraction, and DFT calculations [67]. A short nitroxide biradical O=S$(OR_6)_2$ (**BS**) has been studied by EPR spectroscopy in liquid and frozen toluene, ethanol, and ionic liquid solutions. Variations of the intramolecular dynamics and conformational transitions in **BS** as a function of temperature in the wide range of 240–420 K, polarity and the ionic strength were characterized by the changes in the isotropic ^{14}N hyperfine splitting constant *a*, values of the exchange integral $|J|$, and the empirical parameter γ_3, the ratio between conformations with slow and fast transitions, and thermodynamic parameters of the conformational rearrangements were calculated [67].

The obtained EPR results were compared with the X-ray structural data and quantum chemical calculations of the geometry and intramolecular transitions of biradical **BS**. Three different groups of conformers with large exchange coupling, with $|J| \ll a$, and with an intermediate value of $|J|$ revealed from DFT calculations were also observed experimentally by EPR measurements at three effective conformations, "1," "2," and "*E*" with $|J_1| \ll a$, $|J_2| \gg a$, and $|J_E| \approx 10.0$ mT correspondingly, in solutions with high polarity or with rather high ionic strength [67]. It was concluded that conformation "*E*" presents more or less always in both non-polar and polar solutions. There were discussed possible mechanisms of the polarity and the ionic strength effecting on the biradical **BS** behavior in solutions of different types.

Distances r_{NO-NO} were close to the data obtained from X-ray analysis, but were ~1.2 to 1.3 Å longer than values obtained from EPR measurements in frozen solutions. Inclusion of dispersion correction to DFT calculations resulted in r_{NO-NO} values slightly shorter than observed in EPR experiment, but the absolute error is reduced to 0.1–0.5 Å. Changes in r_{NO-NO} distances in **BS**, due to media polarity, were also smaller than the experimentally observed values, but changes in mutual orientations of nitroxide rings caused by introducing polar media were more pronounced. These changes in geometry, according to calculations carried out in [67], could lead to the increase of the isotropic spin-exchange coupling. Although energy differences between various conformers were close to typical errors of DFT calculations, data allowed satisfactorily explained the experimental observations. Relative stability of conformers in polar media differed significantly: In non-polar media, the lowest energy conformer was stabilized by two intramolecular hydrogen bonds between O=S < oxygen atom and 4-H protons of the R_6 rings. This conformer had the highest $|J|$ value and the shortest r_{NO-NO} distance.

In polar media, the most stable conformers have r_{NO-NO} distance very close to measured experimentally in the ionic liquid at low temperature. These conformers have only one short intramolecular O…H bond and thus are stabilized by polar media. The $|J|$ value in these conformers is smaller than in non-polar solvents. Hence, in polar media, a relatively low $|J|$ value should be observed at low temperatures, while at higher temperatures the admixture of that conformer should grow up along with observed $|J|$. Indeed, this behavior of the **BS** biradical in solutions has been observed in [67]. Note that in polar media, this conformer had a larger $|J|$ value than in non-polar media due to slight changes in a mutual orientation of the nitroxide rings.

5.5 Applications of Biradicals

Numerous publications, original papers, books, and reviews, devoted to fundamental and practical applications of nitroxide mono- and polyradicals [10, 12, 83, 84] (and references therein). The main interest to biradicals as spin probes and labels, in the area of structural biology, for controlling processes of radical polymerization, in biomedicine, etc. Such studies started only in one–two years after opening chemical reactions without involving the unpaired electrons in several countries at once. At the first stage, researchers paid attention to the qualitative characterization of changes in the system based on measuring hfs constants and rotational correlation time τ_c. Quantitative studies in EPR of nitroxides started from measuring distances r_{NO-NO} between unpaired electrons in biradicals and obtained results were extrapolated to spin-labeled proteins, enzymes [85–88], and later to nucleic acids and biomembranes.

As an example, the influence of the ionic strength and pH of the solution on intramolecular spin dynamics has been studied [89. 90] for the biradicals containing the ionizable groups $-COO^-$ or $-NH^+(CH_3)_2$. The changes in the ESR spectra are discussed within the framework of the model of rapid exchange and it is suggested that biradicals of this kind can be used for the study of pH-dependent conformational changes in membranes and other biological media. This work was developed in [70].

Very important direction of further study was devoted to correct determination of $r_{NO\text{-}NO}$ at long distances using various approaches and advantages of EPR spectroscopy [91–98]. A lot of papers were published concerning binitroxide-mediated radical polymerization: biradical initiation and controlling of polymerization processes [22, 99–102].

At last, nitroxide biradicals containing disulfide group in the bridge were used as specific spin probes in in vivo EPR spectroscopy and imaging application for measuring pH, pO_2, redox status, and concentrations of phosphate and glutathione in tumor microenvironment and in tumor-bearing mice tissues [103–106].

References

1. E.G. Rozantsev, V.A. Golubev, M.B. Neiman, First kinetically stable individual iminoxyl biradical. Izv. AN SSSR, Ser. Khim., No. 2, 393–394 (1965)
2. E.G. Rozantsev, V.A. Golubev, M.B. Neiman, Y.V. Kokhanov, On new stable iminoxyl biradicals. Izv. AN SSSR, Ser. Khim., No. 2, 572–573 (1965)
3. R. Briere, R.-M. Dupeyre, H. Lemaire, C. Morat, A. Rassat, P. Rey, Nitroxydes: XVIIBiradicaux stables du type nitroxides. Bull. Soc. chim. France 3290–3297 (1965)
4. A.L. Buchachenko, V.A. Golubev, A.A. Medzhidov, E.G. Rozantsev, EPR spectra of biradicals with weak exchange coupling. Theor. Exper. Khimia **1**, 249–253 (1965)
5. E.G. Rozantzev, M.B. Neiman, Organic radical reactions involving no free valence. Tetrahedron **20**(1), 131–137 (1964)
6. E.G. Rozantzev, On free organic radicals with a hydroxy group. Izv. AN SSSR, Ser. Khim., No. 12, 2187–2191 (1964)
7. E.G. Rozantzev, *Free Nitroxyl Radicals* (Plenum Press, New York, 1970)
8. A. Rassat, Application of electron spin resonance to conformational analysis. Pure Appl. Chem. **25**, 623–634 (1971)
9. A.L. Buchachenko, A.M. Vasserman, *Stable Radicals* (Khimia, Moscow, 1973)
10. L.J. Berliner (ed.), *Spin Labeling: Theory and Applications* (Academic Press, New York, 1976)
11. V.N. Parmon, A.I. Kokorin, G.M. Zhidomirov, Conformational structure of Nitroxide biradicals. Use of biradicals as spin probes. Russ. J. Struct. Chem. **18**, 104–147 (1977)
12. V.N. Parmon, A.I. Kokorin, G.M. Zhidomirov, *Stable Biradicals* (Nauka, Moscow, 1980)
13. L.B. Volodarsky (ed.), *Imidazoline Nitroxide. Synthesis, Properties, Applications*, vol. 1, 2 (CRC Press, Boca Raton, 1988)
14. A. Rassat, Magnetic properties of nitroxide multiradicals. Pure Appl. Chern. **62**, 223–227 (1990)
15. A.I. Kokorin, Regularities of the spin exchange coupling through a bridge in nitroxide biradicals. Appl. Magn. Reson. **26**(1–2), 253–274 (2004)
16. M. Abe, Diradicals. Chem. Rev. **113**, 7011–7088 (2013)
17. M. Baumgarten, High spin organic molecules, in *World Scientific Reference on Spin in Organics*, vol. 4, ed. by J.S. Miller (World Sci. Publ. Co., Singapore, 2017), pp. 1–93
18. P. Michon, A. Rassat, Nitroxides. LXIX. 1,4-Bis(4',4'-dimethyloxazolidine-3'-oxyl)cyclohexane structure determination by electron spin resonance and nuclear magnetic resonance. J. Am. Chem. Soc. **97**(4), 696–700 (1975)

19. J. Michon, A. Rassat, Nitroxides. LIX. Rotational correlation time determination of nitroxide biradical application to solvation studies. J. Am. Chem. Soc. **96**(2), 335–337 (1974)
20. Y. Liao, C. Xie, P.M. Lahti, R.T. Weber, J. Jiang, D.P. Barr, 3,5-Di-tert-butyl-3¢-(N-tert-butyl-N-aminoxy)-4-oxybiphenyl: a heterospin diradical with temperature dependent behavior. J. Org. Chem. **64**, 5176–5182 (1999)
21. J. Fujita, M. Tanaka, H. Suemune, N. Koga, K. Matsuda, H. Iwamura, Antiferromagnetic exchange interaction among the three spins placed in an isosceles triangular configuration in 2,4-dimethoxy-1,3,5-benzenetriyltris(N-tert-butyl nitroxide). J. Am. Chem. Soc. **118**, 9347–9351 (1996)
22. J. Ruehl, N.L. Hill, E.D. Walter, G. Millhauser, R. Braslau, A proximal bisnitroxide initiator: studies in low-temperature nitroxide-mediated polymerizations. Macromolecules **41**, 1972–1982 (2008)
23. T. Iida, J. Ohshita, T. Uemura, H. Fukuoka, N. Ohta, K. Komaguchi, Y. Itagaki, M. Shiotani, S. Yamanaka, A. Kunai, Spin-spin interaction between phenyl nitroxides through the σ-π system. Silicon Chem. **1**, 383–389 (2002)
24. S. Hase, D. Shiomi, K. Satob, T. Takui, Phenol-substituted nitronyl nitroxide biradicals with a triplet ($S \sim 1$) ground stat. J. Mater. Chem. **11**, 756–760 (2001)
25. Y. Liu, F.A. Villamena, A. Rockenbauer, J.L. Zweier, Trityl-nitroxide biradicals as unique molecular probes for the simultaneous measurement of redox status and oxygenation. Chem. Commun. **46**, 628–630 (2010)
26. R. Chiarelli, A. Rassat, P. Rey, Ferromagnetic interactions in a crystalline nitroxide biradical: 1,3,5,7-tetramethyl-2,6-diazaadamantane N,N'-dioxyl. J. Chem. Soc. Chem. Commun. 1081–1082 (1992)
27. D.A. Shultz, A.K. Boal, H. Lee, G.T. Farmer, Structure-property relationships in trimethylenemethane-type biradicals. 2. Synthesis and EPR spectral characterization of dinitroxide biradicals. J. Org. Chem. **64**, 4386–4396 (1999)
28. S. Hase, D. Shiomi, K. Sato, T. Takui, Magnetic properties of phenol-substituted nitronyl nitroxide biradicals as building blocks of organic salt ferrimagnets. Polyhedron **20**, 1403–1409 (2001)
29. A. Caneschi, P. Chiesi, L. David, F. Ferraro, D. Gatteschi, R. Sessoli, Crystal structure and magnetic properties of two nitronyl nitroxide biradicals and of their Copper(II) complexes. Inorg. Chem. **32**, 1445–1453 (1993)
30. D. Wang, Y. Ma, A.I. Kokorin, M. Baumgarten, Temperature dependent intramolecular spin coupling interactions of a flexible bridged nitronyl nitroxide biradical in solution. J. Phys. Chem. A **122**(3), 574–581 (2018)
31. M. Tanaka, K. Matsuda, T. Itoh, H. Iwamura, Syntheses and magnetic properties of stable organic triradicals with quartet ground states consisting of different nitroxide radicals. J. Am. Chem. Soc. **120**, 7168–7173 (1998)
32. W. Zhai, Y. Feng, H. Liu, A. Rockenbauer, D. Mance, S. Li, Y. Song, M. Baldus, Y. Liu, Diastereoisomers of l-proline-linked trityl-nitroxide biradicals: synthesis and effect of chiral configurations on exchange interactions. Chem. Sci. **9**(19), 4381–4391 (2018)
33. Y. Liu, F.A. Villamena, Y. Song, J. Sun, A. Rockenbauer, J.L. Zweier, Synthesis of ^{14}N- and ^{15}N-labeled trityl-nitroxide biradicals with strong spin-spin interaction and improved sensitivity to redox status and oxygen. J. Org. Chem. **75**(22), 7796–7802 (2010)
34. Y. Liu, F.A. Villamena, A. Rockenbauer, Y. Song, J.L. Zweier, Structural factors controlling the spin-spin exchange coupling: EPR spectroscopic studies of highly asymmetric trityl-nitroxide biradicals. J. Am. Chem. Soc. **135**, 2350–2356 (2013)
35. T. Xiaoli, S. Yuguang, L. Huiqiang, Z. Qinwen, A. Rockenbauer, Supramolecular host-guest interation of trityl-nitroxide biradicals with cyclodextrins: modulation of spin-spin interaction and redox sensitivity. Org. Biomol. Chem. **14**, 1694–1701 (2016)
36. G.R. Luckhurst, Alternating linewidths. A novel relaxation process in the electron resonance of biradicals. Mol. Phys. **10**, 543–550 (1966)
37. S.H. Glarum, J.H. Marshall, Spin exchange in nitroxide biradicals. J. Chem. Phys. **47**, 1374–1379 (1967)

38. H. Lemaire, Nitroxydes XX. Résonance paramagnétique électronique d'un biradical nitroxyde; détermination du signe de l'échange. J. Chim. Phys. **64**, 559–571 (1967)
39. J.H. Marshall, Spin exchange in the carboxylic acid dimer of nitroxide monoradical. J. Chem. Phys. **54**, 2762–2763 (1971)
40. A. Nakajima, H. Ohya-Nishiguchi, Y. Deguchi, Magnetic properties of some iminoxyl polyradicals. III. Exchange interaction in iminoxyl biradicals. Bull. Chem. Soc. Japan **45**, 713–716 (1972)
41. V.N. Parmon, A.I. Kokorin, G.M. Zhidomirov, K.I. Zamaraev, Evidence for slow exchange in ESR spectra of nitroxide biradicals. Mol. Phys. **26**, 1565–1569 (1973)
42. V.N. Parmon, G.M. Zhidomirov, Calculation of the E.S.R. spectrum shape of the dynamic biradical system. Mol. Phys. **27**, 367–375 (1974)
43. O.I. Gromov, E.N. Golubeva, V.N. Khrustalev, T. Kálai, K. Hideg, A.I. Kokorin, EPR, the X-ray structure and DFT calculations of the nitroxide biradical with one acetylene group in the bridge. Appl. Magn. Reson. **45**(10), 981–992 (2014)
44. A.I. Kokorin, O.I. Gromov, T. Kálai, K. Hideg, Peculiarities of spin exchange in nitroxide biradicals containing two *para*-phenylene groups in the bridge. EPR investigation and DFT calculations. Appl. Magn. Reson. **47**(11), 1283–1293 (2016)
45. A.I. Kokorin, O.I. Gromov, P.V. Dorovatovskii, V.A. Lazarenko, V.N. Khrustalev, K. Hideg, T. Kálai, The structure and internal dynamics of R6-p-C6H4-R6 biradical: EPR, X-ray crystallography and DFT calculations. Appl. Magn. Reson. **50**(1–3), 425–439 (2019)
46. A.I. Kokorin, V.N. Parmon, A.A. Shubin, *Atlas of Anisotropic EPR Spectra of Nitroxide Biradicals* (Nauka, Moscow, 1984)
47. A.I. Kokorin, E.N. Golubeva, B. Mladenova, V.A. Tran, T. Kálai, K. Hideg, G. Grampp, Behaviour of nitroxide biradicals with acetylene bridges in organic solvents and ionic liquids. Appl. Magn. Reson. **44**(9), 1041–1051 (2013)
48. P.V. Shasmev, K.M. Salikhov, Spin polarization and exchange interaction of the multyelectron systems. Teor. Eksper. Khim. **9**, 291–299 (1973)
49. G.M. Nedlin, To the theory of the exchange interaction. Fiz. Tverd. Tela (Leningrad) **15**, 3048–3052 (1973)
50. S.Ya. Umanskiy, E.N. Golubeva, B.N. Plakhutin, Combined calculation method of weak exchange interactions in biradicals. Russ. Chemi. Bull., Intern. Ed. **62**(7), 1511–1518 (2013)
51. S.Ya. Umanskii, Weak exchange interactions in biradicals: a pseudopotential for unpaired electrons and an asymptotic methods for calculating the exchange integral. Russ. J. Phys. Chem. B **9**(1), 1–8 (2015)
52. D.A. Shultz, R.M. Fico Jr., H. Lee, J.W. Kampf, K. Kirschbaum, A.A. Pinkerton, P.D. Boyle, Mechanisms of exchange modulation in trimethylenemethane-type biradicals: The roles of conformation and spin density. J. Am. Chem. Soc. **125**, 15426–15432 (2003)
53. C. Riplinger, J.P.Y. Kao, G.M. Rosen, V. Kathirvelu, G.R. Eaton, S.S. Eaton, A. Kutateladze, F. Neese, Interaction of radical pairs through-bond and through-space: Scope and limitations of the point-dipole approximation in electron paramagnetic resonance spectroscopy. J. Am. Chem. Soc. **131**, 10092–10106 (2009)
54. A.I. Kokorin, V.A. Tran, K. Rasmussen, G. Grampp, Effect of solvent nature on spin exchange in rigid nitroxide biradicals. Appl. Magn. Reson. **30**, 35–42 (2006)
55. A. Nakajima, J. Yamauchi, Magnetic interactions in TEMPAD biradical. Bull. Inst. Chem. Res. Kyoto Univ. **54**(4), 234–247 (1976)
56. J. Lajzerowicz, Molecular Structures of nitroxides. Ch. 6, in: *Spin Labeling. Theory and Applications*, L.J. Berliner (ed.), (Academic Press, New York, 1976)
57. R.P. Shibaeva, Structure of organic paramagnetics of nitroxide radicals. Zh. Strukt. Khimii **16**, 330–348 (1975)
58. A.I. Kokorin, V.N. Khrustalev, E.N. Golubeva, The structure and EPR behavior of short nitroxide biradicals containing sulfur atom in the bridge. Appl. Magn. Reson. **45**(4), 397–409 (2014)
59. A.I. Kokorin, V.N. Khrustalev, O.I. Gromov, The structure and EPR behavior of nitroxide biradical containing phosphorus atom in the bridge. Appl. Magn. Reson. **46**(12), 1429–1442 (2015)

60. F. Neese, *ORCA—an ab initio, density functional and semiempirical program package. Version 2.8–20* (Max-Planck Institute for Bioinorganic Chemistry, Mülheim an der Ruhr, 2010)
61. V. Barone, I. Cacelli, P. Cimino, A. Ferretti, S. Monti, G. Prampolini, Magnetic interactions in phenyl-bridged nitroxide diradicals: Conformational effects by multireference and broken symmetry DFT approaches. J. Phys. Chem. A **113**, 15150–15155 (2009)
62. C. Boilleau, N. Suaud, R. Bastardis, N. Guihéry, J.P. Malrieu, Possible use of DFT approaches for the determination of double exchange interactions. Theor. Chem. Acc. **126**(3), 231–241 (2010)
63. M.F. Ottaviani, A. Modelli, O. Zeika, S. Jockusch, A. Moscatelli, N.J. Turro, EPR analysis and DFT computations of a series of polynitroxides. J. Phys. Chem. A **116**, 174–184 (2012)
64. A.I. Kokorin, R.B. Zaripov, O.I. Gromov, A.A. Sukhanov, T. Kálai, É. Lamperth, K. Hideg, Spin density distribution in a nitroxide biradical containing ^{13}C-enriched acetylene groups in the bridge: DFT calculations and EPR investigation. Appl. Magn. Reson. **47**(9), 1057–1067 (2016)
65. T.H. Dunning Jr., Gaussian basis sets for use in correlated molecular calculations. I. The atoms boron through neon and hydrogen. J. Chem. Phys. **90**, 1007–1017 (1989)
66. V. Barone, P. Cimino, E. Stendardo, Development and validation of the B3LYP/N07D computational model for structural parameter and magnetic tensors of large free radicals. J. Chem. Theor. Comput. **4**, 751–764 (2008)
67. A.I. Kokorin, B.Y. Mladenova-Kattnig, O.I. Gromov, A.A. Shubin, R.B. Zaripov, G. Grampp, Influence of polarity and ionic strength on intramolecular spin exchange in a short nitroxide biradical containing sulphur atom in the bridge. Appl. Magn. Reson. **49**, 1059–1073 (2018)
68. A.V. Bogdanov, A.K. Vorobiev, Orientation order and rotation mobility of nitroxide biradicals determined by quantitative simulation of EPR spectra. Phys. Chem. Chem. Phys. **18**(45), 31144–31153 (2016)
69. A.I. Kokorin, R.B. Zaripov, O.I. Gromov, K. Hideg, T. Kálai, Tailored nitroxide radicals and biradical containing ^{13}C enriched acetylene groups: ENDOR and DFT investigation. Appl. Magn. Reson. **49**, 137–149 (2018)
70. F. Neese, The ORCA program system. Wires Comput. Mol. Sci. **2**, 73–78 (2012)
71. S. Sinnecker, A. Rajendran, A. Klamt, M. Diedenhofen, F. Neese, Calculation of solvent shifts on electronic g-tensors with the conductor-like screening model (COSMO) and its self-consistent generalization to real solvents (Direct COSMO-RS). J. Phys. Chem. A **110**, 2235–2245 (2006)
72. Y.N. Molin, K.M. Salikhov, K.I. Zamaraev, *Spin Exchange* (Springer, Berlin, 1980)
73. G. Grampp, S. Landgraf, I.A. Grigor'ev, A.B. Shapiro, A.I. Kokorin, Conformational dynamics of some short-chain biradicals in solutions. Appl. Magn. Reson. **19**(2), 187–196 (2000)
74. V.A. Tran, K. Rasmussen, G. Grampp, A.I. Kokorin, The solvent effect on spin exchange in long-chain nitroxide biradicals. Appl. Magn. Reson. **32**(3), 395–406 (2007)
75. V.N. Parmon, A.I. Kokorin, G.M. Zhidomirov, K.I. Zamaraev, On the mechanism of spin exchange in long-chain nitroxide biradicals. Mol. Phys. **30**, 695–701 (1975)
76. G. Ionita, G.A. Vorobieva, V. Chechik, A.I. Kokorin, Intramolecular spin exchange in flexible PEG-based nitroxide biradicals in aqueous solutions. Appl. Magn. Reson. **46**, 251–260 (2015)
77. B. Mladenova-Kattnig, G. Grampp, A.I. Kokorin, Influence of pressure on intramolecular dynamics in a long-chain flexible nitroxide biradical. Appl. Magn. Reson. **46**, 1359–1366 (2015)
78. S.V. Kozlov, A.I. Kokorin, A.B. Shapiro, E.G. Rozantsev, Chained nitroxide biradicals—a model for investigating of oligomers in solutions. Vysokomol. Soed. Ser. B **23**, 322–327 (1981)
79. V.A. Tran, A.I. Kokorin, G. Grampp, K. Rasmussen, Features of spin exchange in nitroxide biradicals in the ionic liquid bmimPF_6. Appl. Magn. Reson. **35**(3), 389–398 (2009)
80. A.I. Kokorin, B. Mladenova, E.N. Golubeva, G. Grampp, Behavior of short nitroxide biradical in room temperature ionic liquids. Appl. Magn. Reson. **41**, 353–362 (2011)

81. A.I. Kokorin, Peculiarities of intramolecular motions, in ionic liquids, in *Ionic Liquids. Theory, Properties, New Applications*, ed. by A.I. Kokorin (InTech Publ, Rijeka, 2011), pp. 183–200
82. A. Weber, O. Schiemann, B. Bode, T.F. Prisner, PELDOR at S-and X-band frequencies and the separation of exchange coupling from dipolar coupling. J. Magn. Reson. **157**, 277–285 (2002)
83. G.I. Likhtenshtein, *The Method of Spin Labeling in Molecular Biology* (Nauka, Moscow, 1974); *Spin labeling methods in molecular biology* (Wiley, New York, 1976)
84. G.I. Likhtenstein (ed.), *Nitroxides: Applications in Chemistry, Biomedicine, and Materials Science* (Wiley VCH, New York, 2008)
85. A.V. Kulikov, G.I. Likhtenstein, E.G. Rozantsev, V.I. Suskina, A.B. Shapiro, On possible determination of distances between functional groups of protein by the method of spin labels. Biofizika **17**(1), 42–48 (1972)
86. A.I. Kokorin, K.I. Zamaraev, G.L. Grigoryan, V.P. Ivanov, E.G. Rozantsev, Measurement of the distances between the paramagnetic centres in solid solutions of nitroxide radicals, biradicals and spin-labeled proteins. Biofizika **17**(1), 34–41 (1972)
87. A.V. Kulikov, Evaluation of the distance between spins of the spin label and paramagnetic centre in spin-labeled proteins from the parameters of saturation curve of EPR spectra of labels at 77 K. Russ. Molek. Biol. **10**(1), 132–141 (1976)
88. A.I. Kokorin, Forty years of the d_1/d parameter, in *Nitroxides: Theory, Experiment and Applications*, ed. by A.I. Kokorin (InTech Publ, Rijeka, 2012), pp. 113–164
89. P. Ferruti, D. Gill, M.P. Klein, M. Calvin, Correlation between conformation and pairwise spin exchange in flexible biradicals in solution. Control of conformation by pH-dependent ionic forces. J. Am. Chem. Soc. **91**, 7765–7766 (1969)
90. P. Ferruti, D. Gill, M.P. Klein, H.H. Wang, G. Entine, M. Calvin, Synthesis of mono-, di-, and polynitroxides. Classification of electron spin resonance spectra of flexible ninitroxides dissolved in liquids and glasses. J. Am. Chem. Soc. **92**, 3704–3713 (1970)
91. P.P. Borbat, J.H. Freed, Double quantum ESR and distance measurements, in *Biological Magnetic Resonance*, vol. **19**, *Distance Measurements in Biological Systems by EPR*, ed. by L.J. Berliner, S.S. Eaton, G.R. Eaton (Kluwer Acad./Plenum Pub, New York, 2000)
92. P. Gajula, S. Milikisyants, H.-J. Steinhoff, M. Huber, A short note on orientation selection in the DEER experiments on a native cofactor and a spin label in the reaction center of *Rhodobacter sphaeroides*. Appl. Magn. Reson. **31**, 99–104 (2007)
93. A. Savitsky, A.A. Dubinskii, H. Zimmermann, W. Lubitz, K. Moebius, High-field dipolar electron paramagnetic resonance (EPR) spectroscopy of nitroxide biradicals for determining three-dimensional structures of biomacromolecules in disordered solids. J. Phys. Chem. B **115**, 11950–11963 (2011)
94. M. Drescher, G. Jeschke (eds.), *EPR Spectroscopy: Applications in Chemistry and Biology* (Springer, Berlin, Heidelberg, 2012)
95. Y.D. Tsvetkov, Notroxide radicals in pulsing double electron-electron resonance spectroscopy (PELDOR). Russ. J. Strukt. Khimii **54**, S46–S75 (2013)
96. G. Jeschke, The contribution of modern EPR to structural biology. Emerg. Topics Life Sci. **2**, 9–18 (2018)
97. Y.D. Tsvetkov, M.K. Bowman, Y.A. Grishin, *Pulsed Electron-Electron Double Resonance: Nanoscale Distance Measurement in the Biological, Materials and Chemical Sciences* (Springer, Cham, 2019)
98. S. Pribitzer, L.F. Ibánez, C. Gmeiner, I. Ritsch, D. Klose, M. Sajid, M. Hülsmann, A. Godt, G. Jeschke, Two-dimensional distance correlation maps from pulsed triple electron resonance (TRIER) on proteins with three paramagnetic centers. Appl. Magn. Reson. **49**, 1253–1279 (2018)
99. W. Huang, R. Chiarelli, B. Charleux, A. Rassat, J.-P. Vairon, Unique behavior of nitroxide biradicals in the controlled radical polymerization of styrene. Macromolecules **35**, 2305–2317 (2002)
100. K. Pietrasik, O. Swiatkowska, A. Kaim, New difunctional mediators based on 4-amino-TEMPO derivatives for controlling free radical polymerization of styrene. Polimery **55**, 812–816 (2010)

101. A. Kaim, J. Szydlowska, K. Pietrasik, The effect of the spacer length on binitroxide mediated radical polymerization of styrene. Macromol. Res. **19**, 1041–1047 (2011)
102. J. Nicolas, Y. Guillaneuf, C. Lefay, D. Bertin, D. Gigmes, B. Charleux, Nitroxide-mediated polymerization. Progr. Polym. Sci. **38**, 63–235 (2013)
103. V.V. Khramtsov, I.A. Grigor'ev, D.J. Luried, M.A. Foster, J.L. Zweier, P. Kuppusamy, Spin pH and SH probes: enhancing functionality of EPR-based techniques. Spectroscopy **18**, 213–225 (2004)
104. G.I. Roshchupkina, A.A. Bobko, A. Bratasz, V.A. Reznikov, P. Kuppusamy, V.V. Khramtsov, In vivo EPR measurement of glutathione in tumor-bearing mice using improved disulfide biradical probe. Free Rad. Biol. Med. **45**, 312–320 (2008)
105. V.V. Khramtsov, In vivo molecular EPR-based spectroscopy and imaging of tumor microenvironment and redox using functional paramagnetic probes. Antiox. Redox Signaling **28**, 1365–1377 (2018)
106. A.A. Bobko, T.D. Eubank, B. Driesschaert, V.V. Khramtsov, In-vivo EPR assessment of pH, pO_2, redox status and concentrations of phosphate and glutathione in tumor microenvironment. J. Vis. Exp. **133**, e56624 (2018)

Chapter 6
Fluorophore–Nitroxide (Profluorescent Nitroxide) Probes

6.1 Introduction

Over the last decades, scientists have faced growing requirements in novel methods of fast and sensitive analysis of antioxidant status of biological systems, redox probing and radical trapping, biological molecules' analysis, investigation of molecular dynamics, and convenient models for studies of photophysical and photochemical processes. This chapter presents a review on the use of tethered nitroxide–fluorophore molecules as probes of redox status, antioxidant activity, oxidative stress, and free radical reaction. In addition, these supermolecules have been proved to serve as tools for the study of molecular dynamics, intermolecular fluorescence quenching and electron transfer mechanism, photoswitching material and analytic reagents. Keeping all properties of spin and fluorescent probes, the dual fluorophore–nitroxide compounds (FNO•) possess important new advantages.

An idea to combine chromophore and nitroxide in one molecule for the study of the probe mobility was designed in 1965 by Jost and Griffith [1]. A supermolecule dansyl 2,2,5,5-Tetramethyl-3-amino-pyrrolidine-1-oxyl) was prepared. Then in liquids, the rotational relaxation time of the probe fluorophore fragment was measured with polarization technique, while an apparent correlation time of the nitroxide segment was estimated by EPR. This approach was developed in works of group of Likhtenstein in which the dual probes were used for the study of molecular dynamics of liquids and albumin in a wide range of temperature [2–10].

In pioneering work of Likhtenshtein with colleagues [3], three fundamental effects were first demonstrated in the dual supermolecules; namely, (1) the nitroxide fragment is a strong quencher of the fluorescence, (2) the radical photoreduction can lead to the decay of the EPR signal and the drastic increase of the fluorescence intensity, and (3) the photoreduction kinetics strongly depends on molecular dynamics of environment. Therefore, any chemical or photo-reduction event of the fragment to a corresponding hydroxylamine derivative, oxidation of the nitroxide fragment, or addition of an active radical yield would result in a decrease of electron spin resonance (ESR) signal that would be accompanied by an increase in fluorescence. These

G. I. Likhtenshtein, *Nitroxides*, Springer Series in Materials Science 292,
https://doi.org/10.1007/978-3-030-34822-9_6

effects form the basis of application of double probes for the quantitative study of redox reactions, molecular dynamics of objects of interest, and the establishment of factors affected on an electronic transfer.

The next principle step was a series of excellent papers by Blough et al. [11, 12] in which the potential of these tethered, optically switching molecules as potent probes of radical was realized. Blough and Simpson were the first to show that biologically relevant reductants, such as ascorbic acid, could be detected in this way using profluorescent naphthalene–nitroxide [13]. Works of the Likhtenshtein and Blough groups have paved the way for thorough investigations of numerous radical reactions in chemistry, biology, and materials science. Oxidative stress, polymer production and degradation, environment pollution are areas of "reign" of free radicals. Another fruitful avenue of the dual fluorophore–nitroxide compounds application appeared to be the analysis of nitric oxide, superoxide, vitamins, and metal ions. Dual probes also are suitable "training areas" for quantitative study factors, such as local molecular dynamics and micropolarity, on intermolecular fluorescence quenching and electron transfer. In addition, the new magnetic materials in which magnetic properties can be controlled by optical stimuli were developed by using photochromic derivatives as photofunctional units and nitroxide radical as spin sources.

The organic synthetic chemistry allows playing with the chemical structure of the dual molecules of different absorption, fluorescence, and ESR spectra, and of redox and spin properties with variety bridges (spacers) tethered to the chromophore and nitroxide segments. Dual probes are described in the literature as dual fluorophore–nitroxide, fluorescence–nitroxide, profluorescent nitroxide, prefluorescent nitroxide, double (spin and fluorescence) sensors, and fluorophore–nitronyl probes. Fundamentals and progress in the area have been discussed in the length in numerous papers and reviews (see, e.g., [10, 14–17]) and are briefly deal within this chapter.

6.2 Structure and Synthesis of Dual Fluorophore–Nitroxide Compounds

Examples of various chemical structures of fluorescence–nitroxide compounds are shown in Fig. 6.1.

Structures of numerous profluorescent nitroxide compounds were described in papers [18–37]. Several examples of the profluorescent nitroxide synthesis are given below.

A range of varying chromophore nitroxide free radicals and their non-radical methoxyamine analogs were synthesized by the reaction of dansyl chloride with the appropriate amino nitroxide in dichloromethane in the presence of base to give the desired dansyl-linked nitroxides [18]. The methyl ether analogs were obtained using Fenton chemistry by the reaction of the nitroxides with methyl radicals generated from dimethyl sulfoxide and hydrogen peroxide.

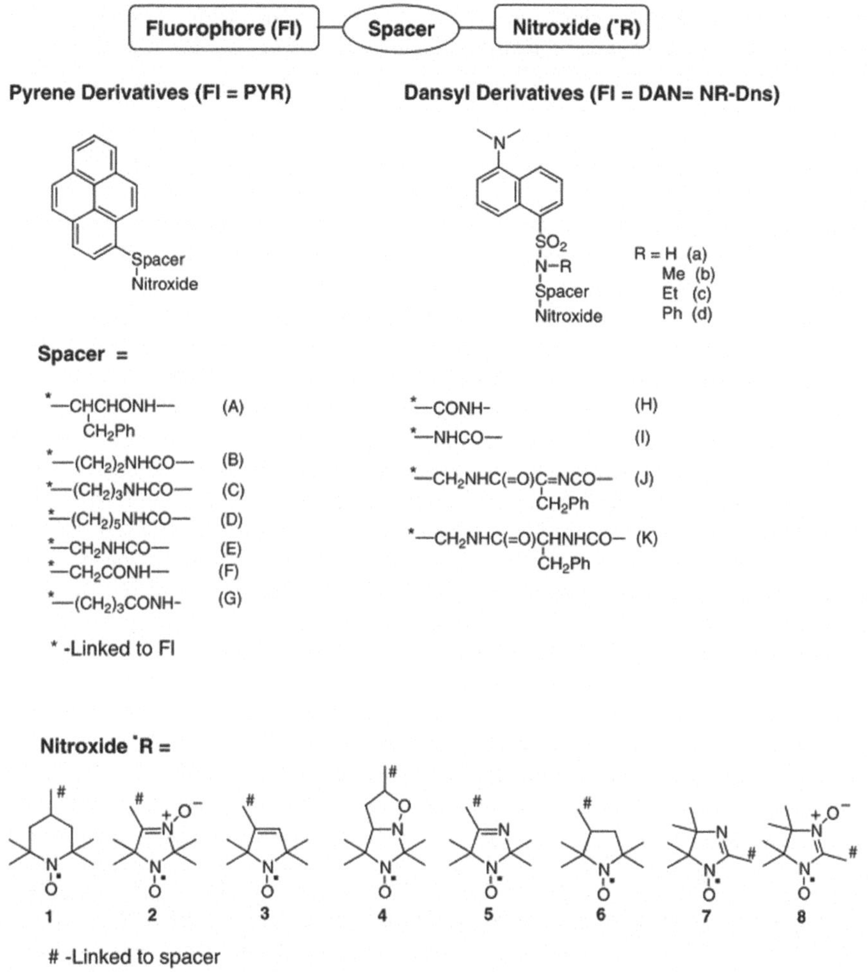

Fig. 6.1 Structure of pyrene and dansyl nitroxides (private communication from Drs. V. V. Martin and A. Weis, Lipitek International, Inc.)

A series of novel frameworks acetylene-substituted isoindoline nitroxide analogs of

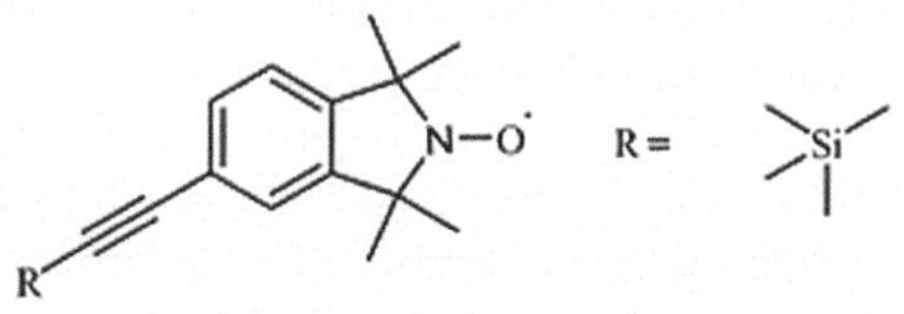

were prepared via palladium-catalyzed copper-free Sonogashira coupling [19]. The coupling gave acetylene-substituted isoindoline nitroxides via 5-iodo-1,1,3,3-tetramethylisoindolin-2-yloxyl,**3**, as an intermediate. Subsequent reaction of the deprotected ethynyl nitroxide with iodinated polyaromatics furnished novel profluorescent aromatic nitroxides with extended conjugation. Ethyne- and butadiyne-linked nitroxide dimers were also be synthesized by this cross-coupling methodology. Design and synthesis of profluorescent polymers [20].

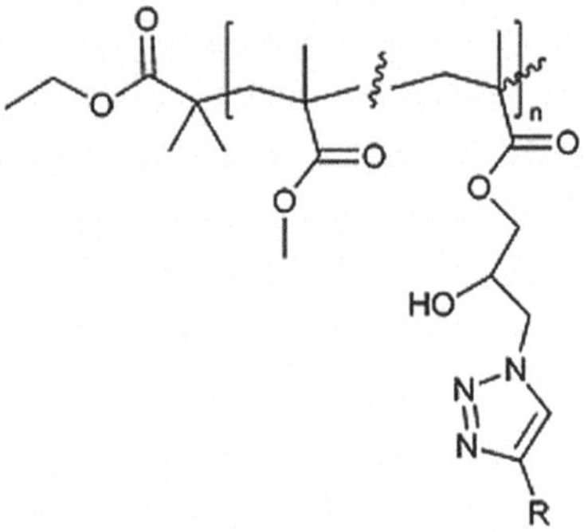

with R

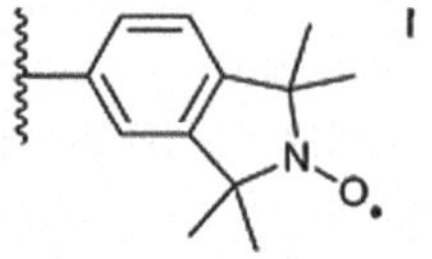

included the following stages: (1) the incorporation of an epoxide as a pendant functionality on a polymer backbone synthesized using atom transfer radical polymerization (ATRP), (2) subsequent nucleophilic ring opening with sodium azide gave hydroxyl and azide functionality within a three-bond radius, and (3) the independent attachment of fluorophore and nitroxide groups in close proximity, giving rise to a profluorescent polymer utilizing orthogonal coupling chemistry. Efficient fluorescence switch-on was observed when the materials were exposed to a model reductant or carbon-centered radical.

Dual probe (**1**) having six-membered piperidine nitroxide fragment was synthesized by treatment of the commercial amino nitroxide with dansyl chloride in pyridine (Fig. 6.2) [26]. Probe (**2**) was derived from the following reaction sequences general for nitronyl nitroxides and imino nitroxide preparation: (1) Condensation of bis-hydroxylamine (**5**) with pyrene 1-aldehyde (**6**) gave 1,3-dihydroxyimidazoline (**7**); (2) the intermediate oxidation by air in the presence of Cu catalyst to give a mixture of nitronyl nitroxide (**9**) and imino nitroxide(**2**); (3) formation of deoxygenated compound (**2**) by dehydration of the intermediate (**7**) into 1-hydroxy-2-imidazoline (**8**) and (4) oxidation into imino nitroxide (**2**).

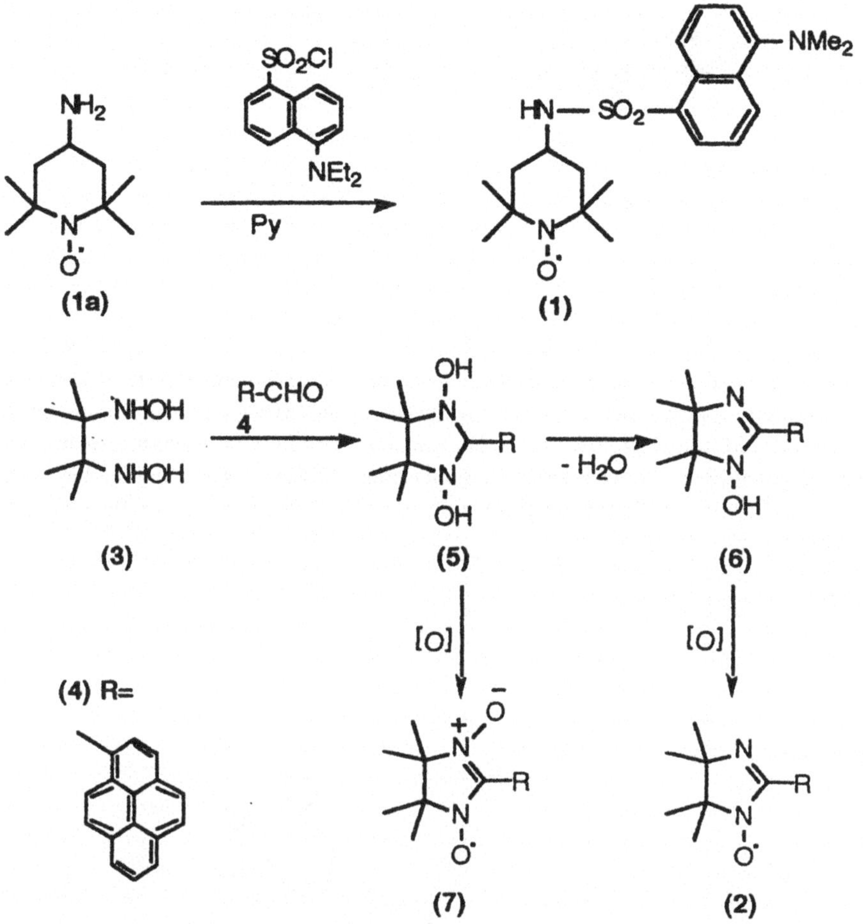

Fig. 6.2 Synthesis of profluorescent nitroxides [26]

Kieber and Blough solved the problem of labile linkages within profluorescent nitroxides by using a two-step process, whereby a simple nitroxide, a water-soluble amino nitroxide (3-(aminomethyl)-2,2,5,5-tetramethyl-l-pyrrolidinyloxy radical, 3-AMP, was prepared. This hybrid compound was firstly utilized to trap carbon-centered radicals [27].

The reaction of dibenzocyclooctyne to its corresponding triazole product with azide **1** in 4:1 dichloromethane/methanol for 32 h gave the novel profluorescent nitroxide probe

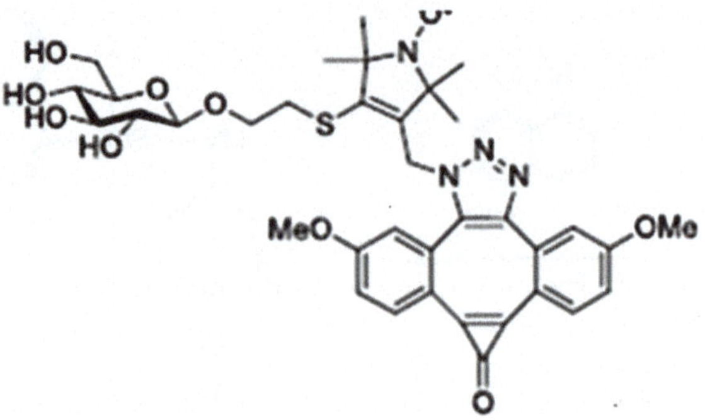

Benzaldehyde-containing comonomer, 4-formylphenyl methacrylate, which was subsequently copolymerized with methyl methacrylate (MMA) utilizing the versatile reversible addition–fragmentation chain-transfer (RAFT) polymerization technique was prepared [28]. The benzaldehyde-containing copolymer was combined as with 4-amino TEMPO in the presence of a drying agent (Na_2SO_4) overnight giving a polymer

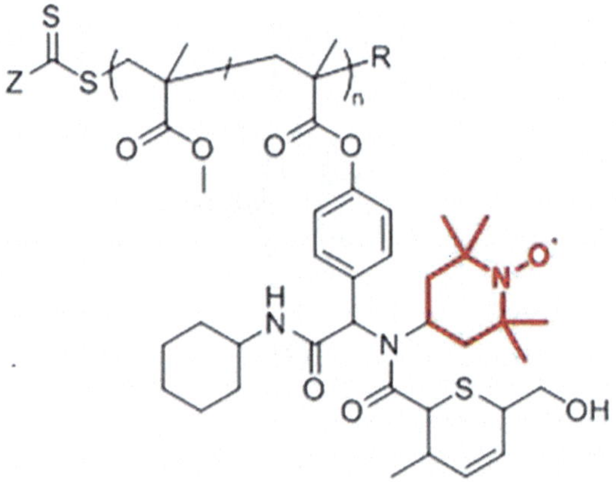

The synthesis of a new class of molecules which are hybrids of long-lived tetramethylisoindolinoxyl (TMIO) radicals and the pyrido[1,2-a]benzimidazole (PyrImid) scaffold was reported [29]. Non-covalent binding of nitroxide radicals to an abasic site in DNA and RNA duplexes at temperatures ranging from 0 to −30 °C was evaluated. Synthesis of PyrImid-TMIO hybrids

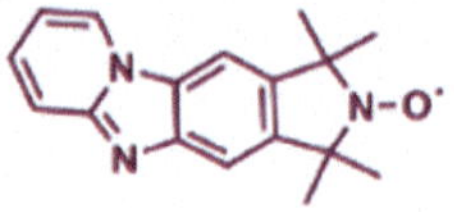

was described in detail.

The synthesis of new profluorescent nitroxides based on 2,5-disubstituted-1,3,4-oxadiazoles as fluorescent moieties and the 2,2,6,6-tetramethylpiperidine-N-oxyl radical (TEMPO) as paramagnetic probes was performed [30]. A linear response to reducing agents, i.e., sodium ascorbate, monitored by fluorescence spectroscopy, suggests the possibility of using the synthesized compounds as potent active probes in the detection of various analytes of interest. To make novel nanomaterial spin probes, three fullerene isoindoline nitroxide N-methyl-3,4-fulleropyrrolidine-2-spiro-50-(10,10, 30, 30 tetramethylisoindolin-20-yloxyl), (C60- (TMIO)m, and C70-(TMIO)n) were synthesized by the covalent bonding of 5-formyl-1,1,3,3-tetramethyl isoindolin-2-yloxyl to the fullerenes C60.

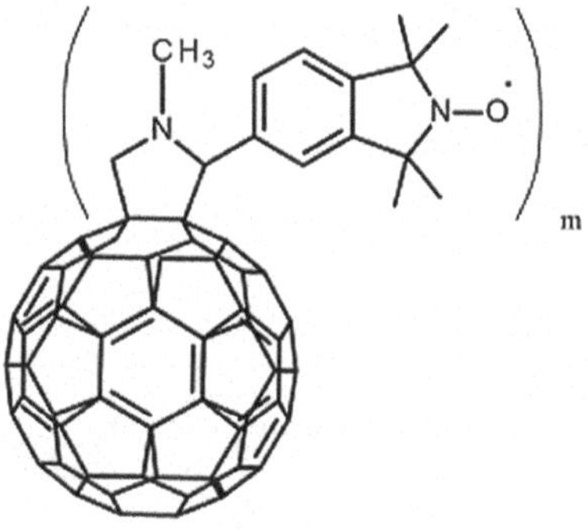

Emission ability of fullerenes has been well documented. For example, at room temperature, C60 and C70 in organic solvents expose fluorescence spectra. These fullerenes modified with nitroxides can be hydrophobic dual probes for biomembranes [31]. In the work of Braslau group [32], reaction of a catechol with a B-alkyl boronic acid derivative to form a B-alkyl boronate ester in situ was carried out. In the presence of a profluorescent nitroxide, a highly fluorescent N-alkoxyamine was formed (Fig. 6.3). The experiments indicated that the generation of a fluorescent signal would be a positive indicator for the presence of the catechol functionality. The addition of two equivalents of orange-colored TEMPO to an NMR tube containing one equivalent of the preformed B-n-butylcatecholboronate ester resulted in the dissipation of the orange color and formation of the fluorescent scN-nbutoxyamine of TEMPO 5.

A series of fluorescent poly(*N*-isopropylacrylamide)s (poly(NIPAM)s) have been synthesized via reversible addition–fragmentation chain-transfer (RAFT) polymerization from a functionalized chain transfer agent (CTA) bearing either dialkoxynaphthalene or dialkoxyphenylene moieties [33].

Exchange reactions between the isoindoline profluorescent nitroxide 1,1,3,3-tetramethyldibenzoisoindolin-2-yloxyl (TMDBIO) and a fluorescent silence TEMPO capped polystyrene (PS) were used for t-synthesis of polymer:

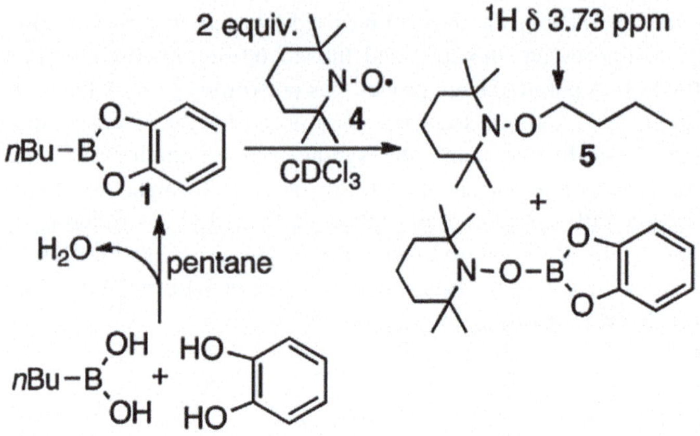

Fig. 6.3 Generation of the N-n-Butoxyamine 5 from TEMPO 4 using preformed B-n-Butylcatecholboronate sster [32]

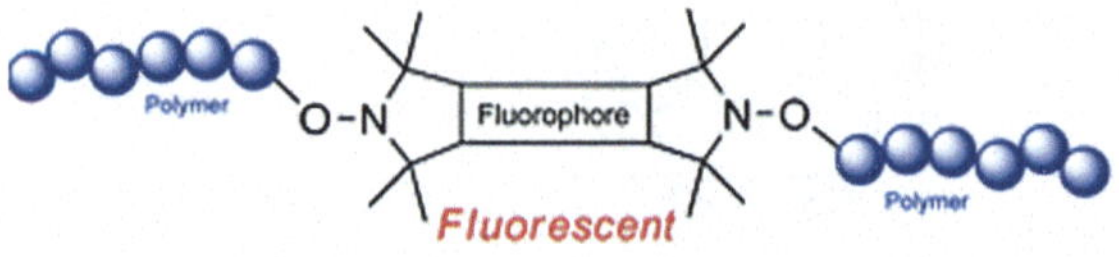

Specifically, the polymer was synthesized via NMP using α,α'-azoisobutyronitrile, as the initiator and 2,2,6,6-tetramethylpiperidine-1-oxyl (TEMPO) as the control agent, and was described as PS–TEMPO. The scope of this study was expanded by exploiting a di-nitroxide 9,10-bis(5-[1,1,3,3-tetramethylisoindolin-2-yloxy])anthracene (BTMIOA) as a connector between two polymer chains forming PS–nitroxide–PS systems.

The synthesis of optically distinct BODIPY-based profluorescent probes bearing *meso*- and β-substituted isoindoline nitroxides and their corresponding methoxyamine derivatives was performed [34]. The probes strongly suppressed fluorescence ($\lambda_{em} = 570$–603 nm) which was revealed upon reduction or reaction with free radicals. To extend the wavelength of fluorescence emission of the BODIPY toward the NIR region to generate profluorescent nitroxides suitable for use in biological systems, dual compounds with the nitroxide unit incorporated through an alkyne linker at the β position of the BODIPY core, for example,

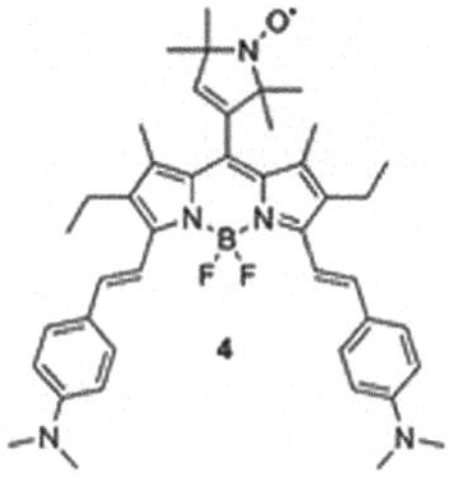

were synthesized.

Polyaromatic profluorescent mono- and bis-isoindoline nitroxides linked to napthalimide and perylene diimide structural cores

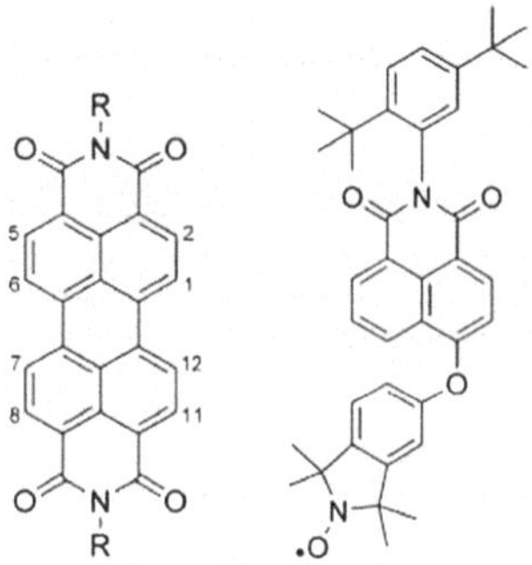

FORM 6.10

were synthesized by treating bromo-naphthalimide with 5-hydroxy-2-methoxy-1,1,3,3-tetramethylisoindoline in a base-assisted nucleophilic phenoxide substitution reaction by heating with KOH in DMF to give substituted naphthalimide [35]. Then nitroxide was obtained by oxidation of methoxyamine with *m*CPBA in a Cope-type elimination process. (A novel synthetic avenue for the preparation of profluorescent nitroxides via nitrile imine-mediated tetrazole-ene cycloaddition (NITEC) was introduced [36]. The NITEC involves two steps: (1) under irradiated with UV light, a nitrile imine is formed via a first-order reaction with the release of nitrogen, and (2) the in situ generated nitrile imine subsequently undergoes a rapid cycloaddition with the olefin. Figure 6.4 schematically illustrates the synthesis of profluorescent nitroxides.

A bifunctional stilbene-nitroxide label (BFL1) was synthesized (Fig. 6.5) [37]. The synthesized dual stilbene–nitroxide probe was covalently immobilized onto the surface of a quartz plate as an eventual sensor for ascorbic acid and microviscosity (Fig. 6.6).

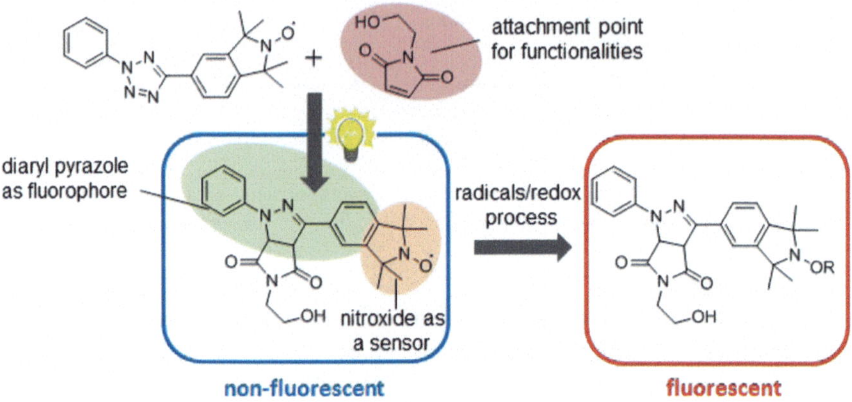

Fig. 6.4 Schematic presentation of the NITEC of diaryl tetrazole and maleimide [36]

6.2.1 Profluorescent Nitroxides as Redox Probes

As was first demonstrated in [3] in a fluorescence–nitroxide supermolecule,

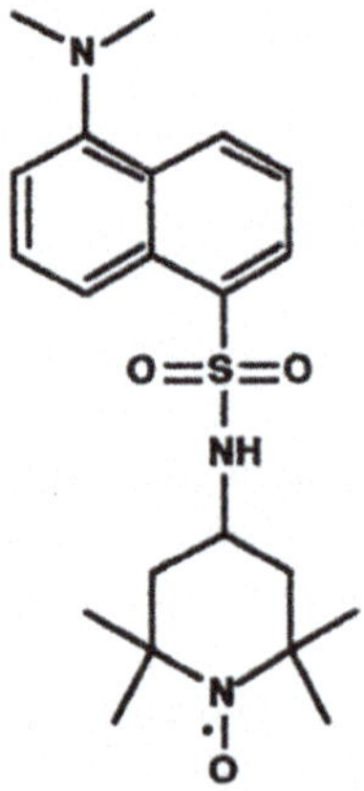

FN I

the nitroxide serves as a strong intramolecular quencher of the fluorescence from the chromophore fragment and the fluorescence is restored during the nitroxide photoreduction Later, it was shown that in such molecules, fluorescence is restored via radical scavenging to the alkoxyamine or redox processes to the hydroxylamine or oxoammonium cation [13]. Therefore, chemical reduction of the nitroxide fragment by antioxidant would result in a decrease of electron spin resonance (ESR) signal and accompanied with a rise of the fluorescence intensity and can be used in antioxidant analysis. A series of the dual FNO molecules including **FN I** and FN II

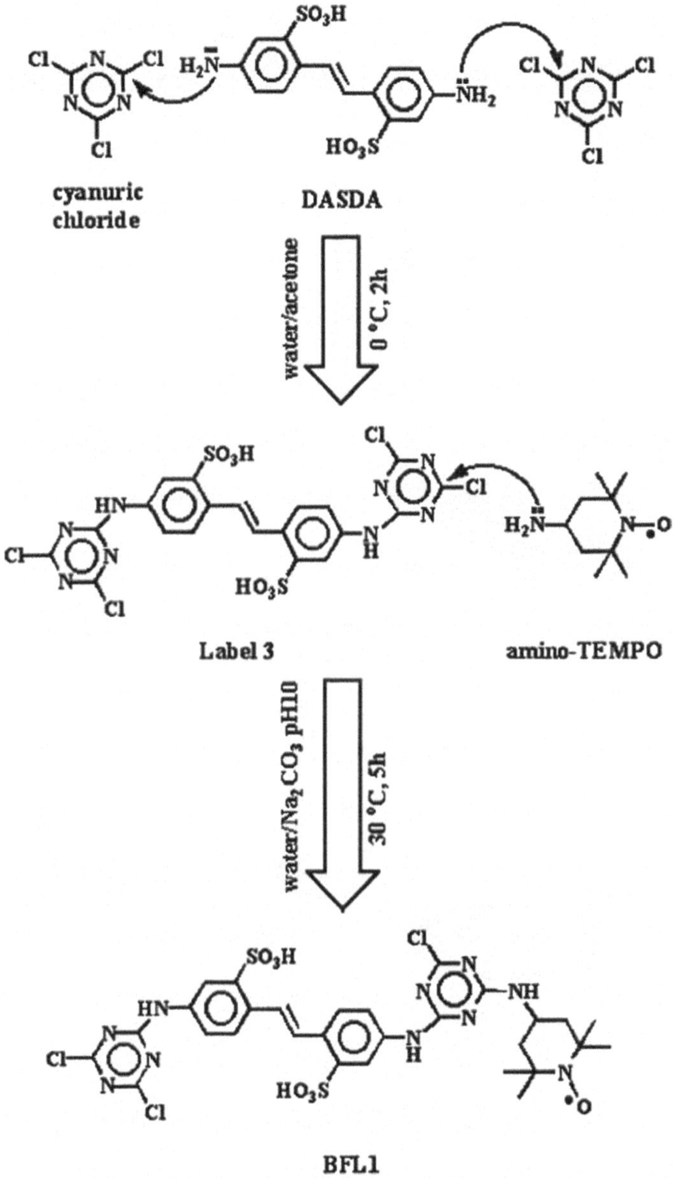

Fig. 6.5 Synthesis of the dual stilben-nitroxide probe BFL1 [37]

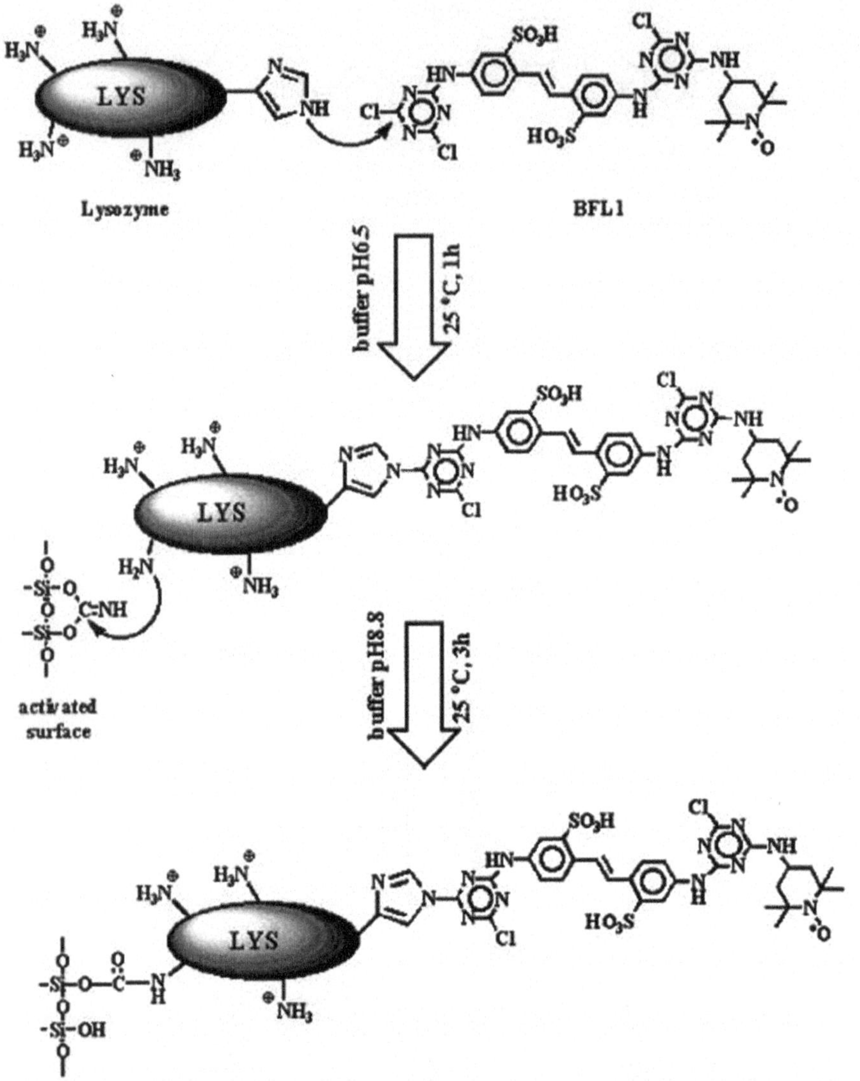

Fig. 6.6 Immobilization of the dual stilbene–nitroxide probe onto the surface of a quartz plate via lysosome [37]

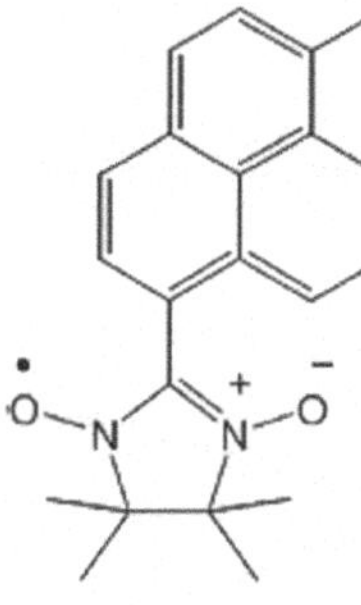

FNO• I

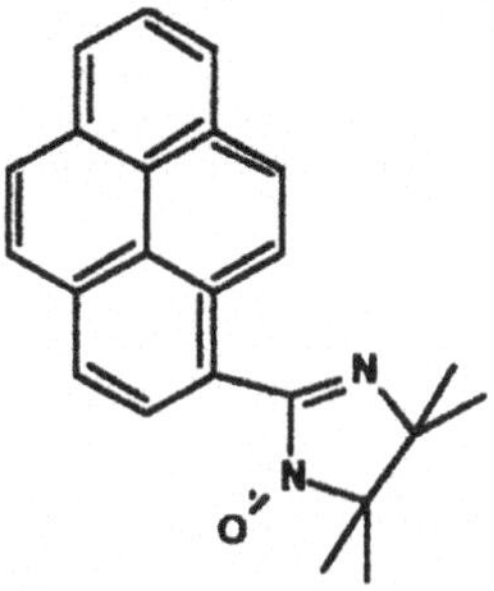

FNO II

have been used first in the analysis of antioxidants of different reducing capacities. On the basis of the abovementioned unique properties of the dual molecules, a method for the quantitative analysis of vitamin C in biological and chemical liquids has been proposed and implemented [26, 38–41]. In the presence of ascorbic acid, an increase in the fluorescent intensity and a decay of the EPR signal of the dual probe DT occurred with the same rate constant (Figs. 6.7 and 6.8). By performing a series of pseudo-first-order reactions between the dual molecule and ascorbic acid and by consequently plotting the rate constants (k_{app}) versus the ascorbic acid concentrations, the calibration curves for the vitamin C analysis were obtained. In the fluorescence and ESR measurements (25 C, phosphate buffer), the slope of the calibration line corresponded to the second-order constant $k = 7\ M^{-1}\ s^{-1}$ for FNO I. Nitroxide FNO II displayed a faster kinetics with $k = 51\ M^{-1}\ s^{-1}$. The observed order of reactivity for the probes is in agreement with oxidation potentials of their nitroxide fragments, expected from the inductive effects of the substituents. The kinetics of this series has been described in a very good approximation by the model of pseudo-first-order process with apparent rate constant k_{app}. The linear dependencies of k_{app} on ascorbic acid concentration obtained either by ESR or fluorescence measurements are shown in Fig. 6.8. The proposed method was applied to the determination of ascorbic acid in mandarin beverage.

Effect of bovine serum albumin (BSA) on the kinetics of ascorbate oxidation was evaluated [39]. The oxidation of ascorbic acid by the fluorophore–nitroxide, probe dansyl piperidine–nitroxide, was monitored by steady-state fluorescence and electron paramagnetic resonance. Experiments showed that the probe reduction by ascorbic acid was accelerated with the increase of the BSA. The influence of BSA on the rate is attributed to the adsorption of both ascorbate and the probe to BSA. Adsorption of ascorbate to BSA was confirmed by NMR relaxation experiments. In the presence of BSA, the autoxidation of ascorbate at different pH values is significantly slowed down. This effect was explained in terms of the electrostatic interaction between the ascorbate anion and the BSA molecule.

The fluorescent probe, R2c, consisted of silicon phthalocyanine and two 2,2,6,6-tetramethyl-1-piperidinyloxy radicals, and is encapsulated by the dimer of bovine serum albumin was used for detecting ascorbic acid [42]. Due to this encapsulation, the **R2c**–BSA complex was prevented from reacting with various redox species in

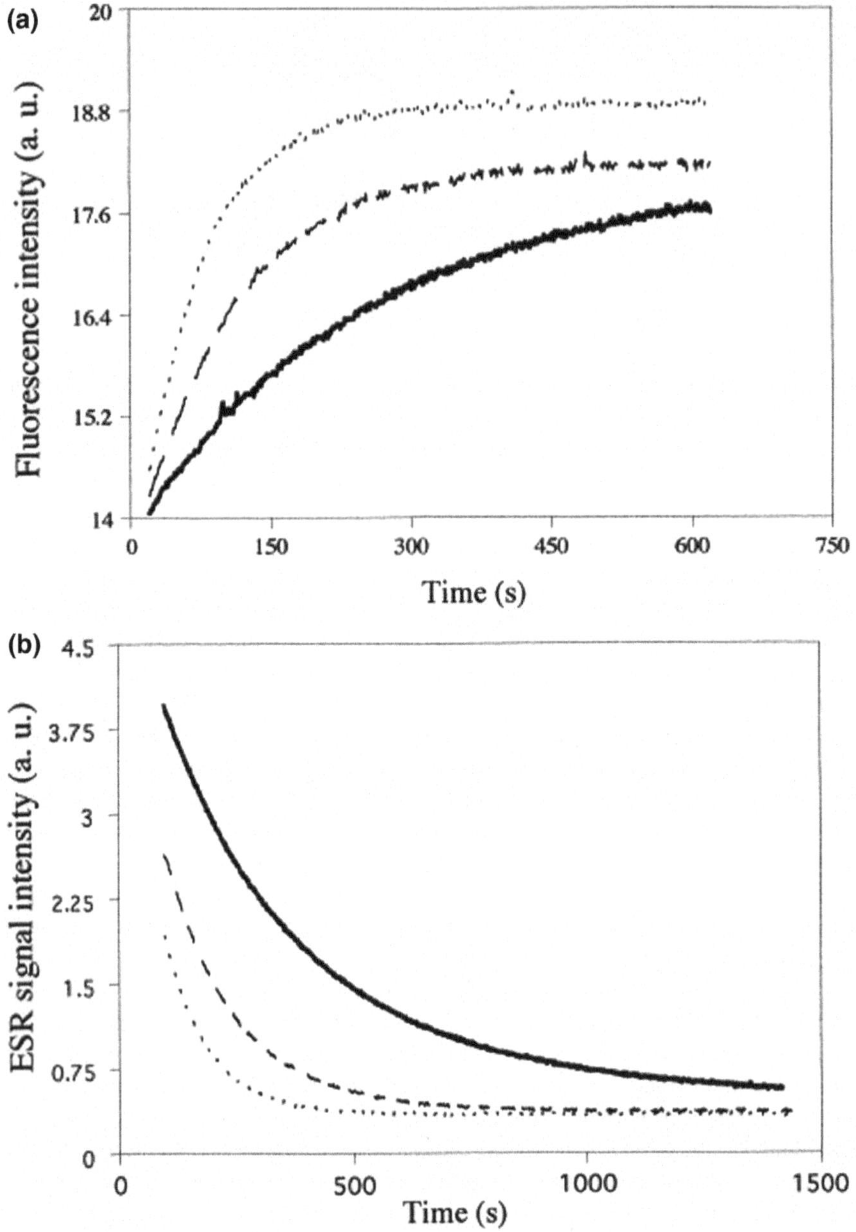

Fig. 6.7 Fluorescence enhancement (**a**) and ESR decay (**b**) of the probe (**1**) caused by excess of ascorbic acid. Solid line, 0.1 mM; dashed line, 0.4 mM; dotted line, 0.8 mM. Acquisition parameters of fluorescence: excitation slit 1 nm, emission slit 16 nm, voltage 480 V, *T*5258C, phosphate buffer [26]

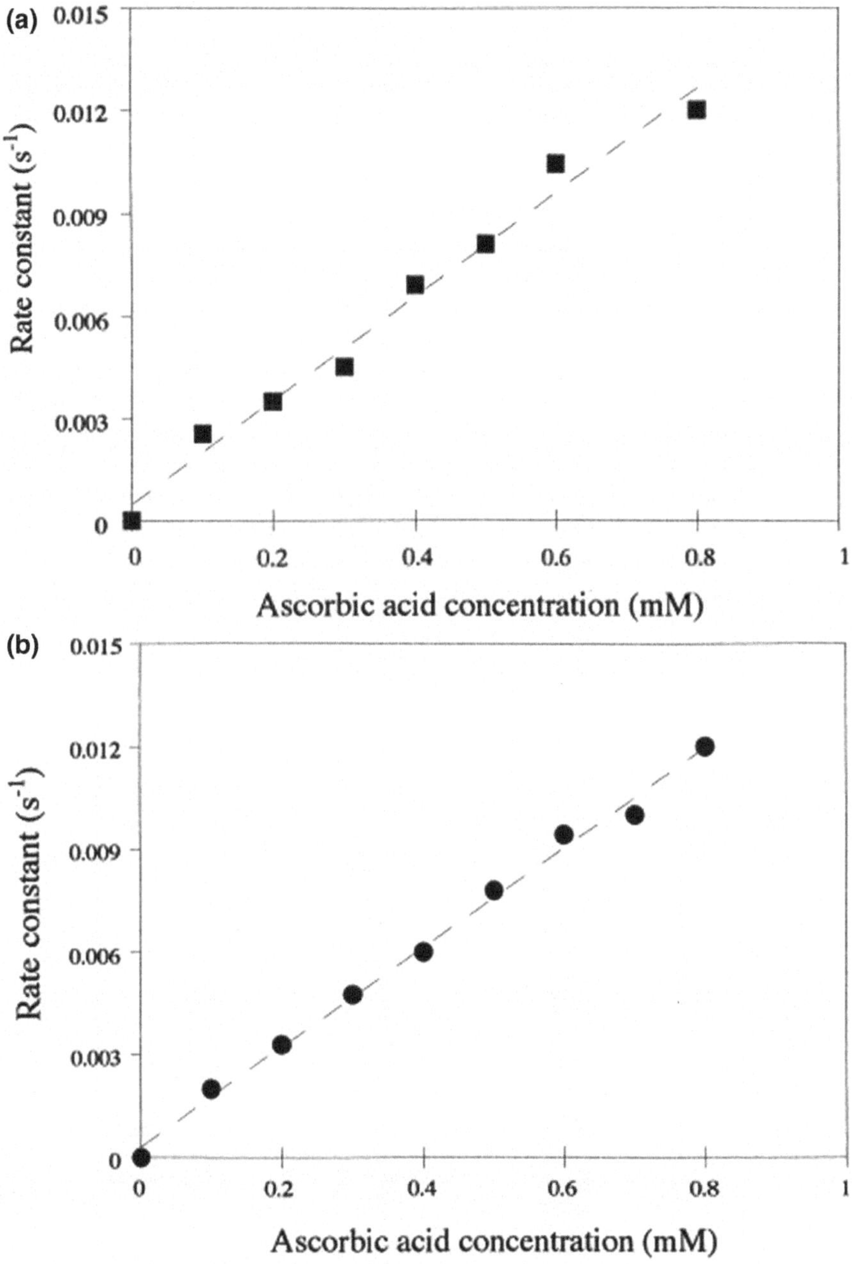

Fig. 6.8 Dependence of rate constant of reduction for the nitroxide FNO **I** on ascorbic acid concentration (25C, phosphate buffer pH 7), **a** fluorescence, **b** ESR [26]

biological, but reacted selectively with ascorbic acid. The complex fluorescence of after ascorbic acid addition depended on the pH and fluorescence intensity increased in the order pH 6 < pH 5 < pH 4 < pH 3, but decreased at pH 2 compared with that at pH 3 and also decreased with increasing temperature after ascorbic acid addition. These effects were explained in terms of the relationship between the shielding effects of BSA and the folding ↔ unfolding structural changes by the change in relative proportions of α-helix and β-sheet in BSA.

A nitronyl nitroxide radical was covalently linked to an organic fluorophore, pyrene, and was used to detect superoxide radicals and to quantitatively estimate the antioxidant activity of biological compounds of different antioxidant redox potential [43]. This approach is also based on the phenomenon of intramolecular fluorescence quenching of the pyrene fragment by the nitroxide in the dual fluorophore–nitronyl (FNO• II). The nitroxide fragment of the dual non-fluorescence molecule can be reduced by a variety of free radicals (e.g., superoxide) and antioxidants (e.g., ascorbic acid, quercetin, galangin). The reduction of FNO is accompanied by a drastic increase in fluorescence intensity (up to 2000) (Fig. 6.9) and a corresponding decrease of the EPR signal. The fluorescence technique, which has been developed, allows the detection of antioxidants at the submicromolar concentration scale, while the EPR and light absorption techniques are limited to several micromolars (Fig. 6.10).

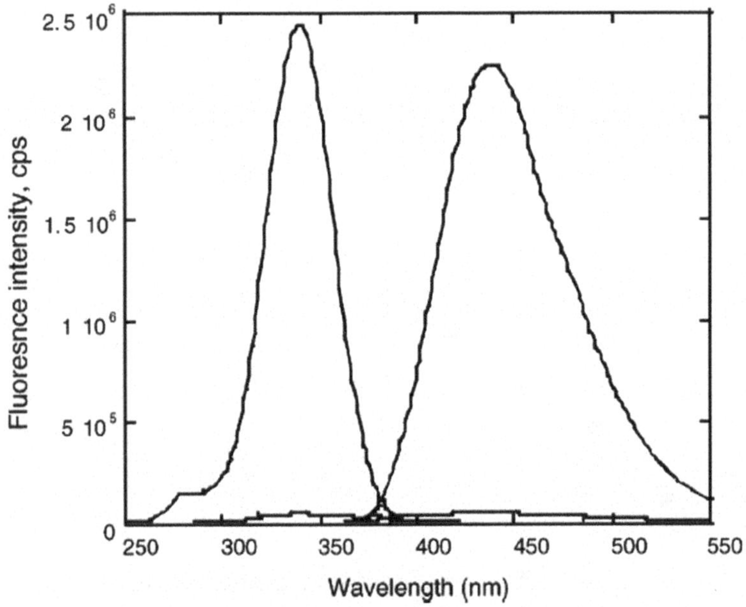

Fig. 6.9 Fluorescence emission and excitation spectra of the FNO• II (5×10^{-5} M) before and after reduction by quercetin (5×10^{-7} M) in PBS (pH 7.4, $T = 300$ K) [43]

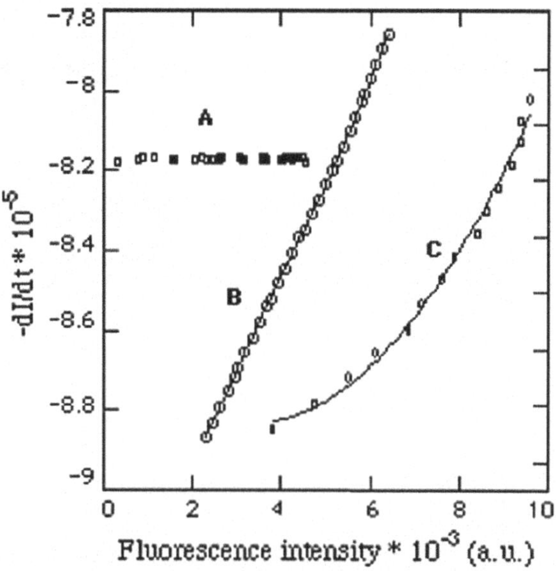

Fig. 6.10 FNO fluorescence increases rate versus fluorescence intensity: **A** zero-order reduction rate of FNO (5×10^{-5} M) by superoxide ($\omega i = 7 \times 10^{-8}$ M/min); **B** pseudo-first-order reduction rate of FNO (5×10^{-5} M) by ascorbic acid (5×10^{-6} M); and **C** second-order reduction rate of FNO (5×10^{-5} M) by ascorbic acid (5×10^{-5} M). The fluorescence kinetics was measured in the presence of catalase (280 U/ml) in PBS (pH 7.4, $T = 300$ K [43]

A scheme of the chemical reaction upon addition of reducing agent to the dual fluorescence–nitronyl probe was suggested (Fig. 6.11) This method has also been applied to the determination of hydrogen peroxide in submicromolar concentrations.

A novel nitronyl nitroxide derivative of

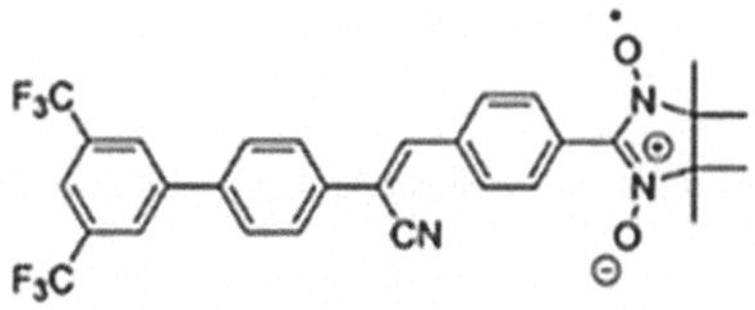

showed 260-fold fluorescence turn-on and diminished electron spin resonance signal upon ascorbic (AA) addition [44]. The probe could detect AA over a broad concentration range from 1 μM to 2 mM. It was reported that the fluorescence of a polyamidoamine (PAMAM) dendrimer-entrapped gold nanocluster Au_8 cluster is quenched by the paramagnetic nitroxide radical with a combination of static and dynamic quenching processes [45]. Combined fluorescence and electron paramagnetic resonance studies formed a basis for the development of a turn-on fluorescence probe for sensing in AA in living cells. Fluorescence of negatively charged blue fluorescent carbon dots (CDs) was found to be efficiently quenched by the cationic 4-amino-2,2,6,6-tetramethylpiperidine-*N*-oxide free radical (4-AT) in solution presumably by electron transfer mechanism [46]. CDs tethered with TEMPO exhibits sensitive fluorescence and ESR bimodal response toward ascorbic acid (AA) at the μM level.

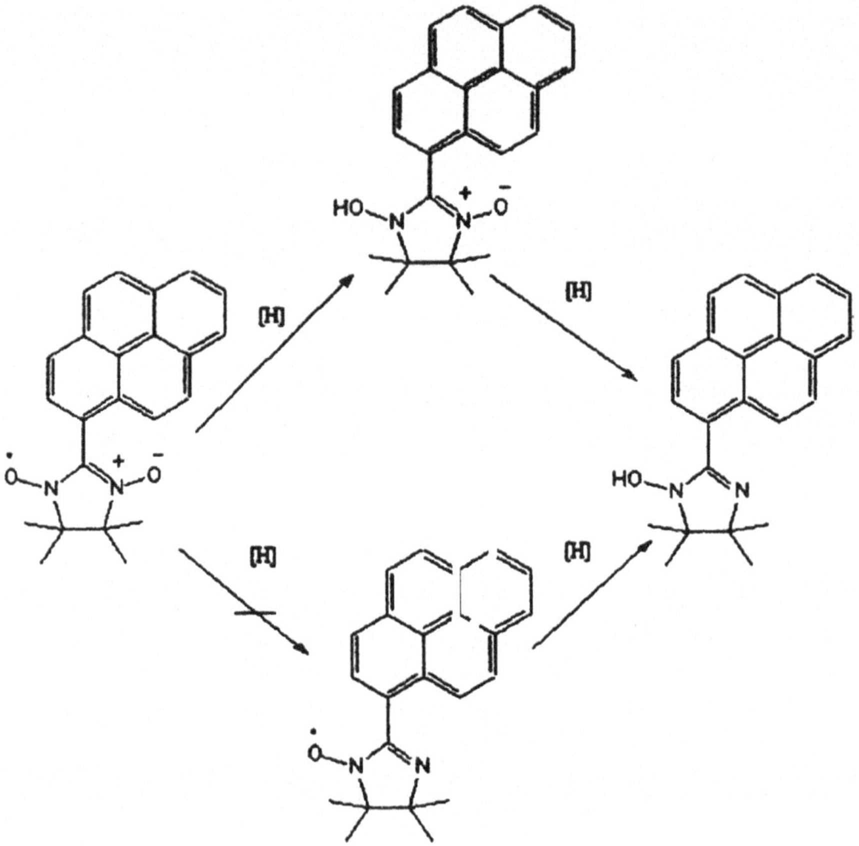

Fig. 6.11 Suggested mechanism the chemical reaction upon addition of reducing agent to the dual fluorescence–nitronyl probe [43]

Six fluorescein-nitroxide radical hybrid compounds were synthesized, and their physical and chemical properties were characterized by fluorescence and EPR spectroscopies [25]. Fluorescence intensities (around 520 nm) of compound

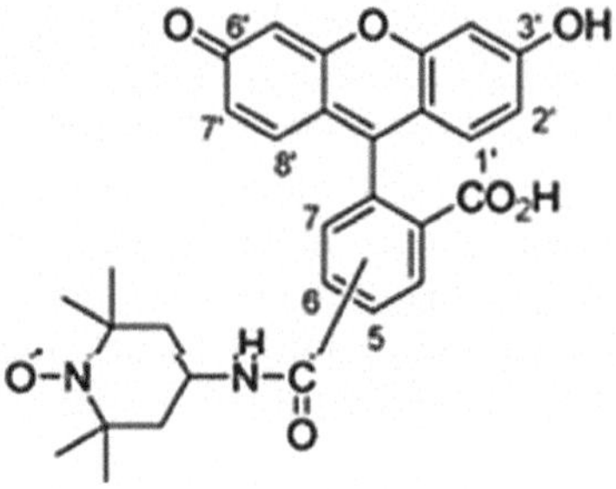

after reduction by ascorbic acid of the radical increased from 1.43- to 3.21-folds.

A rhodamine nitroxide probe (R-NOc), combining rhodamine fluorophore with a 2,2,6,6-tetramethylpiperidinyl-1-oxy (TEMPO) receptor unit, was introduced to probe glutathionyl radicals (GSc) with high sensitivity and selectivity [47]. The reaction between GSc and nitroxides had low energy barriers, being radical–radical reactions, and had rate constants near the diffusion limit (>10^8 dm^3 mol^{-1} s^{-1}) [48]. In this work, horseradish peroxidase (HRP)-catalyzed and metal-catalyzed oxidation systems were selected as the model of simulating the generation of GSc. It was found that the metal-catalyzed system had the same experimental results with the HRP-catalyzed system, demonstrating the strong oxidant ability of the hydroxyl radical (•OH) to initiate toxic GSc. By using combined high-performance liquid chromatography (HPLC) detection and MS analysis, it was shown that the R-NOc was converted into fluorescent secondary amine derivative (R-NH).

Synthesis of novel profluorescent nitroxides, 2,5-disubstituted-1,3,4-oxadiazole–TEMPO hybrid analogs of

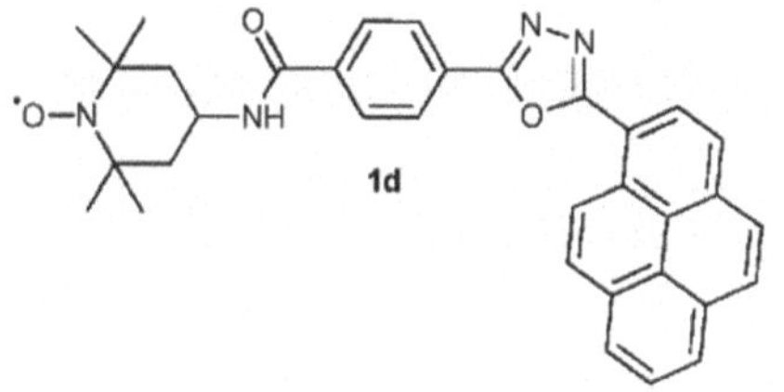

as dual luminescent-paramagnetic active probes for ascorbic acid assay was performed [49]. These nitroxides in solutions (5 × 10^{-5} M in 15% DMSO in HEPES, 0.01 M, pH = 7.51) with increasing amounts of sodium ascorbate (between 0 and 2 equivalents) and measurement of the fluorescence emission at λ_{em} = 388 nm (λ_{ex} = 295 nm) after incubation for 30 min showed a linear response up to the stoichiometric amount of the reducing agent (0.5 equivalents).

A method of the measurement of ascorbic acid and microviscosity of media developed in [37] is based on the use of stibene-photochrome-fluorescence-spin probes (**BFL1**) immobilized onto the surface of a quartz plate as an eventual sensor (Fig. 6.6). In such a hybrid compound, the nitroxide moiety quenches the fluorescence of the fluorophore (stilbene moiety). The reduction of nitroxide segment by an ascorbic acid causes a rise of fluorescence of the fluorophore. The rate constant of the stilbene fragment photoisomerization in such systems is dependent upon the viscosity of the media. Therefore, this probe was applied for the parallel determination of the antioxidant status and measurement of micro- and macroviscosity of the media. Such an approach made it possible to measure the concentration of ascorbic acid in solution in a range of (1–9) × 10^{-4} M and the viscosity of a medium in the range 1–500 cP (Figs. 6.12 and 6.13).

The hybrid compounds comprised of five nitroxides with Ibuprofen were synthesized [50]. The rate constants in reduction reaction with 200-fold excess of ascorbic acid were determined in order of 1.0–6 $M^{-1}s^{-1}$). Two sterically shielded nitroxides

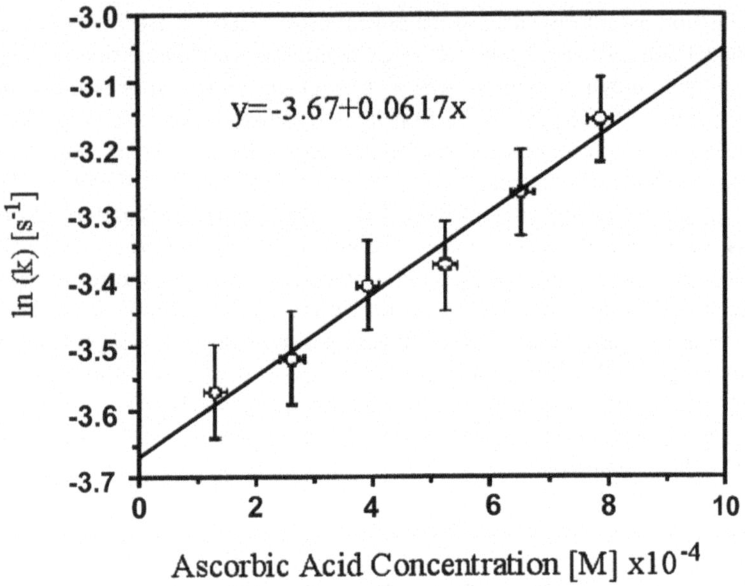

Fig. 6.12 Logarithmic dependence of the nitroxide moiety reduction rate for the immobilized stibene-photochrome-fluorescence-spin probe (BFL1) molecule versus the ascorbic acid concentration in aqueous solution. $T = 298$ K, PBS pH 7.4 [37]

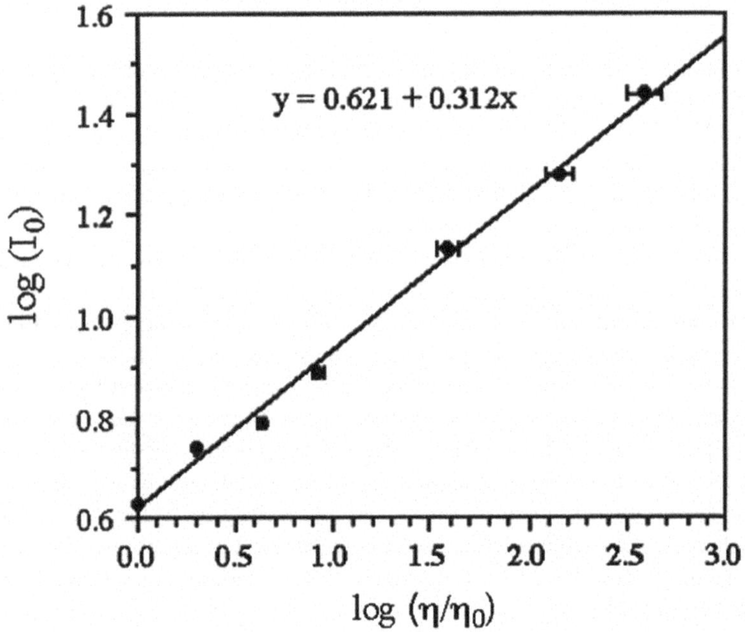

Fig. 6.13 Logarithm of initial intensity of the immobilized stibene-photochrome-fluorescence-spin probe versus logarithm of viscosity of aqueous–glycerol solutions [37]

reacted with AA markedly slower. In the reaction with the more reactive methyl radicals, produced by 200-fold excess of Fenton's reagent, the reduction rates were in the following increasing order from 2.0 to 5.0 $M^{-1}s^{-1}$.

To increase the reactivity and selectivity toward the detection of ascorbic, the dual nitroxide probe 15-((9-(Ethylimino)-10-methyl-9*H*benzo[*a*]phenoxazin-5-yl)amino)-3,11-dioxa-7-azadispiro hexadecan-7-yloxyl (Nile-DiPy)

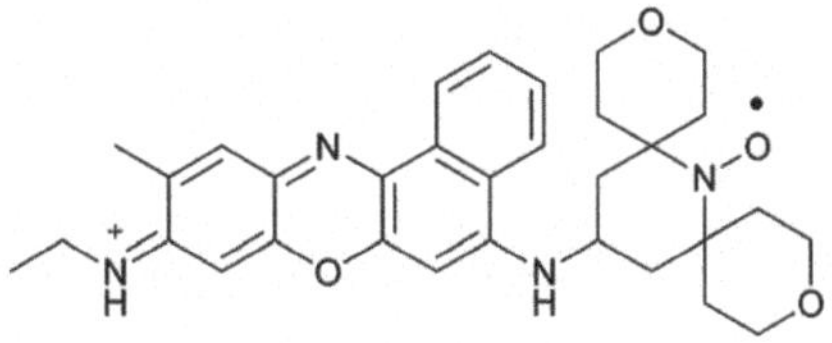

Nile-DiPy

was synthesized [51–53]. This fluorophore–nitroxide probe rapidly reacted with ascorbic acid and showed in parallel fluorescence enhancement in PBS at pH 7.4, containing 5% (v/v) DMSO. In the presence of ascorbic acid, the fluorescence intensity of Nile-DiPy increased in a dose-dependent manner concentration of ascorbic acid (0.13–8.0 mM). The second-order rate constant for the reaction of Nile-DiPy and of Nile-TEMPO with ascorbic acid was calculated as 246 M^{-1} s^{-1} and 17.4 M^{-1} s^{-1}, respectively. The kinetic isotope effect (KIE) for the detection of ascorbic acid was determined to be 9.8, indicating that Nile-DiPy reacts with ascorbic acid to produce the fluorescent Nile-DiPy-H via hydrogen atom transfer. The limit of detection (LOD) of this fluorometric method was estimated to be 9.7 nM. The application of Naph-DiPy nitroxide for the measurement only ascorbic acid in the plasma of osteogenic disorder Shionogi rats when fed an ascorbic acid-deficient diet was reported. [51–53]. In this condition, other biological reductants, including uric acid, glutathione, NADH, catechin, 2,2,5,7,8-pentamethyl-6-chromanol (HPMC) as a vitamin E model compound, and uric acid, did not react with the nitroxide. The usefulness of Naph-DiPy nitroxide for the measurement of ascorbic acid in the plasma of streptozotocin-induced diabetic animals was confirmed.

A sensitive and selective sensor for ascorbic acid detection based on the recovered fluorescence of NAPS-NO (*N*-propyl-triethoxysilane-4-(4-ylamino-1-oxy-2,2,6,6-tetramethylpiperdine)- naphthalimide) probe was designed [54]. Over a wide range from 80 nM to 50 μM, a good linear relationship between the fluorescence intensity and the concentration of ascorbic acid was found and the detection limit was estimated to be as low as 20 nM. The use of NAPS-NO for the measurement of ASA in human blood serum and determination of the concentration of AA in HEK 293 cell lysate was demonstrated. In addition, confocal laser scanning microscopy experiments showed that this chemosensor is cell permeable and can be used as a fluorescent probe for monitoring ascorbic acid in living cells.

The kinetics of reduction of the dual fluorescence–nitroxide probe R*, 5-dimethylaminonaphthalene-1-sulfonyl-4-amino-2,2,6,6-tetramethyl-1-piperidine-oxyl by human blood and its components were studied using the EPR technique [40].

The results indicated that (1) R* is adsorbed to the outer surface of the membrane and does not penetrate into the erythrocytes, (2) the observed first-order rate of disappearance of the nitroxide radical k is k(blood) > k(eryth) > k(plasma) and k(blood) approximately = k(eryth) + k(plasma), (3) the erythrocytes catalyze the reduction of R* by ascorbate, (4) the rate of reduction of the radical is high though it does not penetrate the cells, (5) there is an efficient electron transfer route through the cell membrane, and (6) ascorbate is the only natural reducing agent in blood that reacts with R*. The study points out that R* is a suitable spin label for measuring the reduction kinetics and antioxidant capacity in blood and other biological liquids as expressed by reduction by ascorbate.

There are controversial reports in the literature regarding the total antioxidant capacity of neonate blood [41]. In order to investigate directly the antioxidant properties in the newborn's and adults blood, a method of dual fluorescence–nitroxide probes was applied to investigate antioxidant status of on vein blood samples taken from 38 newborn's and 40 healthy adults using the probe R* and EPR spectroscopy. It was found that the reduction rates of R* by ascorbate in neonate's whole blood are significantly higher ($P < 0.001$) than the reduction rates of R* in adult's whole blood. The authors concluded that newborn's blood has significantly higher ability to deal with oxidative stress, caused by R, in comparison with adult blood and the system that responds to the recycling of ascorbate is more efficient in neonate blood than in adult's blood.

Perylenebisimide-linked nitroxide (PBILN) was employed as a profluorescent reagent, which permits for the selective determination of ascorbic acid in a stream [55]. Under optimized conditions, a good linear relationship between the concentration of ascorbic acid, fluorescence peak height in the concentration range from 0.5 to 10 μmol L^{-1} was found, and the detection limit ($S/N = 3$) was 0.28 μmol L^{-1}. The proposed method was applied to the determination of ascorbic acid in several soft drink beverages.

A novel, highly sensitive, and appropriately selective fluorescent probe consisting of silicon phthalocyanine (SiPc) and two 2,2,6,6-tetramethyl-1-piperidinyloxy (TEMPO) radicals, R2c, encapsulated in dimeric bovine serum albumin was employed for imaged ascorbic acid intravenously injected into a mice [56]. Ascorbic acid intravenously injected into mice was efficiently transported to the liver, heart, lung, and cholecyst. The proposed dual probes obeyed to principle requirements: (1) The excitation and fluorescence wavelengths should be >650 nm to penetrate deeply into living tissues and to avoid overlap with natural fluorescent species, and (2) nitroxide radicals should be shielded from biological redox-active species before injection of ascorbic acid, but should then efficiently react with ascorbic acid after injection. The changes in fluorescence before and after injection of ascorbic acid were dependent on the organs, as shown in Fig. 6.14. Eight minutes after ascorbic acid injection, the ratios of fluorescence intensity significantly increased at several organs, such as the liver, heart, and lung.

A range of novel, biostable, isoindoline nitroxide-based antioxidants, profluorescent nitroxide ME-TRN, and nitroxide DCTEIO and CTMIO

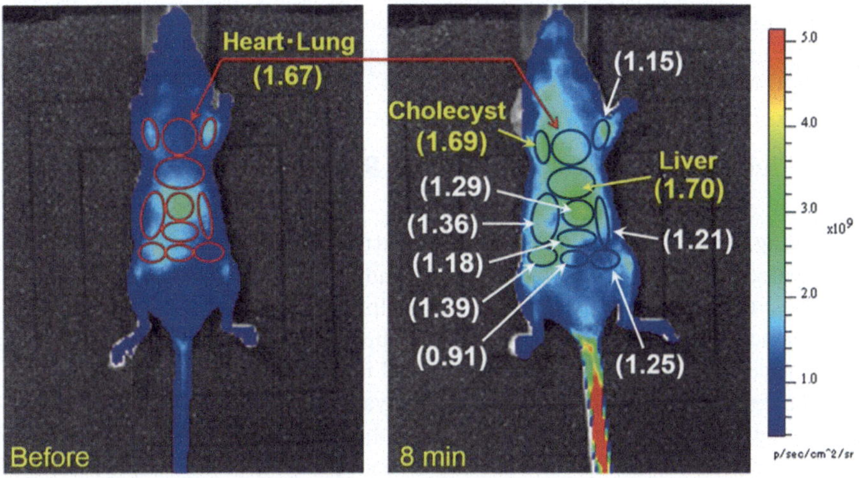

Fig. 6.14 Ratio of fluorescence intensity of imaging. Ratios were calculated using fluorescence intensities before and after ascorbic acid injection (8 min) [56]

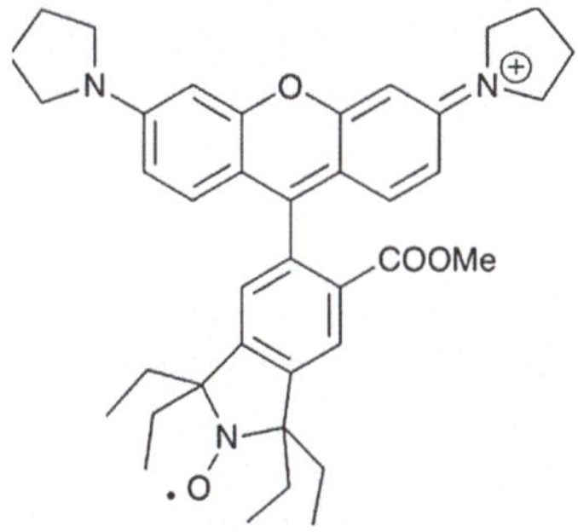

has been prepared and used for real-time quantification of oxidative stress and the protective effect of nitroxide antioxidants [57]. The efficiency of these compounds as antioxidant therapies in reducing ROS both in vivo (the rat eye model) and in vitro (661 W photoreceptor cells) was compared with the established antioxidant resveratrol. By assessing changes in fluorescence intensity of profluorescent nitroxide ME-TRN, in the rat retina in vivo, the ability of antioxidant therapy to ameliorate ROS production and reverse of the accumulation of ROS after complete acute ischemia followed by reperfusion (*I*/*R*) were evaluated. The novel agents' capacity to prevent ROS-mediated metabolic dysfunction in the 661 W photoreceptor cell induced by the oxidant, tert-butyl hydroperoxide, was investigated. As an example, changes in fluorescence of reduced ME-TRN probe in induced by tert-butyl hydroperoxide ischemia/reperfusion injury (*I*/*R*) treated rat eyes upon antioxidant administration were quantificated in a vivo fundus imaging. The nitroxide compounds DCTEIO and CTMIO successfully reversed the effects of ROS upon ME-TRN probe fluorescence. Thus, the reversible properties of the ME-TRN probe, which can detect and quantify dynamic alterations in the cellular redox status in vivo and in real time, in response to sequential pro- and antioxidant stimuli were demonstrated.

6.3 Fluorescence Detection of Free Radicals

Pioneering experiments of Blough group opened the way to the fluorescence detection of radical scavenging reactions in which the paramagnetic nitroxide-substituted naphthalene, 4-(1-napthoyloxy)-2,2,6,6-tetramethylpiperidine-l-oxyl (I), is converted to a diamagnetic N-alkoxy derivative [58]. The latter showed the fluorescence quantum yield in acetonitrile and hexane 55 and 30-fold is higher, respectively, than those of paramagnetic analog. In the experiments, 2-Cyanopropyl free radicals were generated by the thermal decomposition of azobisisobutyronitrile (AIBN) in cyclohexane or in acetonitrile containing I and the fluorescence intensity of the sample increased proportionally to the decrease in its ESR signal intensity.

The profluorescent nitroxide, 1,1,3,3-tetramethyldibenzo[*e*,*g*]isoindolin-2-yloxyl (TMDBIO), was investigated as a probe for the formation of polymer alkyl radicals during the thermo-oxidative degradation of unstabilised polypropylene [59]. A very low fluorescence quantum yield of the dual probe was drastically enhanced after reaction with alkyl radicals during the polymer oxidation. Importantly, the trapping of polymer alkyl radicals during the "induction period" at 120 C was observed, when it is not possible to detect changes in the polymer using either chemiluminescence or infrared spectroscopy.

A sensitive method for the detection of glutathionyl radical ($GS^{\bullet}$) based on their specific interaction with Ac-TEMPO, a non-fluorescent conjugate of fluorogenic acridine with paramagnetic nitroxide TEMPO was developed [60]. The method was employed for detection radicals generated through phenoxyl radical recycling by peroxidase. During reaction of Ac-TEMPO with $GS^{\bullet}$, TEMPO EPR signals decayed and acridine fluorescence concurrently increased. Using combined HPLC and mass spectrometry, it was determined that 90% of the Ac-TEMPO was converted into fluorescent acridine (Ac)–piperidine.

The preparation of nitroxide-containing polymer coupled to a fluorophore

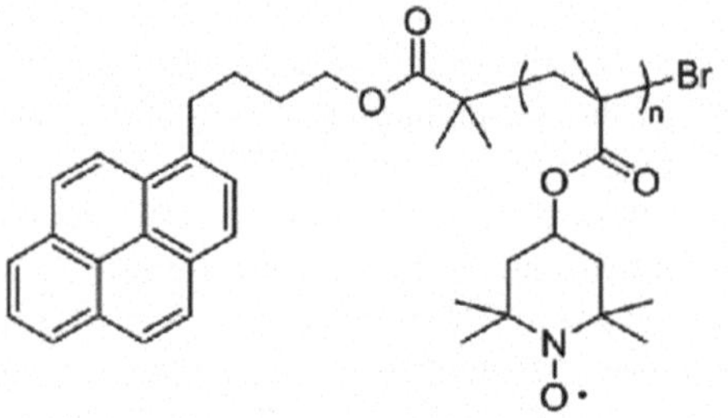

was reported, and its reduction with pentafluorophenylhydrazine was examined [17]. The fluorescence switch-on kinetics and radical concentrations were monitored by fluorescence and ESR spectroscopy. The polymeric compound displayed significant delays in reduction and fluorescence switch-on and higher turn-on ratios than their single-nitroxide counterparts.

A range of profluorescent nitroxides analogs of

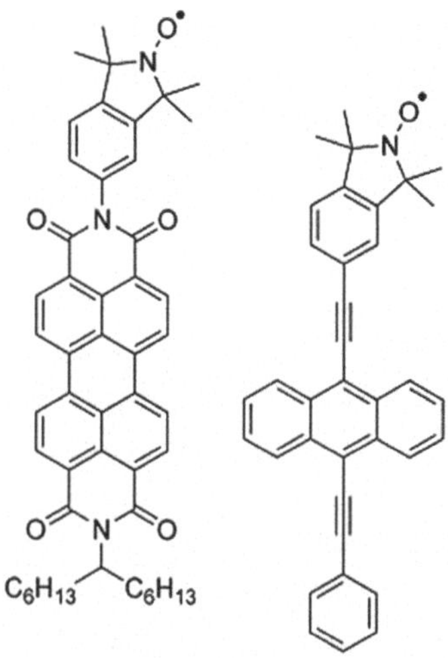

were tested as probes to monitor photo-induced radical-mediated damage in film polymer materials [21]. The most stable and sensitive probe of the perylenediimide (PFNs) tested was an alkyne-linked PFN, with naphthalimide and 9,10-bis(phenylethynyl)anthranene-based versions giving lower stability and sensitivity.. It was also determined that the nitroxide-containing compounds have higher stability in the photo-oxidative environment over their non-radical analogs and the alkyne-linked perylenediimide PFN has a switch on ability when exposed to photo-induced radicals. It is important that the dual probes provided information about the early stages of the film's degree of degradation, justifying the value of the PFN technique as a tool to give deeper insight into the oxidation induction period then that IR and EPR spectroscopies.

6.4 Photophysical and Photochemical Properties of Fluorescence–Nitroxide

6.4.1 Dual Compounds for Photoswitching Magnetic Materials

Considerable studies have recently been made by using photochromic derivatives as photofunctional units and nitroxide radical 1 as spin source. On this base, the new magnetic materials in which magnetic properties can be controlled by optical stimuli were developed [61–72]. The introduction of one or more radical spins to a photochromic fluorophore would give a photofunctional magnetic material [61]. A series of anthracene derivatives with nitroxide radicals were prepared as a photochrome-nitroxide system and as a spin coupler unit [61] and references cited. A photochromic reaction based on photodimerization of anthracene chromophore and heat-mediated dissociation of the photo-dimer was investigated in details. For example, an azobenzene derivative carrying two nitronyl nitroxide (NN) radicals was prepared, and in a frozen toluene at 10 K, its ESR spectra at cryogenic temperature were found to differ before and after irradiation.

Several *trans*-azobenzene derivatives carrying a nitroxide (aminoxyl) radical analogs

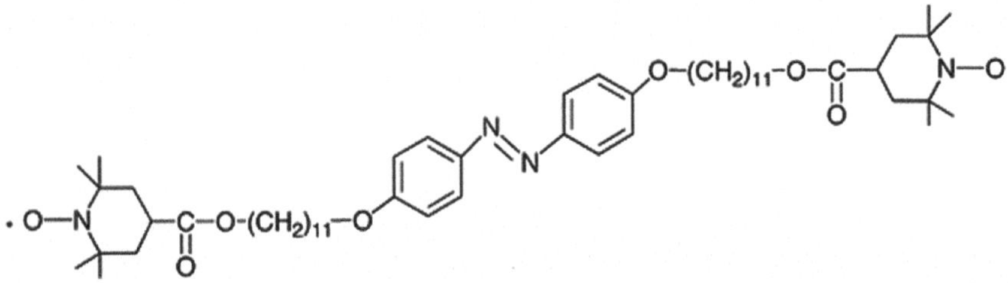

were synthesized, and their photoisomerization reactions to the corresponding *cis*-isomers were investigated [65]. These compounds, isolated as relatively stable solid substances, were isomerized by irradiation in solution to the corresponding *cis*-isomers. For example, an antiferromagnetic interaction model in the *trans*-isomer carrying TEMPO substituent with a large antiferromagnetic exchange interaction of $J = 47.6$ K was observed, while a weak ferromagnetic one of Curie–Weiss (CW) model with Weiss temperature $h = +0.11$ K was found in the corresponding *cis*-isomer.

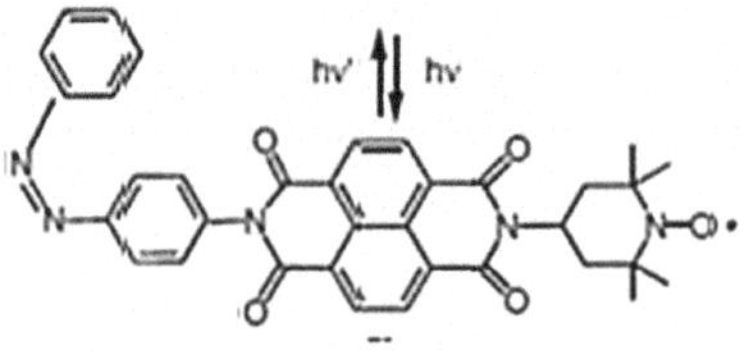

The unsubstituted derivatives at the *para* position were found to show photoisomerizations by irradiation to give the corresponding *cis*-isomers. The change of intermolecular magnetic interactions was observed by the structural change for each photochromic couple. A naphthalenediimide derivative carrying TEMPO radical and azobenzene substituent was found to show photo-responsive magnetic as well as FET properties upon illumination applying magnetic field [66]. Specifically, azobenzene derivatives had shown reversible photochromism in solution as well as in thin film. The change of FET behavior has been revealed by the photoisomerization together with the change of magnetic properties. The magnetic properties of both photoisomers were found to show antiferromagnetic interactions of CW behavior. Namely, the magnitude of Weiss temperature of the compounds turns out from 1.13 K to 2.41, indicating the existence of weaker spin–spin interactions in the former isomer.

Several spiropyrans with TEMPO-substituents and naphthopyrans with TEMPO- or PROXYL-substituents

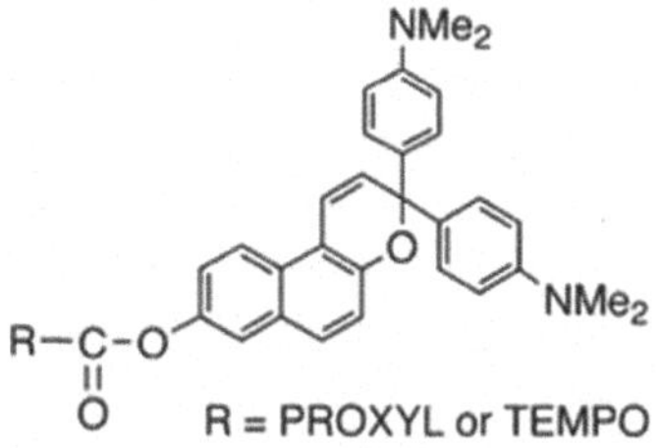

were prepared, and they showed photochromic behaviors in solution giving the respective stable open-formed merocyanines [67]. For example, naphthopyran derivatives with nitroxide substituents gave by irradiation the corresponding open-formed isomers. The product treatment with SiO_2 as a catalyst changed its back to the starting closed-formed naphthopyrans. The low-temperature region of spiropyran derivatives was found to be ferromagnetic with CW behavior ($h = 0.38$ K), whereas the interaction in the corresponding merocyanine was antiferromagnetic with CW behavior (0.82 K).

A couple of naphthalenediimide derivatives carrying a nitroxide radical and an azobenzene group were prepared, and photochromic property of compound with p-dimethylamino-substituent in solution were observed [68]. Syntheses, electronic structures in the ground state, unique photoexcited states, and spin alignment were reported for biradical 1 [69].

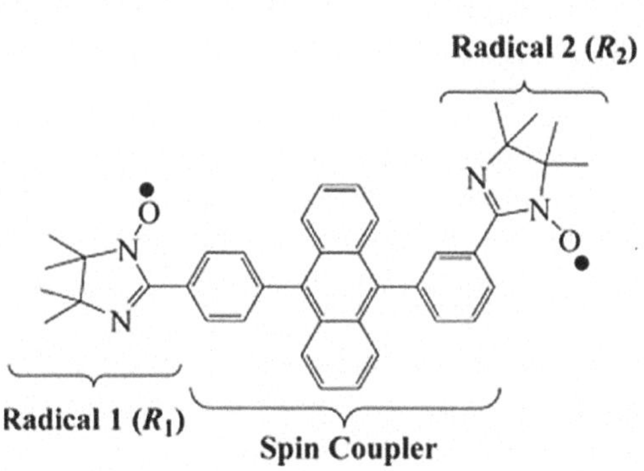

In the biradical electronic ground state, two radical moieties connected by an aromatic bridge in the diamagnetic state interact very weakly. On photo-irradiation, a novel lowest photoexcited state with the intermediate spin ($S = 1$), arising from four unpaired electrons with low-lying quintet ($S = 2$) photoexcited state, was detected using time-resolved ESR (TRESR). The triplet state has a unique electronic structure, the D value of which is reduced by antiferromagnetic spin alignment between two radical spins through the excited triplet spin coupler. The general theoretical predictions of the spin alignment and the reduction of the fine-structure splitting of the triplet bis(radical) systems were also formulated.

The synthesis of two nitroxide-based diradicals connected to a 2,3,4,5-tetraphenylsilole (TPS) unit, especially designed to present high-spin photo-excited states, was reported [70]. In ground state, the diradical displays weak intramolecular antiferromagnetic interactions (J/kB ≈ -1 K). In photo-excited high-spin states, the nature of the magnetic exchange coupling between two dangling radical spins through the spin coupler changes from antiferromagnetic to ferromagnetic after photoexcitation. Photo-physical properties of 4-(phenylazo)-benzoyl-2,2,6,6-tetramethylpiperidine-1-oxyl radical (AzO-TEMPO) and of the 4-(2-thienylazo)-benzoyl-2,2,6,6-tetramethylpiperidine-1-oxyl radical (ThiO-TEMPO) before and after their grafting to two polyethylene matrices (a copolymer ethylene/α-olefin (*co*-EO) and a high-density polyethylene (HDPE) were characterized by UV–Vis, FT-IR, and EPR spectroscopy [72]. The structural (FT-IR analysis), the thermal (thermal gravimetric analysis and EPR), and the photo-physical (UV–vis) properties of the RO-TEMPO derivatives before and after their grafting were evaluated.

Two approaches to surface modification based on the use TEMPO and its derivatives are discussed [73]. The first relies on the immobilization of TEMPO moieties on the surface of various materials including silicon wafers, silica particles, organic polymers as well as diverse nanomaterials. The second approach utilizes TEMPO and its derivatives for the grafting of polymer chains and polymer brushes' formation on flat and nanostructure surfaces via nitroxide-mediated radical polymerization (NMRP).

6.4.2 Photophysical Effects in Paramagnetic Complexes Bearing Nitroxides

Dual fluorophore–nitroxide compounds have been proved to be convenient "training area" for investigation of mechanism of photophysical and photochemical processes and factor affected on these processes that are guided with the electron exchange interactions with paramagnetic species [16, 73]. Quenching mechanism of the excited chromophore by a nitroxide originates from changes in the spin multiplicity of the electronic states and is guided with the electron exchange interactions with paramagnetic species.

In a chromophore-nitroxide supermolecules (Fig. 6.15), because of the unpaired electron spin of the doublet nitroxide radical (NR), the singlet ground (S_0) state and the lowest excited singlet (S_1) state of the chromophore become the doublet (D_0 and D_n, respectively) (Fig. 6.16) [16]. On the other hand, for a dual compound consisting a chromophore in the excited triplet state (T_1) and nitroxide, the lowest excited doublet (D_1) and quartet (QA_1) states are generated by an interaction between the NR and the T_1 chromophore. In such a case, the spin-forbidden transitions of

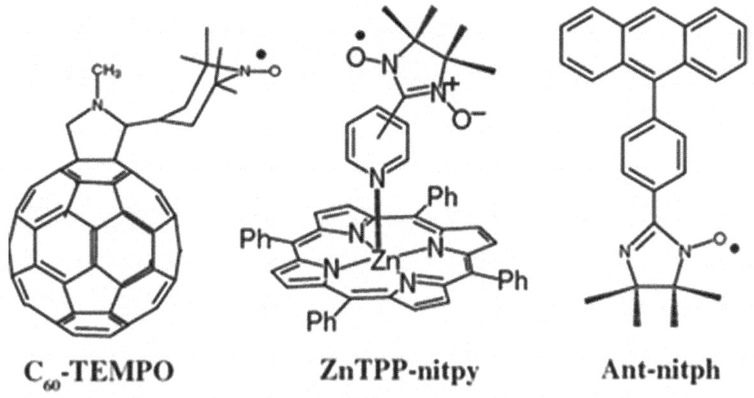

Fig. 6.15 Typical examples of the formation of excited multiplet states [16]

the chromophore, i.e., $S_1 \rightarrow T_1$ and $T_1 \rightarrow S_0$, partially transform into the $D_n \rightarrow D_1$ and $D_1 \rightarrow D_0$ transitions, respectively. These doublet states (Dn, D_1, and D_0) have the same spin multiplicity, and the lifetimes of the excited state can be very short as comparable to those of the $S_1 \rightarrow S_0$ transition. Thus, fluorescence quenching was interpreted by the intersystem crossing (ISC) enhancement resulting from transitions between states having the same spin multiplicity.

The photophysical properties of silicon phthalocyanine (SiPc) covalently linked to one or two NR have been studied in detail by fluorescence, transient absorption, and TRESR spectroscopies [16, 73]. The electronic states of R0, R1, and R2 are depicted in Fig. 6.16. Experiments indicated that (1) quantum yield (Φ_F) decreases in the order R0 (0.57) > R1c (0.21)_R1a (0.16) > R2c (0.012) > R2a (0.0036); (2) the quantum yield of 3SiPc* (FTSiPc) increases in the order R0 (0.34) < R1a (0.54)_R1c (0.59) < R2c (0.67); (3) the fluorescence decays of R1c and R2c (R1c = 42 ps and 4.7 ns, R2c < 30 ps and 4.7 ns) are faster than that (6.8 ns) of R0; and (4) the S_1 and T_1 energies of R_0 were found to be 1.47×10^4 and 8.90×10^3 cm^{-1}, respectively; and (5) rate constants of the $D_1 \rightarrow D_0$ transitions are 3.5×10^6, 2.1×10^6, 3.9×10^5 and 2.0×10^5 s^{-1} for R_{1a}, R_{1b}, R_{1c} and R_{1d}, respectively.

As shown in Fig. 6.17, in the case of the Dn $\rightarrow D_1$ and $D_1 \rightarrow D_0$ transitions, the electron exchange process is dominant. These transitions occur along with the electron exchange process between the SiPc and NR moieties, and the decay rate of the $D_1 \rightarrow D_0$ transition is well correlated with the magnitude of the electron exchange interaction between the T_1 chromophore and the doublet NR.

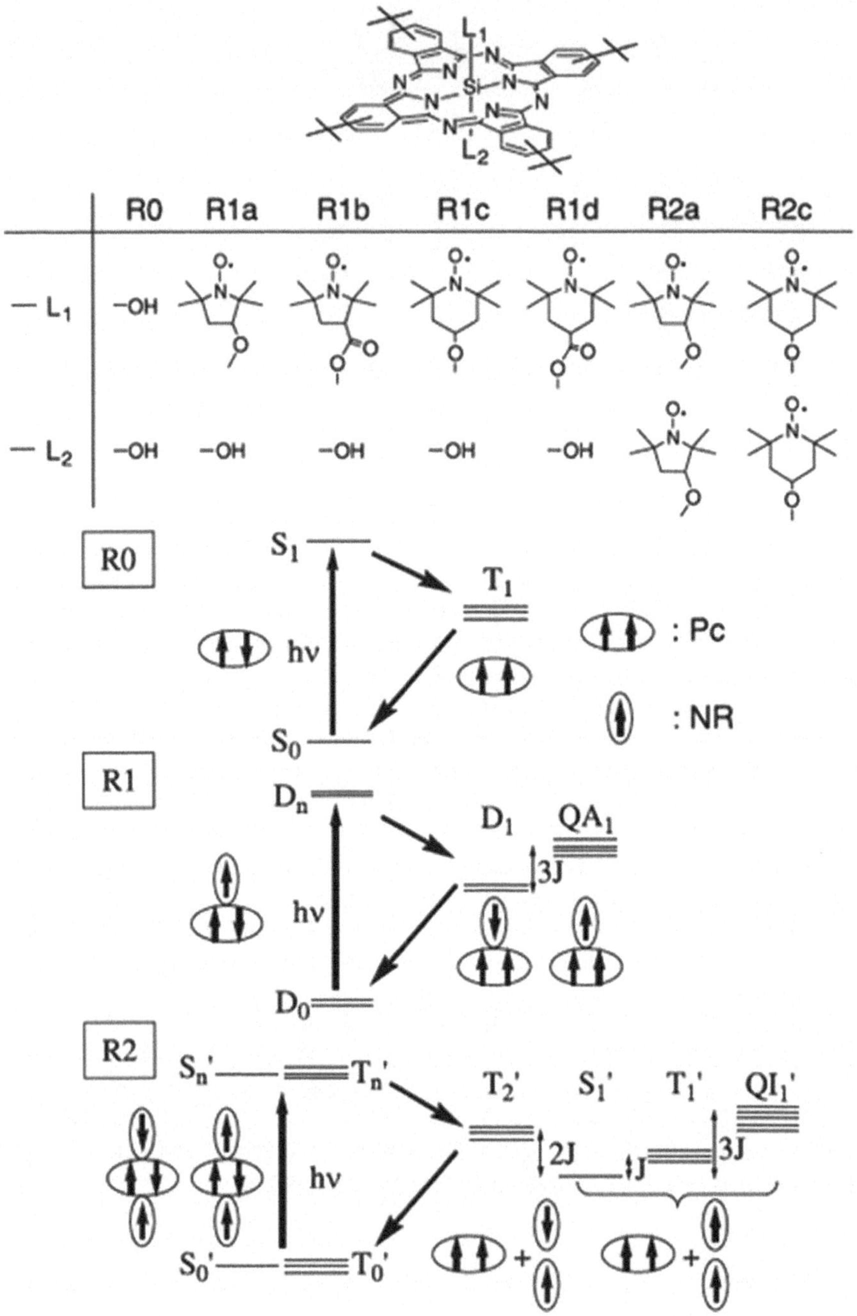

Fig. 6.16 Molecular structures and electronic states of SiPc covalently linked to NRs [16]

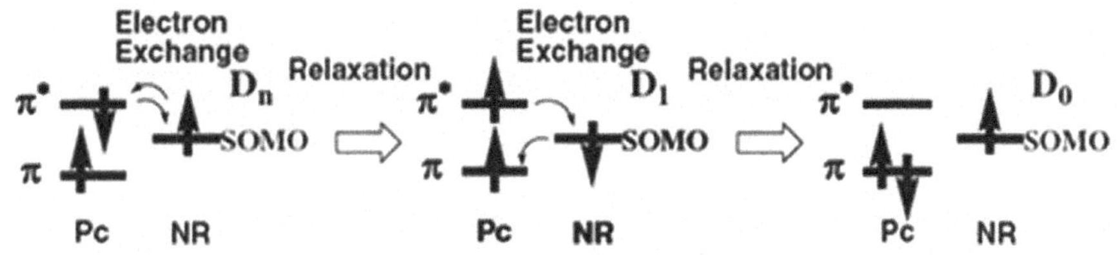

Fig. 6.17 Dn fi D1 and D1 fi D0 transitions via electron exchange processes [7]

6.4.3 Factors Affected on Intramolecular Fluorescence Quenching, Electron Transfer, and Photoreduction in Dual Compounds

One of the principal advances of the dual probes is an ability to monitor molecular dynamics of a media of interest in wide range temperatures and correlation times, which can be affected on chemical and physical reactions including electron transfer [1–10]. To establish the effect of molecular dynamics of media on electron transfer in a donor–acceptor pair, super-slow molecular dynamic of o liquids (ethanol, water–glycerol), and albumen were investigated at temperature region 30–100 K [2]. The following parameters related to dynamical properties of the probes, two nitroxides and a dual fluorophore–nitroxide, were measured: (1) the intensity of phosphorescence, (2) the relaxation shift of phosphorescence spectra, (3) the amplitude of EPR spectra, and (4) the width of EPR spectra central lines. Experiments with the use of the probes revealed various dynamic processes from the correlation time $\tau_c = 10^2$ s^{-1} (30 K) to $\tau_c = 10^{-4}$ s^{-1} (95 K). Obtained temperature dependences are in tune with correspondent data on the heat capacity and the proton NMR spectra width.

In pioneering work [3], it was first demonstrated that the nitroxide fragment is a strong quencher of the fluorescence and acts as an acceptor in the radical photoreduction, which strongly depends on molecular dynamics of environment. Specifically, irradiation of the chromophore segment of dansyl-TEMPO in a glassy liquid (glycerol 75%, water 20%, ethanol 5%) invoked producing the hydroxylamine derivative accompanying a decay of the nitroxide ESR signal and parallel eightfold increase in fluorescence. Both processes run with the same rate constant k_{red} under identical conditions with the quantum yield 8×10^{-2} (Fig. 6.18). The k_{red} values drastically increased when the temperature increased starting from 210 K (Fig. 6.19). The k_{red} increase correlated with an animation of the nanosecond relaxation dynamics in media monitored by the fluorescence (relaxation shift) and ESR techniques (change superfine splitting A_{ZZ}), while the rate constant of the intramolecular fluorescence quenching k_q was found to be temperature independent.

In order to establish a mechanism of intramolecular fluorescence quenching (IFQ) and photo-reduction of the nitroxide segment in the dual molecules, a series of c dansyl-nitroxides of different structures and flexibility of the spacer group, and different redox potentials of nitroxide (private communication from Drs. V.V. Martin

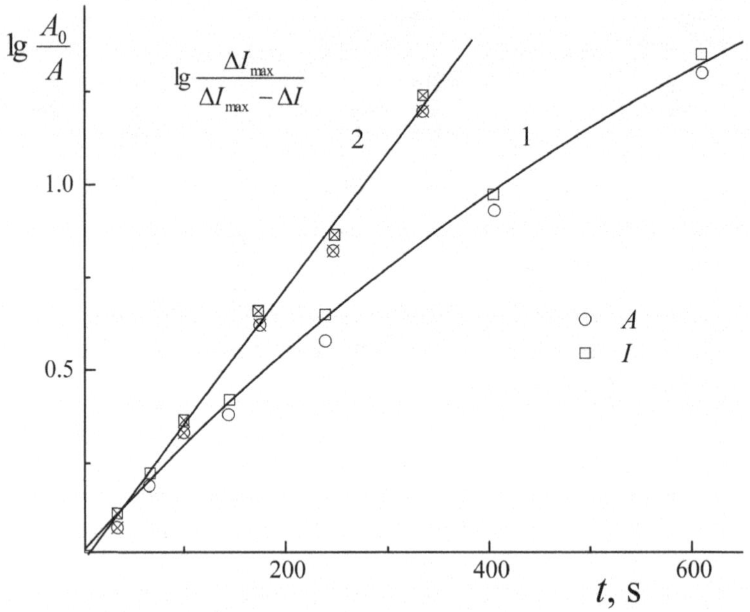

Fig. 6.18 Kinetics of change of the amplitude of nitroxide fragment EPR spectra (A) and intensity of fluorescence of the fluorophore segment (I) during photochemical reaction in the present (1) and absent (2) of oxygen in solvent (glycerol 75%, water 20%, ethanol 5%) [3]

and A. Weis, Lipitek International, Inc. (Fig. 6.1)) were synthesized and investigated [4–10, 74]. The following quantitative characteristics of the processes were obtained: (1) the rate constants of reversible electron transfer monitored by the time-resolved picosecond fluorescence technique (k_q), (2) fluorescence quantum yields of the probe fluorophore segment before and after the nitroxide moiety reduction (ϕ_{fl}) paramagnetic, and (3) the rate constant of photoreduction (k_{red}). As seen in Fig. 6.20, the positive correlation between the rate constant of the nitroxide fragment photoreduction k_{red} and the equilibrium constant K_{eq} for the chemical exchange reaction between different nitroxides depending on the nitroxide redox potential takes place [16]. On the basis of these data, two mechanisms of intermolecular quenching were proposed: the major mechanism, intersystem crossing (ISC) and the minor mechanism, irreversible intramolecular electron transfer (ET) from the excited singlet of the fluorophore (donor D) to nitroxide (acceptor A) followed by fluorophore segment regeneration and hydroxylamine formation. The latter mechanism is responsible for photoreduction. Nevertheless, the k_q values are not dependent on K_{eq}. In such a case, the fluorescence quenching occurs by the ISC mechanism or/and by reversible ET guided by the vibration modes [75, 76].

Blough et al. [12] suggested that the fluorescence quenching in a series of dual compounds arises through electron exchange, which causes relaxation of the singlet state to the triplet.

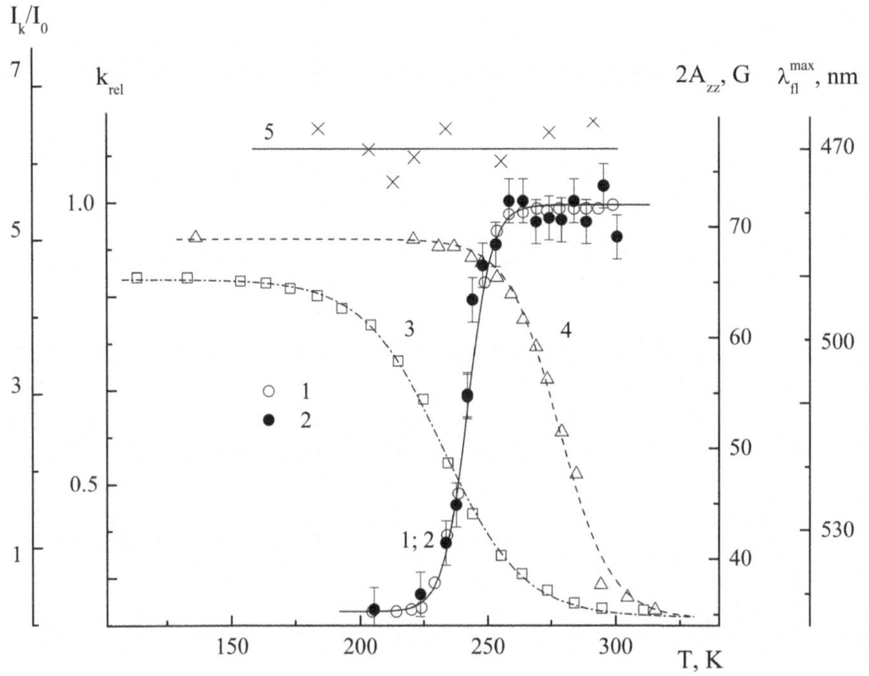

Fig. 6.19 Temperature dependences of relative rate constant of the photochemical reduction of the probe nitroxide fragment (k_{rel}) and the spectral parameters characterized dynamical state of the probe and the solvent: 1 and 2—the k_{rel} values obtained by the fluorescence and EPR techniques, respectively, 3—λ_{fl} is related to the fluorophore emission; 4—2Azz that is the nitroxide EPR superfine splitting of the nitroxide EPR spectra [3]

With a aim to established detail mechanism of quenching in fluorophore–nitroxide in a protein, a donor–acceptor hybrid molecule (DA) consisting of a 1-dimethylaminonaphthalene-5-sulfonate group in an excited singlet state (donor) and a nitroxide radical (acceptor) was incorporated into a hydrophobic cavity of bovine serum albumin (BSA) (Fig. 6.21) [4–10]. The kinetics of reversible intramolecular electron transfer (ET) from the donor to the acceptor and the nitroxide segment photoreduction, the micropolarity, and the intramolecular dynamics in the vicinity of the donor and the acceptor groups were monitored by fluorescence and ESR techniques in the temperature range from 77 to 300 K. The Arrhenius dependence of the reversible ET constant (k_{ET}) was found to be nonlinear and the apparent activation energy (E_{app}) changes from $E_{app} = 0$ eV (at $T = 77$–100 K) to 0.25 eV (near $T = 298$ K). The temperature region of the E_{app} increase (100–240 K) was close to the temperature of the increase of relaxation shift of the donor fluorescent spectra. Using the Marcus-Levich [78, 79] model of electron transfer in polar media, on the basis of the obtained kinetic data and of the data of the micropolarity in the vicinity of the donor and the acceptor groups, values of standard Gibbs energy, the reorganization energy, the resonance integral and the Franck–Condon factor for reversible electron transfer were

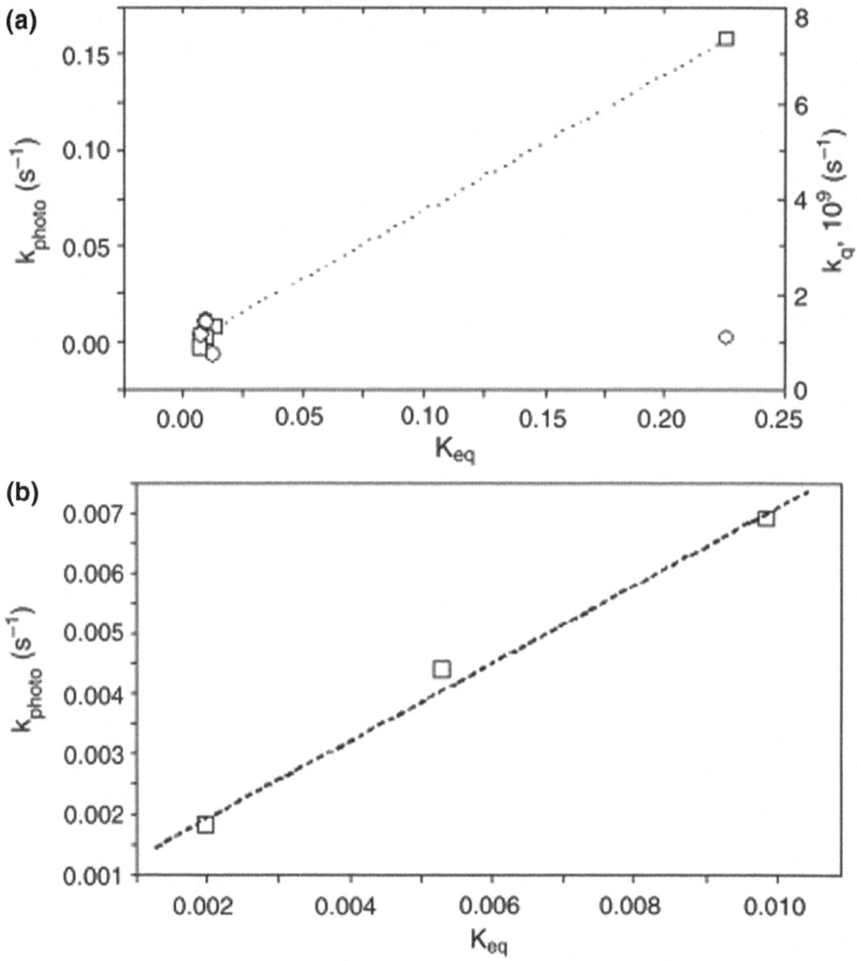

Fig. 6.20 Dependence of rate constant of photoreduction, kphoto, on redox power of the nitroxide fragment, K_{eq} (squares), contrary to independence of rate constant of IFQ, k_q (circles) [16]

estimated. The key role of molecular dynamics in the vicinity of the donor and the acceptor groups in the nitroxide photoreduction was stressed.

In order to model the effects of protein on the ET in **FNO•**, two doxyl-nitroxide probes FN1 and FN2 were incorporated in bovine (BSA) and human serum albumin (HSA) (Fig. 6.21) [8, 9]. The photoreduction of the nitroxide fragment was monitored by ESR, and fluorescence quenching was measured by steady-state and picosecond time-resolved techniques. The same groups allow to investigate the factors affecting the ET, namely the molecular dynamics and micropolarity of the medium in the vicinity of the donor (by fluorescence technique) and acceptor (by ESR) moieties.

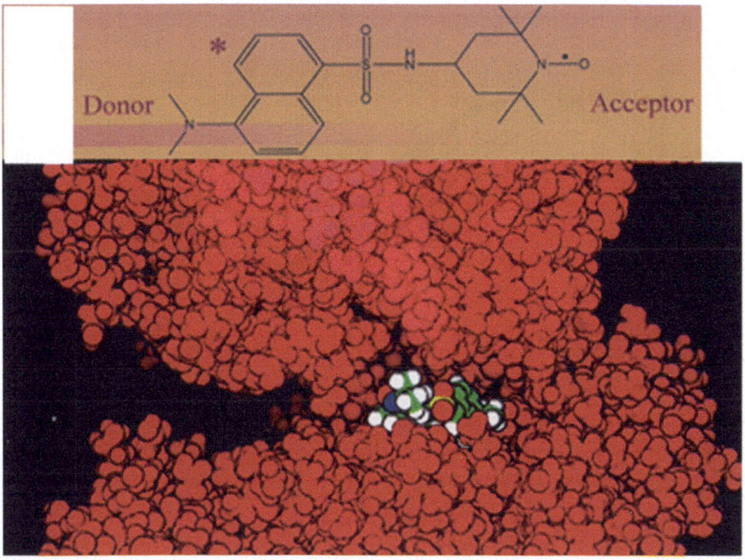

Fig. 6.21 Dual probe corporated in hydrophobic packet of bovin serum albumin

The measurements of fluorescence polarization and ESR at ambient temperature, and by direct monitoring of relaxation dynamics of the protein-binding site around the dansyl moiety of the dual fluorophore–nitroxide probe FN2, using picosecond fluorescence time-resolved technique (Fig. 6.22), led to conclusion that the nanosecond dynamics of the protein medium is one of the decisive factors affecting the photoreduction and the light energy conversion in the given system. Such an intramolecular flexibility of the protein makes it possible to stabilize products of the reactions of the oxidized donor D^+ and acceptor A^- due to interactions with surrounding dipoles of protein, thus providing favorable thermodynamics for these reactions.

Comparison of results of modeling of the torsion angle rotation in the probe FN1 with the experimental data on the nitroxide segment dynamics in protein provided additional information about the mechanism of the protein molecular "breating" [8, 9]. According to computer simulation in vacuum, a single C-N bond twisting around the piperidine ring undergoes minimal steric hindrances as compared to that of the dansyl fragment. The following values of the twisting correlation time τ_c were theoretically estimated (the twisting angles is shown in brackets): 3×10^{-10} s (170°–220°), 10^{-10} s (220°–275°), and 10^{-8} s (50°–150°). Two first values are significantly lower than $\tau_c = 4 \times 10^{-9}$ s for the probe attached to BSA (EPR data) and especially for the rotation correlation time of the BSA molecule as a whole (8×10^{-8} s). From these findings, an important conclusion can be formulated; that is, the experimental correlation time of the probe nitroxide fragment can be attributed to its motion relative the protein globule, reflecting molecular "breathing" of the protein in the vicinity of the fragment (Fig. 6.23).

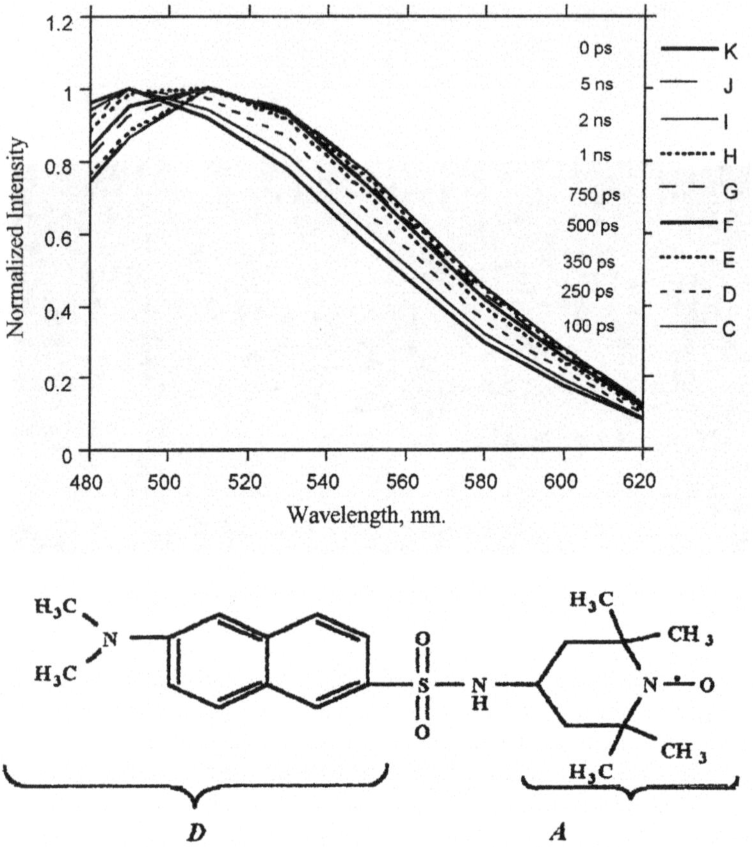

Fig. 6.22 Time-resolved fluorescence spectra of dansyl-TEMPO dual probe in HSA [77]

For analysis of experimental data obtained in the investigation of system, a Marcus–Levich formula for non-adiabatic electron transfer in a polar media was utilized (6.1) [78, 79]:

$$k_{\mathrm{ET}} = \frac{2\pi V^2}{h\sqrt{4\pi\lambda k_{\mathrm{B}}T}}\exp\left[-\frac{(\lambda+\Delta G_0)^2}{4\lambda k_{\mathrm{B}}T}\right] \quad (6.1)$$

where ΔG_0 is the driving force of the value of the process, standard Gibbs free energy, λ is the reorganization energy defined as energy for ET without replacement of the nuclear frame, and V is the electronic coupling or the resonance integral. Based on experimental data on the local apparent dielectric constant ε_0 in the vicinity of the donor dansyl groups (fluorescence technique) and around the acceptor nitroxide segments (ESR), the following parameters of the Marcus–Levich theory were estimated for the ET at $T = 300$ K [4, 5, 9]: Gibbs energy $\Delta G_0 = -1.7$ eV, reorganization energy $\lfloor\lambda = 0.9$ eV, and free activation energy $\Delta G^{\#} \approx 0.25$ eV. Such a set of parameters is

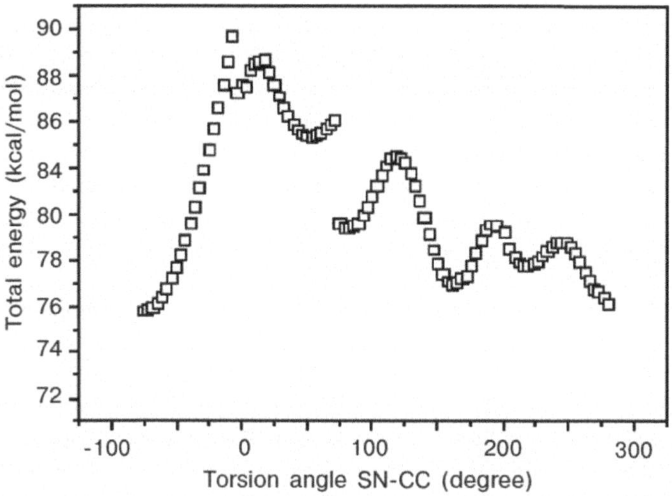

Fig. 6.23 Intramolecular rotation energy versus the torsion angle SN-CC for X in vacuum [9]

related to the inverted Marcus region. In this condition, the ET under thermodynamically equilibrium conditions can occur with the rate substantially lower ($k_{ET} \approx 10^5$ s^{-1}) than the experimental value $k_q \approx 1 \times 10^9$ s^{-1}. Therefore, the fast primary ET in the system under investigation occurs at the expense of vibrational stabilization of the photoseparated ionic pair (D^+A^-) and/or at the expense of partial stabilization due to the fast polar relaxation modes [75, 76].

The discussed photochemical reactions in the dual molecules may be considered as the light energy transfer processes, which meet the main requirements for any efficient light energy conversion system, that is, favorable thermodynamics, optimum orbital overlap of the donor and acceptor groups, and nanosecond dynamics in vicinity of the donor–acceptor pair [3–10].

The photochemical reaction between the fluorophore and the nitroxide radical in dual compounds

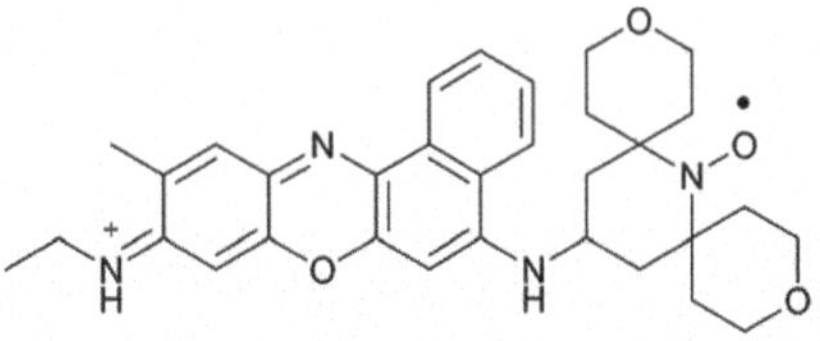

Nile-DiPy

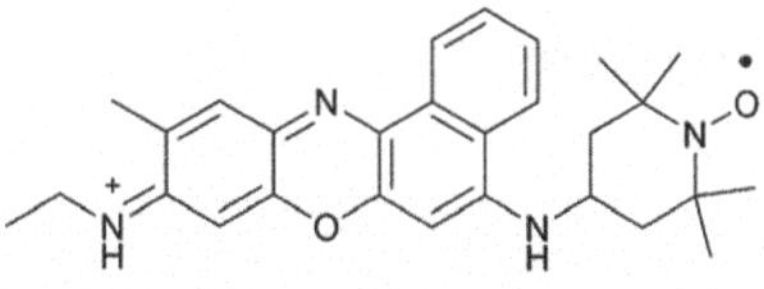

Nile-TEMPO

was investigated by optical, electrochemical, and femtosecond laser-induced transient absorption spectroscopy experiments [50]. Redox potentials of Nile-DiPy in MeOH containing 0.10 M TBAPF6 were measured by cyclic voltammetry and the second harmonic alternating current voltammetry SHACV at 298 K (vs. Fc/Fc^+). The following values of redox potentials were found: E_{red} (Nile-Dipy) = 0.521 eV, E_{ox} (Nitroxide) = 0.786 eV, E_{red} (Nile-TEMPO) = 0.519, E_{ox} (Nitroxide) = 0.797 eV. These values were allowed to compose energy diagram for photo-induced reversible electron transfer in Nile-DiPy. The femtosecond transient spectra of Nile-DiPy in an N2-saturated aqueous solution containing 5% DMSO displayed the very fast rise in broad transient absorption and emission around 530 and 645 nm after the selective excitation of Nile blue at 627 nm. The decay rate constants of the singlet excited state of Nile-DiPy and Nile-TEMPO were found to be as (4.9×10^9 s^{-1}) and (4.5×10^9 s^{-1}), respectively.

All in all, the dual compounds, keeping all properties of chromophore, fluorescence, and nitroxide spin probe, at the same time have gained new principal advantages. The most important new properties of chromophore–nitroxide (CN) compounds are its ability to: (1) obtain spatial and temporal information about both fluorescence and ESR behavior of the probe from the same specific part of the system under interest; (2) investigate micropolarity and molecular dynamics of media in the vicinity of fluorophore segment by the fluorescence technique and nitroxide segment by ESR; (3) monitor redox and spin trapping processes in systems of any optical density by ESR technique and of low optical density by fluorescence which is two to three orders more sensitive than ESR and optical absorption spectroscopy; (4) construct a variety of organic dual magnetic systems with photofunctionality; (5) analyze metal ions using both fluorescence and ESR techniques; (6) investigate mechanisms of intramolecular photochemical and photophysical processes in multispin systems in particular, (7) form the basis for the development of novel methods for the analysis of antioxidants and for real-time monitoring of important biochemical processes such as NO, superoxide, and reactive radical generation; (8) compose novel mulitspin and photoswitching systems; and (9) give a clue to the access of bland new organic photomagnetic materials such as novel mulitspin and photoswitching systems.

Thus, there are strong evidences to believe that the use of dual molecules will continue to be one of the most effective tools for solving excited and complicated problems in photochemistry, biochemistry, and chemistry of materials.

References

1. P.C. Jost, O.H. Griffith, The spin-labeling technique. Methods Enzym. **49**, 369–418 (1978)
2. I.G. Likhtenstein, V.R. Bogatyrenko, A.V. Kulikov, K. Hideg, H.O. Hankovsky, N.V. Lukoianov, A.I. Kotelnikov, B.S. Tanaschelchuk, Study of superflow motion in solid solutions by the method of physical probes. Dokl. Akad. Nauk. SSSR **253**, 481–484 (1980)
3. I.M. Bystryak, G.I. Likhtenshtein, A.I. Kotelnikov, O.H. Hankovsky, K. Hideg, The influence of the molecular dynamics of the solvent on the photoreduction of nitroxy-radicals. Russ. J. Phys. Chem. **60**, 1679–1983 (1986)
4. V.R. Vogel, E.T. Rubtsova, G.I. Likhtenshtein, K. Hideg, Factors affecting photoinduced electron transfer in a donor- acceptor pair (D-A) incorporated into bovine serum albumin. J. Photochem. Photobiol. A, Chem. **83**, 229–236 (1994)

5. E.T. Rubtsova, V.R. Vogel, G.I. Likhtenstein, D.V. Khudjakov and A.I. Kotelnikov Influence of molecular dynamics of protein matrix on the photoinduced electron transfer kinetics. Biofizika 38, 222–227 (1993)
6. G.I. Likhtenstein, Novel fluorescent methods for biotechnological and biomedical sensoring: assessing antioxidants, reactive radicals, NO dynamics, immunoassay, and biomembranes fluidity. Appl. Biochem. Biotechnol. **152**, 135–155 (2009)
7. G.I. Likhtenshtein, *Solar Energy Conversion. Chemical Aspects* (Wiley-VCH, Weinheim, 2012)
8. G.I. Likhtenshtein, Factors affecting light energy conversion in dual fluorophore-nitroxide molecules in solution and a protein. Pure Appl. Chem. **80**(10), 2125–2139 (2008)
9. G.I. Likhtenshtein, D. Pines, E. Pines, V. Khutorsky, Effect of a protein on the light energy conversion in dual fluorophore-nitroxide probes studied by ESR and fluorescence spectroscopy. Appl. Magn. Reson. **35**(3), 459–472 (2009)
10. G.I. Likhtenshtein, *Electron Spin in Chemistry and Biology: Fundamentals, Methods, Reactions Mechanisms, Magnetic Phenomena, Structure Investigation* (Springer, Berlin, 2016)
11. N.V. Blough, D.J. Simpson, Chemically mediated fluorescence yield switching in nitroxide-fluorophore adducts, optical sensors of radical/redox reactions. J. Am. Chem. Soc. **110**, 1915–1917 (1988)
12. J.A. Green, D.J. Simpson, G. Zhou, P.S. Ho, N.V. Blough, J. Am. Chem. Soc. **112**, 7337 (1990)
13. N.V. Blough, D.J. Simpson, Chemically mediated fluorescence yield switching in nitroxide-fluorophore adducts: optical sensors of radical/redox reactions. J. Am. Chem. Soc. **110**, 1915–1917 (1988)
14. A.G. Coman, C.C. Paraschivescu, A. Paun, A. Diac, N.D. Hadade, L. Jouffret, A. Gautier, M. Matache, P. Ionita, Synthesis of novel profluorescent nitroxides as dual luminescent-paramagnetic active probes. New J. Chem. **41**, 7472 (2017)
15. J.P. Blinco, A.C.K.E. Fairfull-Smith, B.J. Morrow, S.E. Bottle, Profluorescent nitroxides as sensitive probes of oxidative change and free radical reactions. Aust. J. Chem. **64**, 373–389 (2011)
16. G.I. Likhtenstein, K. Ishii, S. Nakatsuji, Dual chromophore-nitroxides: novel molecular probes, photochemical and photophysical models and magnetic materials. Photochem. Photobiol. **83**, 871–881 (2007)
17. K.A. Hansen, K.E. Fairfull-Smith, S.E. Bottle, J.P. Blinco, Development of a redox-responsive polymeric profluorescent probe. Macromol. Chem. Phys. **217**(20), 2330–2340 (2016)
18. H.-Y. Ahn, K.E. Fairfull-Smith, B.J. Morrow, V. Lussini, B. Kim, M.V. Bondar, S.E. Bottle, K.D. Belfield, Two-photon fluorescence microscopy imaging of cellular oxidative stress using profluorescent nitroxides. J. Am. Chem. Soc. **134**, 4721–4730 (2012)
19. D.J. Keddie, K.E. Fairfull-Smith, S.E. Bottle, The palladium-catalysed copper-free Sonogashira coupling of isoindoline nitroxides: a convenient route to robust profluorescent carbon–carbon frameworks. Org. Biomol. Chem. **6**, 3135–3143 (2008)
20. E.M. Simpson, Z.D. Ristovski, S.E. Bottle, K.E. Fairfull-Smith, J.P. Blinco, Modular design of profluorescent polymer. Sensors. Polym. Chem. **6**, 2962 (2015)
21. V.C. Lussini, J.M. Colwell, K.E. Fairfull-Smith, S.E. Bottle, Profluorescent nitroxide sensors for monitoring photo-induced degradation in polymer films. Sens. Actuat. B Chem. **241**, 199–209 (2017)
22. B.A. Chalmers, S. Saha, T. Nguyen, J. McMurtrie, S.T. Sigurdsson, S.E. Bottle, K-S. Masters, TMIO-PyrImid hybrids are profluorescent, site-directed spin labels for nucleic acids. Org. Lett. **16**, 5528–5531 (2014)
23. M. Barzegar, A. Oliaa, A. Zavras, C.H. Schiesser, S.A. Alexandera, Blue 'turn-on' fluorescent probes for the direct detection of free radicals and nitric oxide in *Pseudomonas aeruginosa* biofilms. Org. Biomol. Chem. **14**, 2272–2281 (2016)
24. S. Sato, S. Endo, Y. Kurokawa, M. Yamaguchi, A. Nagai, T. Ito, T. Ogata, Synthesis and fluorescence properties of six fluorescein-nitroxide radical hybrid-compounds. Spectrochim. Acta Part A Mol. Biomol. Spectrosc. **169**, 66–71 (2016)

25. S. Endo, Y. Kurokawa, M. Yamaguchi, A. Nagai, T. Ito, T. Ogata, Synthesis and fluorescence properties of six fluorescein-nitroxide radical hybrid-compounds. Spectrochim. Acta Part A Mol. Biomol. Spectrosc. **169**, 66–71 (2016)
26. E. Lozinskya, V.V. Martin, T.A. Berezina, A.I Shames, A.L. Weis, G.I. Likhtenshtein, Dual fluorophore–nitroxide probes for analysis of vitamin C in biological liquids. J. Biochem. Biophys. Methods **38**, 29–42 (1999)
27. M.M. Haugland, J.E. Lovett, E.A. Anderson, Advances in the synthesis of nitroxide radicals for use in biomolecule spin labeling. Chem. Soc. Rev. **47**, 668–680 (2018)
28. M. Eing, B.T. Tuten, J.P. Blinco, C. Barner-Kowollik, Visible light activation of spin-silenced fluorescence. Chem. Eur. J. **24**, 12246–12249 (2018)
29. B.A. Chalmers, S. Saha, T. Nguyen, J. McMurtrie, S.T. Sigurdsson, S.E. Bottle, K.S. Masters, TMIO-PyrImid hybrids are profluorescent, site-directed spin labels for nucleic acids. Org. Lett. **16**, 5528–5553 (2014)
30. G-P. Yan, B. Zhao, S.E. Bottle, Q. Zhang, J-L. Li Synthesis and properties of fullerene C60 and C70 spin probes containing isoindoline nitroxides. Fuller. Nanotubes Carbon Nanostructures **23**(8), 734–741 (2015)
31. Y. Zhao, Y. Fang, Y. Jiang, Fluorescence study of fullerene in organic solvents at room temperature. Spectrochim. Acta A Mol. Biomol. Spectrosc. **64**(3), 564–567 (2006)
32. R. Braslau, F. Rivera III, E. Lilie, M. Cottman, Urushiol detection using a profluorescent nitroxide. J. Org. Chem. **78**, 238–245 (2013)
33. J.P. Blinco, K.E. Fairfull-Smith, A.S. Micallef, S.E. Bottle, Highly efficient, stoichiometric radical exchange reactions using isoindoline profluorescent nitroxides. Polym. Chem. **1**, 1009 (2010)
34. J.P. Allen, M.C. Pfrunder, J.C. McMurtrie, S.T. Bottle, J.P. Blinco, K.E. Fairfull-Smith, BODIPY-based profluorescent probes containing meso- and β-substituted isoindoline nitroxides. Eur. J. Org. Chem. **2017**(3), 476–483 (2017)
35. V.C. Lussini, K.E. Fairfull-Smith, S.E. Bottle, Polyaromatic profluorescent nitroxide probes with enhanced photostability. Chem. A Eur. J. **21**(50), 8258–8268 (2015)
36. P. Lederhose, N.L. Haworth, K. Thomas, S.E. Bottle, M.L. Coote, C. Barner-Kowollik, J.P. Blinco, Design of redox/radical sensing molecules via nitrile imine-mediated tetrazole-ene cycloaddition (NITEC). J. Org. Chem. **80**, 8009–8017 (2015)
37. P. Parkhomyuk-Ben Arye, N. Strashnikova, G.I. Likhtenshtein, Stibene-photochrome-fluorescence-spin molecules: covalent immobilization on silica plate and application as redox and viscosity probes. J. Biochem. Biophys. Methods **51**, 1–15 (2002)
38. G.I. Likhtenshtein, Dual phluorophore-nitroxides as a tool for analysis of antioxidants, nitric oxide and superoxide in biological liquids and tissues, in *International Conference "Recent Advantagesin Sensoring, Signals and materials"*, November 3–5, 2010, Faro, Portugal, pp. 162–168 (2010)
39. E. Lozinsky, A. Novoselsky, A.I. Shames, O. Saphier, G.I. Likhtenshtein, D. Meyerstein, Biochim. Biophys. Acta Gen. Subj. **1526**(1), 53–60 (2001)
40. O. Saphier, T. Silberstein, A.I. Shames, G.I. Likhtenshtein, E. Maimon, D. Mankuta, M. Mazor, M. Katz, D. Meyerstein, N. Meyerstein, Free Radic. Res. **37**, 301–308 (2003)
41. T. Silberstein, D. Mankuta, A.I. Shames, G.I. Likhtenshtein, D. Meyerstein, N. Meyerstein, O. Saphier, Neonatal blood is more resistant to oxidative stress induced by stable nitroxide radicals than adult blood. Arch. Gynecol. Obstet. **277**(3), 233–237 (2008)
42. T. Yokoi, K. Ishii, Dependence of phthalocyanine-based fluorescence on albumin structure: a fluorescent probe for ascorbic acid. J. Photochem. Photobiol. A **364**, 1–5 (2018)
43. N. Medvedeva, V.V. Martin, A.L. Weis, G.I. Likhtenshten, Dual fluorophore-nitronyl probe for investigation of superoxide dynamics and antioxidant status of biological systems. J. Photochem. Photobiol. A **163**, 45–51 (2004)
44. H. Nam, J.E. Kwon, M.-W. Choi, J. Seo, S. Shin, S. Kim, S.Y. Park, Highly sensitive and selective fluorescent probe for ascorbic acid with a broad detection range through dual-quenching and bimodal action of nitronyl-nitroxide. ACS Sens. **1**(4), 392–398 (2016)

45. C.-P. Liu, T.-H. Wu, C.-Y. Liu, H.-J. Cheng, S.-Y. Lin, Interactions of nitroxide radicals with dendrimer-entrapped Au8-clusters: a fluorescent nanosensor for intracellular imaging of ascorbic acid. J. Mater. Chem. B **3**, 191–197 (2015)
46. F. Lin, D. Pei, W. He, Z. Huang, Y. Huang, X. Guo, Electron transfer quenching by nitroxide radicals of the fluorescence of carbon dots. J. Mater. Chem. **22**, 11801–11807 (2012)
47. H. Yu, L. Cao, F. Li, Q. Wu, Q. Li, S. Wang, Y. Guo, The antioxidant mechanism of nitroxide TEMPO: scavenging with glutathionyl radicals. RSC Adv. **5**, 63655 (2015)
48. M.A. Lam, D.I. Pattison, S.E. Bottle, D.J. Keddie, M.J. Davies, Nitric oxide and nitroxides can act as efficient scavengers of protein-derived free radicals. Chem. Res. Toxicol. **21**, 2111–2119 (2008)
49. A.G. Coman, C.C. Paraschivescu, A. Paun, A. Diac, N.D. Hădade, L. Jouffret, A. Gautier, M. Matache, P. Ionita, Synthesis of novel profluorescent nitroxides as dual luminescent-paramagnetic active probes. New J. Chem. **41**, 7472 (2017)
50. K. Sasaki, T. Ito, H.G. Fujii, S. Sato, Synthesis and reduction kinetics of five Ibuprofen-nitroxides for ascorbic acid and methyl radicals. Chem. Pharm. Bull. **64**(10), 1509–1513 (2016)
51. M. Yamato, T. Yamasaki, F. Mito, K. Yamada, Rapid and convenient detection of ascorbic acid using a fluorescent nitroxide switch. Free Radic. Biol. Med. **53**(11), 2112–2118 (2012)
52. Y. Matsuoka, K. Ohkubo, T. Yamasaki, M. Yamato, H. Ohtabu, T. Shirouzu, S. Fukuzumi, K-i. Yamada, A profluorescent nitroxide probe for ascorbic acid detection and its application to quantitative analysis of diabetic rat plasma. RSC Adv. **6**, 60907–60915 (2016)
53. Y. Matsuoka, M. Yamato, K-i. Yamada, Fluorescence probe for the convenient and sensitive detection of ascorbic acid. J. Clin. Biochem. Nutr. **58**(1), 16–22 (2016)
54. T. Yang, B. Zheng, H. Liang, Y. Wan, J. Du, D. Xiao, A sensitive and selective chemosensor for ascorbic acid based on a fluorescent nitroxide switch. Talanta **132**, 191–196 (2015)
55. T. Maki, N. Soh, K. Nakano, T. Imato, Flow injection fluorometric determination of ascorbic acid using perylenebisimide-linked nitroxide. Talanta **85**, 1730–1733 (2011)
56. T. Yokoi, T. Otani, K. Ishii, In vivo fluorescence bioimaging of ascorbic acid in mice: development of an efficient probe consisting of phthalocyanine, TEMPO, and albumin. Sci. Rep. **8**, 1560 (2018)
57. C.L. Rayner, S.E. Bottle, G.A. Gole, M.S. Ward, N.L. Barnett, Real-time quantification of oxidative stress and the protective effect of nitroxide antioxidants. Neurochem. Int. **92**, 1–12 (2016)
58. J.L. Gerlock, P.J. Zacmanidis, D.R. Bauer, D.J. Simpson, N.V. Blough, I.T. Salmeen, Fluorescence detection of free radicals by nitroxide scavenging. Free Radic. Res. Commun. **10**(1–2), 119 (1990)
59. A.S. Micallef, J.P. Blinco, G.A. George, D.A. Reid, S.E. Bottle, The application of a novel profluorescent nitroxide to monitor thermo-oxidative degradation of polypropylene. Polym. Degrad. Stabil. **89**(3), 427–435 (2005)
60. G.G. Borisenko, I. Martin, Q. Zhao, A.A. Amoscato, V.E. Kagan, Nitroxides scavenge myeloperoxidase-catalyzed thiyl radicals in model systems and in cells. Am. Chem. Soc. **126**, 9221. (2004)
61. S. Nakatsuj, Preparation, reaction and properties of functional nitroxide radicals, in *Nitroxides: Application in Chemistry, Biomedicine, and Materials Science*, ed. by G.I. Likhtenshtein, J. Yamauchi, S. Nakatsuji, A. Smirnov, R. Tamura (Wiley-VCH, Weinhem, 2008)
62. photoswitching of an intramolecular magnetic interaction, K. Matsuda, K. M. Irie Photochromism of diarylethenes with two nitronyl nitroxides. Chemistry **7**, 3466–3473 (2001)
63. K. Matsuda, K.M. Irie, Diaryethene as a photoswitching unit. J. Photochem. Photobiol. C, Photochem. Rev. **5**, 169–182 (2004)
64. S. Nakatsuji, Recent progress toward the exploitation of organic radical compounds with photo-responsive magnetic properties. Chem. Soc. Rev. **33**, 348–353 (2004)
65. S. Nakatsuji, M. Fujino, H. Akutsu, J. Yamada, V.S. Gurman, A.K. Vorobiev, Azobenzene derivatives carrying a nitroxide radical. J. Org. Chem. **72**, 2021–2029 (2007)
66. M. Nakagawa, H. Akutsu, J-i Yamada, M. Karakawa, Y. Aso, S. Fall, T. Heiser, S. Nakatsuji, A spin-carrying naphthalenediimide derivative with azobenzene unit. Chem. Lett. **41**(2), 175–177 (2012)

67. T. Kaneko, T. Amano, H. Akutsu, J. Yamada, S. Nakatsuji, Photochromic radical compounds based on a naphthopyran systems. Org. Lett. **5**, 2127–2129 (2003)
68. M. Nakagawa, H. Akutsu, J-i. Yamada, M. Karakawa, Y. Aso, S. Nakatsuji. Photo-responsive naphthalenediimide derivative carrying a nitroxide and an azobenzene group. Ann. Acad. Roman. Sci. Ser. Chem. Sci. **1**(1) (2014) Online ISSN 2393–2902
69. Y. Teki, T. Toichi, S. Nakajima, π-Topology and spin alignment in unique photoexcited triplet and quintet states arising from four unpaired electrons of an organic spin system. Chem. Eur. J. **12**, 2329–2336 (2006)
70. N. Roques, Ph. Gerbier, Y. Teki, S. Choua, P. Lesniakova, J.-P. Sutter, P. Guionneau, C. Guerin, Towards a better understanding of photo-excited spin alignment processes using silole diradicals. New J. Chem. **65**, 1319–1326 (2006)
71. S. Nakatsuji, Organic multifunctional spin systems based on aminoxyl radicals. Adv. Mater. **13,** 1719–1724 (2001)
72. F. Cicogna, I. Domenichelli, S. Coiai, F. Bellina, M. Lessi, R. Spiniello, E. Passaglia, Structural, thermal and photo-physical data of azo-aromatic TEMPO derivatives before and after their grafting to polyolefins. Data Brief. **6**, 562–570 (2016)
73. S. Takeuchi, K. Ishii, N. Kobayashi, Time-resolved EPR transient absorption studies on phthalocyaninatosilicon covalently linked to two PROXYL radicals. J. Phys. Chem. A **108**, 3276–3280 (2004)
74. E. Lozinsky, A. Shames, G.I. Likhtenshtein, Dual fluorophore-nitroxides: MODELS for investigation of intramolecular quenching and novel redox probes, in *Recent Research Development in Photochemistry and Photobiology*, vol. 2, ed. by S.G. Pandalai (Transworld Research Network, Trivandrum, India, 2000), pp. 41–45
75. S. Efrima, M. Bixon, Vibrational effects in outer-sphere electron-transfer reactions in polar media. Chem. Phys. **13**, 447 (1976)
76. M. Bixon, J. Jortner, in *Advances in Chemical Physics*, vol. 107, Part 1, ed. by J. Jortner, N. Bixon (Wiley, New York, 1999), pp. 35–202
77. G.I. Likhtenshtein, F. Febrario F, Nucci R. Intramolecular dynamics and conformational transitions in proteins studied by biophysical labeling methods. Common and specific features of proteins from thermophylic microorganisms. Spectrochem. Acta Part A. Biomolec. Spectrosc. **56**, 2011–2031 (2000)
78. V.G. Levich, R. Dogonadze, Dokl. Akad. Nauk **78**, 2148 (1959)
79. R.A. Marcus, N. Sutin, Electron transfers in chemistry and biology. Biochim. Biophys. Acta **811**, 265 (1985)

Chapter 7
Nitroxide-Mediated Polymerization

7.1 Introduction

Controlled radical polymerization (CRP) under radical initiation conditions belongs to priority areas in the development of the synthetic chemistry of polymers of the last years [1–16]. Nitroxide-mediated polymerization (NMP) was invented by Solomon [1, 13]. Since this discovery, nitroxide-mediated radical polymerization is a powerful method to synthesize well-defined macromolecular architectures with precisely controlled topologies, compositions, microstructures, and functionalities [3–5].

The most common mechanisms for reversible activation in polymerization reactions are schematically illustrated in Scheme 7.1.

Persistent radical effect (PRE) occurs when two radicals are generated at the same time, at the same rate, and one is more persistent than the other, the self-termination reactions are lowered, leading to an unusually high selectivity for the cross-coupling reaction [10]. The effect has been investigated for the preparation of macromolecules with a narrow molar mass distribution through radical polymerization.

Nitroxide-mediated polymerization is widely applied in industrial polymer syntheses as a method for production of large-tonnage polymers and is employed to manufacture new pigments, sealants, emulsion stabilizers, and block copolymers, etc., with a various set of properties. NMP has also paved an avenue for complex macromolecular architectures (statistical, block, graft) in the fields of nanoscience and nanotechnology [5, 9, 12] and references cited therein. A brief summary of NMP developments in both the patent and open literature during the period of the early

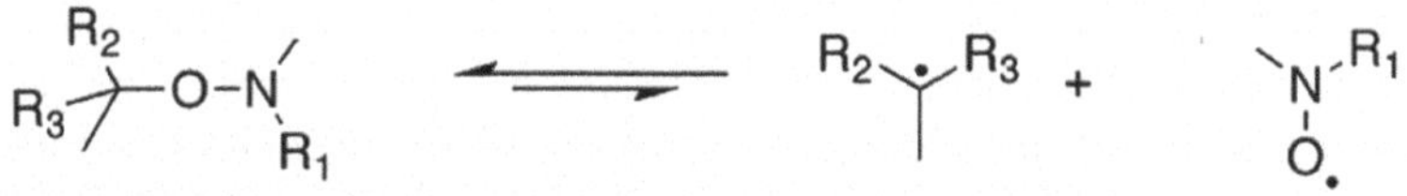

Scheme 7.1 Mechanisms for reversible activation in polymerization reactions [6]

G. I. Likhtenshtein, *Nitroxides*, Springer Series in Materials Science 292,
https://doi.org/10.1007/978-3-030-34822-9_7

1980–2000 was presented in [11]. Various important aspects in this area, such as synthesis of nitroxides and alkoxyamines, fundamentals to applications in materials science, kinetic aspects of NMP, recent developments in NMP, nitroxide-mediated polymerization in dispersed media, NMP of methacrylic esters, complex macromolecular architectures prepared by NMP, surface-initiated NMP, from nanoporous materials to microelectronics, NMP under homogenous conditions, and NMP-derived materials for biomedical applications were presented and discussed in separate chapters of the book [12]. In this chapter, fundamentals and recent developments in the NMP area are briefly summarized and illustrated by particular examples.

7.2 Mechanism of the Nitroxide-Mediated Polymerization

The basic idea of the NMP method (Scheme 7.1) is that the initiation of the "living" free polymerization (LEP) process occurs as a result of the dissociation of a special compound (A-S) into two radicals, active (AR˙) and stable (SR˙) ones. Persisted radical effect (PRE) can also occur for radicals generated from the decomposition of two different initiators affording the active radical and a persistent radical. The active radical involves into the polymer chain initiation and self-recombination reactions. The recombination of a stable radical is thermodynamically forbidden, and, in this case, it can only participate in reversible cross-recombinations with the partner radical (AR˙) and with a polymer radical Rn˙.

The perfect candidates for the role of initiator of LFP were found to be alkoxyamines. These compounds cleave into transient alkyl and persistent aminoxyl radicals which then combine and regenerate the parent compounds. Simultaneously, the alkyl species self-terminate, and this causes a continuous buildup of excess aminoxyl. Hence, the back-reaction to the alkoxyamine (e.g., R_1R_2NOR) accelerates, and the self-termination slows down in time. The cross-recombination of macroradical (Rn˙) with $R_1R_2NO^{\cdot}$ leads to a very low concentration of AR˙, limiting the self-termination reaction of the alkyl radical, and allowing the controlled growth of the macroradical.

The first work on alkoxyamines R_1R_2NOR was reported by Jones and Major in 1927 [17]. The radical reactivity of these compounds was discovered by Kovtun et al. in 1974 [18]. The formation of R_1R_2NOR in recombination of corresponding radicals was also demonstrated.

In NMP, the principle stages are [9]: (1) Initiation can be either performed in a bicomponent system or in a unimolecular initiation process. (2) The nitroxide can then cross-recombine with the propagating active chain to give a nitroxide-terminated radical as dormant species. (3) This macroalkoxyamine can undergo reversible C–ON bond homolysis at elevated reaction temperatures and releases the stable nitroxide radical as well as the active polymer chain. Detail schemes of the reaction mechanism in NMP and initiation mechanism of uni- and bimolecular initiators were presented in review [9]. The development of alkoxyamines as initiators permits controlled homopolymerization of acrylates and acrylamides. Data on

TEMPO derivatives, alkoxyamines including alkoxyamines bearing stereo centers, and light-sensitive alkoxyamines used in DNP were also described in the review.

Steps involved in the TEMPO-mediated nitroxide-mediated polymerization are as follows: (1) The TEMPO/monomer adduct is formed with a source of radicals from a conventional initiator like benzoyl peroxide. (2) With an increase in temperature, the equilibrium is shifted to the addition of monomer, and the radical concentration is controlled by the persistent radical effect. (3) Propagation continues and the growing polymer is capped by TEMPO and the cycle continues. The recent review [5] also provided information on the key components of NMP such as TEMPO derivatives used as nitroxides, alkoxyamines derivatives showing improved bond hydrolysis, alkoxyamines bearing stereo centers, nitroxides and alkoxyamines used for the polymerization (RDRP), of methyl methacrylate (MMA), and functional alkoxyamines.

Spin traps are agents that are commonly used in electron spin resonance spectroscopy (ESR) as a tool to convert transient radicals into a stable form. In polymerizations, spin traps act as radical scavengers, and hence, mostly inhibit chain growth. Nitroxides and nitrons can be radical spin trap scavengers and used in ESR studies to quantify radical initiation [19–22]. A nitrone, for example,

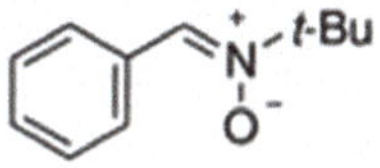

possesses higher affinity toward reacting with oxygen-centered radicals, acrylates, styrenic, and tertiary carbon-centered radicals [20, 21]. In enhanced spin capturing polymerization (ESCP), (macro)radical spin trapping by nitrons is in competition with conventional macroradical chain growth, leading to an effective chain length control of the polymerization, namely trapping macroradicals and shorting the residual polymer chains. The trapped species constitutes a macronitroxide, which by itself is able to trap a further transient radical in a radical recombination event. Thus, the polymer product consists of a coupling product of two polymers, which is bridged by an alkoxyamine functionality in a typical example of the nitroxide-mediated polymerization, the preparation of ABA-type block copolymers

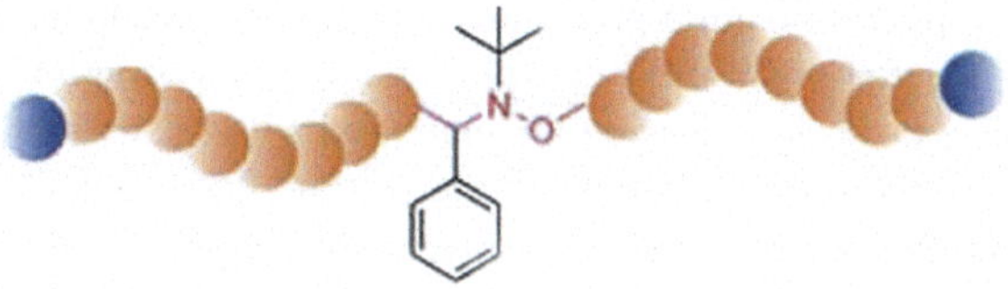

via tandem enhanced spin capturing polymerization (ESCP) [23]. Midchain alkoxyamine functional polystyrenes (M_n = 6200, 12,500, and 19,900 g mol^{-1}) were chain extended with styrene as well as *tert*-butyl acrylate at elevated temperature NMP conditions (T = 110 °C), generating a tandem ESCP-NMP sequence. It was found that the efficiency of the block copolymer formation process decreases with an increasing chain length of the ESCP precursor macromolecules.

Photopolymerization reactions are recognized as powerful tool for modification of coatings, inks, photoresists, dual-cure systems, and others [24, 25]. Nitroxide-mediated photopolymerization (PNMP) is characterized by important advantages, e.g., thin films can be polymerized very rapidly by UV irradiation, and PNMP is an ecological alternative to thermal processes and is able to modify and pattern surfaces.

During PNMP, the photolysis of an alkoxyamine bearing a chromophore, for example,

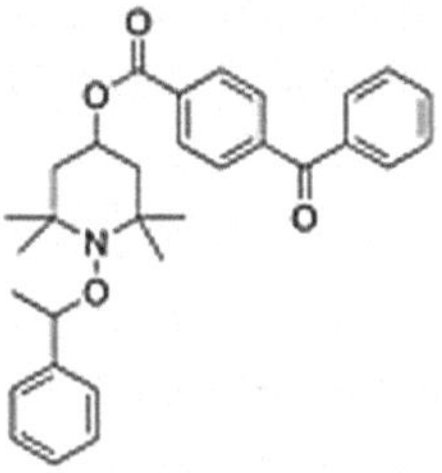

leads to a reversible equilibrium between the alkoxyamine and the generated nitroxide and alkyl radicals, providing a controlled radical photopolymerization [26]. In the alkoxyamines with a closer proximity between the chromophore and the C–O bond, the homolysis of the alkoxyamine is facilitated due to an energy transfer from the chromophore ("antenna") to the weak C–O bond. In work [27], the thermally driven polymerizations were performed via the cleavage of the alkoxyamine functionality; whereas, the photochemically induced polymerizations were carried out either by nitroxide-mediated photopolymerization (NMP2) or by a mechanism, depending on the structure of the light-sensitive alkoxyamine employed. With each alkoxyamine, as initiators of thermally and photoinduced polymerizations, block copolymers were obtained, and the system was applied to the post-modification of polymer coatings for application in patterning and photografting.

The photoradical polymerization of vinyl acetate was performed using 4-methoxy-2,2,6,6-tetramethylpiperidine-1-oxyl as the mediator in the presence of bis(alkylphenyl)iodonium hexafluorophosphate (BAI) [28]. It was found that in this condition: (1) The polymerization proceeds by the living mechanism based on linear increases in the first-order time–conversion and conversion–molecular weight plot. (2) The molecular weight distribution increased with the increasing conversion due to cloudiness of the solution as the polymerization proceeded. (3) The polymerization had a photolatency because the propagation stopped by interruption of the irradiation and was restarted by further irradiation. It was shown [29] that the combination of living radical photopolymerization and lithographic techniques can provide the regiospecificity of graft-polymerized surfaces and precision graft-chain architectures, including their surface graft-chain density, spatio segmental or graded block chains, molecular shape or configuration segmental density. Via such an approach, it was possible to obtain thickness above 100 nm and a linear growth of the thickness versus irradiation time.

7.3 Nitroxide-Mediated Polymerization Kinetics

The complete kinetic scheme is composed of three stages [6, 7, 30–34].

In Scheme 7.2, k_d and $k_{d,ds}$ are the homolysis rate constants of the initiator R_1R_2NOR and of the dormant species, respectively; k_c and $k_{c,ds}$ are the rate constants for the cross-coupling between the alkyl radicals (the initiating one R˙ and the propagating one RMn˙) and the controlling nitroxide $R_1R_2NO^{\cdot}$, respectively; k_{add} is the rate constant of addition of the initiating alkyl radical R˙ onto monomer M; k_p is the propagation rate constant of the propagating radical RMn˙; k_t the self-termination rate constants for the C-centered radicals; k_{dD} is the rate constants for the intramolecular proton transfer (Cope-type elimination) between the initiator and the dormant species; and k_{cD} the rate constant for the intermolecular H-transfer between nitroxides and C-centered radicals.

In comprehensive review [7], principle aspects of the NMP kinetics were described and discussed in separate paragraphs including (1) equations for NMP and the phase

Initiation

$$R^1R^2NOR \longrightarrow R^1R^2NO\bullet + R\bullet \qquad (1)\ k_d$$
$$R^1R^2NO\bullet + R\bullet \longrightarrow R^1R^2NOR \qquad (2)\ k_c$$
$$R\bullet + M \longrightarrow R{-}M_1\bullet \qquad (3)\ k_{add}$$

Propagation

$$R^1R^2NO\bullet + RM_n^\bullet \longrightarrow R^1R^2NOM_nR \qquad (4)\ k_{c,ds}$$
$$R^1R^2NOM_nR \longrightarrow R^1R^2NO\bullet + RM_n^\bullet \qquad (5)\ k_{d,ds}$$
$$RM_n\bullet + nM \longrightarrow RM_n^\bullet \qquad (6)\ k_p$$

Termination

Self-termination

$$R\bullet + RM_n\bullet \longrightarrow \text{dead products} \qquad (7)\ k_{t,ds}$$

Cross side-reactions

$$R^1R^2NOR \longrightarrow R^1R^2NOH + \text{alkene} \qquad (8)\ k_{dD}$$
$$R^1R^2NO\bullet + R\bullet \longrightarrow R^1R^2NOH + \text{alkene} \qquad (9)\ k_{cD}$$

Scheme 7.2 Three-stage scheme for NMP [6]

diagram approach, (2) evidence of the occurrence of PRE in NMP, (3) importance of the k_c, k_d, initiation stage, and side-reactions on the fate of NMP. Data on cross-recombination of nitroxides and alkyl radicals are discussed in Chap. 3. In the review, equations for the PRE dependence on the quasi-equilibrium constant, affording the control, the livingness, and the polymerization time were also presented.

Figure 7.1 illustrates the enthalpic and entropic contributions of the various effects in the NDP homolysis stage.

Effects of the polarity and polarizability on the NDP hemolysis are schematically described in Fig. 7.2.

One of examples of the occurrence of PRE is the decomposition of the TEMPO-based alkoxyamine in monocomponent systems initiated in both bicomponents and

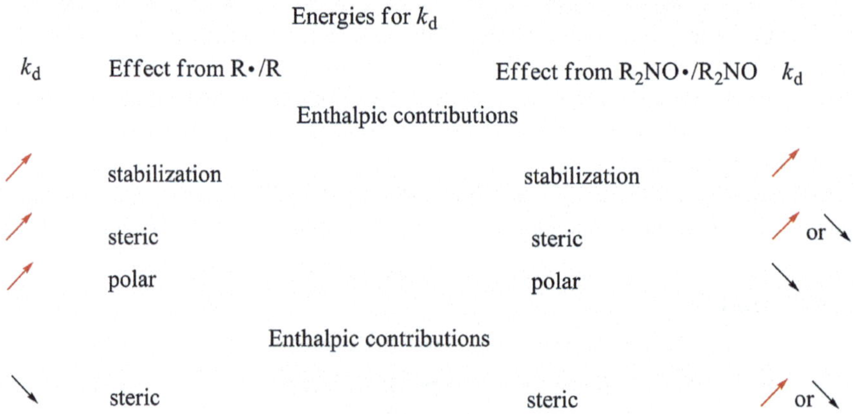

Fig. 7.1 Enthalphic and entropic contributions of the various effects governing k_d values [7]. Private communication from Prof. Elena Bagryanskaya

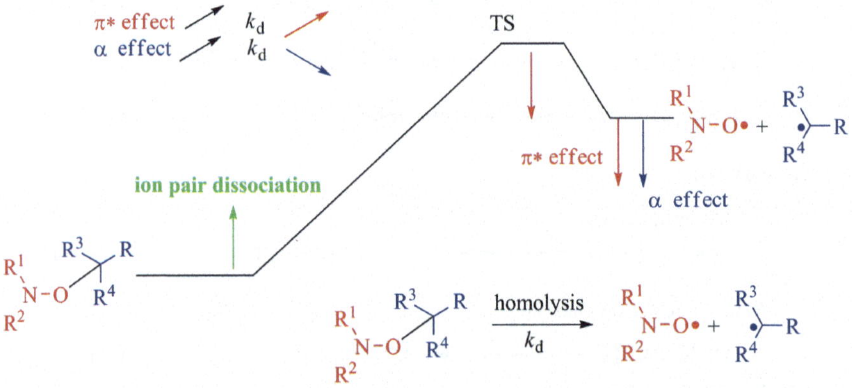

Fig. 7.2 Energy profile for the solvent displaying the effects of the polarity/polarizability of the displaying solvent [7]. Private communication from Prof. Elena Bagryanskaya

monocomponent systems [33]. The following stages are involved in the reaction (1) the C–ON bond hemolysis in the 2-phenyl-2-(2′,2′,6′,6′-tetramethylpiperidine-1′-oxyl)propane (cumyl–TEMPO) 1 afforded an persistent nitroxide and an alkyl radical $PhMe_2C^{\cdot}$ (k_d), (2) the latter either cross-recombined (k_c) in or self-terminated (k_t), and (3) the decay of 1 is accelerated in the presence of a free alkyl radical scavenger galvinoxyl. Solving the kinetic equations predicted the possibility of three different kinetics regimes, that is (1) very short regime I for less than 0.1 s where a transient and a persistent radical are generated at the same rate, (2) regime II lasting around 10^7 s, when the pre-equilibrium is set, and the growth of the nitroxide obeys a $t^{1/3}$ law and the decay of the transient cumyl radical obeys a $t^{-1/3}$ law; and (3) the third regime III which would last 10^7 s with the decays of alkoxyamine 1 and of the cumyl radical 3 obeying a $1/t$ law.

It was shown [34] that in NMP, the concentration of persistent species obeys a $t^{1/3}$ law (7.1) [7]

$$[\mathrm{P}^*] = [\mathrm{PT}]_0^{2/3} \cdot \left(\frac{3k_d^2 \cdot 2k_t}{k_c^2}\right)^{1/3} \cdot t^{1/3} \tag{7.1}$$

where $[PT]_0$ is the initial concentration of alkoxyamine, k_d is the rate constant of alkoxyamine homolysis, k_t is the recombination rate constant of transient radicals T*, and k_c is the recombination rate constant for radical T* and persistent radical P*. In reversible-deactivation radical polymerization (RDRP), the self-termination stage can be quantified by determining the dead chain fraction (DCF) as given in (7.2) [35].

$$\mathrm{DCF} = \frac{[\mathrm{T}]}{[\mathrm{R}-\mathrm{X}]} = \frac{2DP_r k_t[\ln(1-p)]^2}{[\mathrm{M}]_0 k_p^2 t} \tag{7.2}$$

where [T] is the number of dead chains, [R X] is the number of all initiated chains [R X], DP_T is the decreasing degree of polymerization, p is the conversion (p), and k_t and k_p rate coefficients of termination and propagation (k_p). $[M]_0$ is the monomer concentration, and t is the reaction time (t).

Review [35] describes the methods influencing C–ON bond homolysis without alteration of temperature are protonation of functional groups in an alkoxyamine, formation of metal–alkoxyamine complexes, and chemical transformation of alkoxyamines. These approaches can have a significant effect on the homolysis rate constant, by a factor of up to 30, and can shorten the half-lifetime from days to seconds. Importance of the side-reactions on the fate of NMP such as the dimerization and disproportionation of alkyl and polymeric radicals, self-termination reactions of alkyl radicals, the nitroxide self-decomposition, the intramolecular proton transfer reaction, the thermal decomposition of the alkoxyamine and nitroxides, and the presence of an additive was also emphasized in [7]. More details on the development of the kinetic equations and the subsequent consequences are available in papers [2–7, 34, 36].

7.4 Surface and Dispersed Systems

Surface-initiated nitroxyl-mediated polymerization (SI-NMP) on either flat or curved (e.g., nanoparticles) structures has been applied to a diverse range of substrates and applications including the formation of polymer brushes, and the graft modification of nanoparticles (organic and inorganic) and renewable polymers to make new hybrid materials for electronics, optoelectronics, biomedicine, etc. [37–56]. The most significant peculiarity of the surface-initiated nitroxyl-mediated polymerization (SI-NMP) is that the local chain concentrations can be much higher than in solution or bulk [5–9, 12]. In such systems, the initiating and propagating sites are in close proximity, possibly several nanometers apart, and the polymer chain ends remain constrained in closer proximity to each other than occurs in homogeneous NMP. Therefore, the presence of a surface can have a profound impact on the course of polymerization, including kinetics, chain length distribution, and chain end livingness.

Among the numerous inorganic–organic hybrid materials, silica–polymer hybrid materials are the most commonly investigated. Grafting of polymer brushes by SI-NMP has been widely dedicated to surface modification of spherical silica particles exhibiting cylindrical organized mesopores of 2.5–10 nm [37]. Using three alkoxyamine chemical structures on the silicon wafers based on TEMPO, SG1 and TIPNO

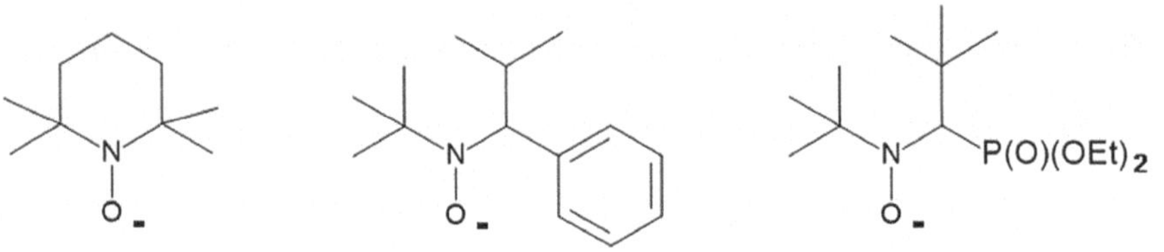

TEMPO **TIPNO** **SG1**

and the surface pressure during Langmuir–Blodgett (LB) deposition, polystyrene brushes with grafting densities of 0.3–1.0 chains nm 2 and stretching values of 40–70% were obtained [38]. Other inorganic surfaces subjected to NMP are clays, Fe_3O_4 (Magnetite), c-Fe_2O_3 (Maghemite), TiO_2 (titanium dioxide), zinc oxide ZnO nanoparticles, CuO (cupric oxide), quantum dots cadmium selenide (CdSe) nanoparticles, polystyrene surfaces, multiwalled carbon nanotubes, graphene oxide, and chitosan. A newly designed diazonium-based isoindoline nitroxide was grafted to a number of carbon allotropes including single and multiwalled carbon nanotubes, graphene, glassy carbon, and carbon fiber [39]. Synthetic pathway for grafting of various carbon allotropes with isoindoline nitroxide-free radicals was descried. Spin-labeled compound

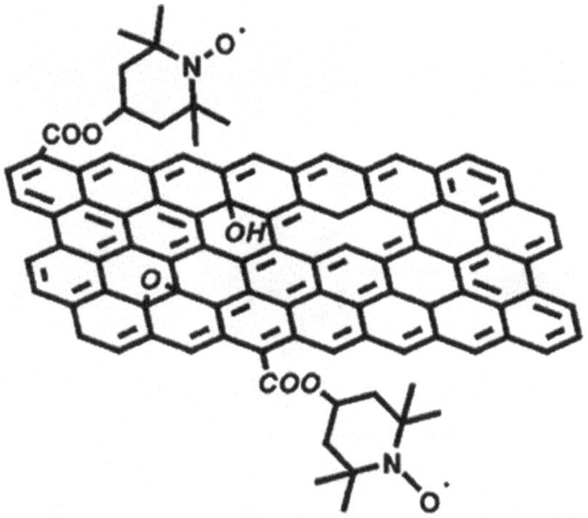

was also synthesized.

To design higher capacity next generation lithium ion batteries, graphene was grafted with 4-hydroxy-TEMPO via conventional esterification of its carboxylic acid groups [40]. The hybrid nitroxide–graphene anode conjugate material possesses superior lithium storage capacity as a result of its dual reservoir of graphene and the reversible redox cycling nitroxide.

Though the majority of polymers produced via a radical mechanism in the industry originates from bulk processes, emulsion and miniemulsion polymerizations together with dispersion polymerization have been also widely studied techniques in reversible-deactivation radical polymerization protocols (RDRP). Implementation of polymerization in a dispersible environment is characterized as the row of features that is especially important at an industrial production. In this process, kinetics is usually faster, and final conversion and degree of polymerization are higher [41–58]. The concentration of radicals in the entire organic phase of a latex is largely superior to that of the corresponding homogeneous system. The final product of such a polymerization in aqueous dispersed media is a latex, suspension of stable nano- or microparticles that found applications as elastomers, adhesives, paints, and in biotechnology.

As an example, using oleic acid/potassium hydroxide and alkoxyamine, particles with diameters less than 20 nm were obtained by conventional radical polymerization and nitroxide-mediated radical polymerization of styrene conducted in miniemulsion SG1 [41]. The method was based on miniemulsion polymerization exploiting combination of the in situ surfactant generation approach whereby the surfactant is formed at the oil–water interface by reaction between an organic acid and a base and by ultrasonication. Microemulsions are formed spontaneously upon TEMPO mixing of oil in water with a very high concentration of surfactant. In emulsions, the initiation can occur in the aqueous phase and leads to the rapid formation of oligomers reaching a critical length for exclusion from the aqueous phase, in combination with appropriate concentrations of nitroxide in both aqueous and organic phases [42].

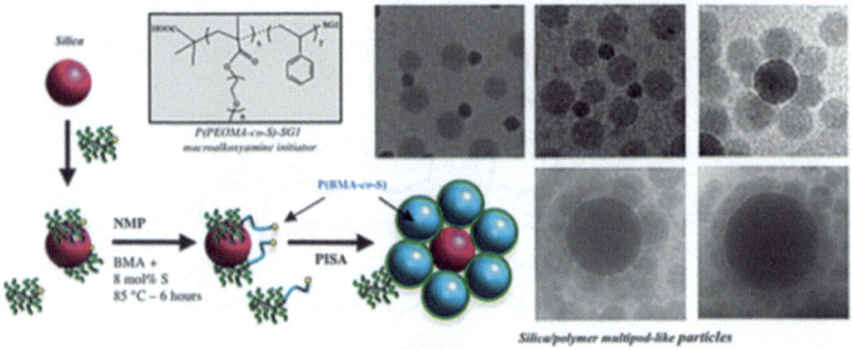

Fig. 7.3 Strategy for the synthesis of hybrid polymer latex/silica particle supracolloidal assemblies and typical examples of obtained morphologies [44]. Reprinted from [44], Copyright 2012 American Chemical Society

Ab initio emulsion polymerizations, in which monomer droplets are present during polymerization, droplet polymerization can be responsible for the formation of large (>1 μm) particles that can lead to coagulum formation. Coagulum-free latexes can be produced using a TEMPO-mediated ab initio emulsion polymerization by suppressing droplet polymerization [42].

Overall strategy for the synthesis of hybrid polymer latex/silica particle supracolloidal assemblies is schematically illustrated in Fig. 7.3 [43].

Dispersion polymerization operated in an organic solvent and yielded large particles (200 nm to 10 μm) was accomplished in the presence of a stabilizer [45]. Diblock copolymers consisting of a poly(acrylic acid) (PAA) segment and a LCST-type poly(*N*,*N*-diethylacrylamide) (PDEAAm) block were obtained by nitroxide-mediated polymerization in aqueous dispersion using a water-soluble macroalkoxyamine. Polymeric nanoparticles, prepared via atom transfer radical polymerization (ATRP BEFOR) and reversible addition–fragmentation chain transfer (RAFT BEFOR), have gained importance as carrier systems for drugs [46, 47]. About 70 imaging scaffolds, micellar systems as nanostructures for biomedical applications, and as component in polymer mixtures to change the rheology of materials were prepared [46]. Polymeric nanoparticles for biomedical delivery applications were also designed in work [47]. The synthesis of micelles from diblock copolymers consisting of a poly(acrylic acid) (PAA) segment and an LCST-type poly(*N*,*N*-diethylacrylamide) (PDEAAm) block, utilized the water soluble of the SG1-capped poly(sodium acrylate) macroalkoxyamine, was performed by nitroxide-mediated aqueous dispersion polymerization. The polymer characterization, physical core cross-linking of pH- and thermoresponsive dynamic diblock copolymer micelles, and proposed mechanism for *PAA-SG1*-initiated DEAAm dispersion polymerization were reported [45].

The generation of block copolymers micelles from biohybrid polymers of poly(n-butyl acrylate) (PnBA) and polystyrene using a TIPNO and SG1-based bifunctional initiator in combination with a controlled polypeptide synthetic technique, such as N-carboxyanhydride (NCA) ring-opening polymerization was achieved [48]. The peptide block, poly(L-glutamate) was used as the hydrophilic component and the formed vesicles and micelles for the transportation of thermolysin and elastase. Gemcitabine, a hydrophilic anticancer drug, was linked to an SG1-based alkoxyamine and served as an initiator for the NMP of isoprene which led to stable nanoparticles of 130–160 nm in diameter with narrow particle size distribution and great colloidal stability [49]. A perfluorinated SG1-capped macroalkoxyamine at 90% conversion was synthesized in supercritical carbon dioxide ($scCO_2$) by MAMA-initiated homogeneous polymerization of heptadecafluorodecyl acrylate and used to initiate the polymerization and simultaneously stabilize the particles. Polymerization of styrene at 110 °C in the presence of a large excess of nitroxide was carried out [51]. TIPNO (2,2,5-trimethyl-4-phenyl-3-azahexane-3-nitroxide)-alkoxyamine was used for nitroxide-mediated precipitation polymerizations of styrene in supercritical carbon dioxide ($scCO_2$) [52]. Under the same conditions, the analogous SG1 (N-tert-butyl-N-(1-diethylphosphono-2,2-dimethylpropyl)nitroxide)-alkoxyamine gave higher rates of polymerization and inferior controlled/living character. All polymerizations proceeded at a similar rate and level of control to the TIPNO system in solution toluene. Suspension polymerization was proved as a route to the synthesis of poly(methacrylic) resins with controlled molecular weight and complex macromolecular architectures [53]. Using 3-(((2-cyanopropan-2-yl)oxy)(cyclohexyl)amino)-2,2-dimethyl-3-phenylpropanenitrile, the synthesis of methacrylic homopolymers by nitroxide-mediated suspension polymerization was carried out. Both methyl methacrylate and *n*-butyl methacrylate were successfully polymerized and showed the narrow molecular weight up to 100,000 g/mol. Dialkoxyamine bearing two carboxylic acid groups was synthesized and used as an initiator for the bulk polymerizations of *n*-butyl acrylate and styrene [54]. Very stable latexes were obtained with small particles and narrow particle size distributions employing the sodium salt counterpart as a water-soluble initiator in the emulsion polymerizations of n-butyl acrylate and styrene via a multistep process. As a result, well-defined poly(n-butyl acrylate) and polystyrene-b-poly(n-butyl acrylate)-b-polystyrene triblock copolymer were synthesized.

In review [55], progress by 2012 in living radical polymerization (LRP) dispersions was presented with a discussion of outstanding issues and challenges as well as the outlook for adoption of LRP dispersions by industry. It was stressed that conducting LRP within nanoparticles enables faster reaction times and will provide better control over the polymer livingness; it also has the potential in the control of the particle mesostructure and microstructure.

Statistical NMP copolymerizations of α-gamma butyrolactone methacrylate, 3-hydroxy-1-adamantyl methacrylate, and 2-methyl 2-adamantyl methacrylate with 5–10 mol% of controlling comonomers (i.e., styrene, p-acetoxystyrene, 2-vinyl naphthalene, acrylonitrile, and pentafluorostyrene) were accomplished [56]. All of the resulting copolymers have relatively low dispersities and show relatively low absorbance at 193 nm. Reported in [57], emulsifier-free emulsion iodine transfer polymerization of styrene with self-assembly nucleation was comprised with the following stages: (1) poly(methacrylic acid) (PMAA) was synthesized in dioxane at 40 °C, (2) dioxane solution of PMAA37-I and styrene were added stepwisely under stirring into an aqueous solution(pH $\sim$ 9), and (3) emulsion was initiated by adding 4,4′-azobis(4-cyanopentanoic acid) at 60 °C. Stable polystyrene emulsion was obtained without coagulation.

The first nitroxide-mediated synthesis of multipod-like silica/polymer latexes by polymerization-induced self-assembly of amphiphilic block copolymers in aqueous emulsion was reported [58]. A water-soluble brush-type PEO-based macroalkoxyamine initiator composed of poly(ethylene oxide) methacrylate and a small amount of styrene was synthesized and adsorbed on the surface of silica particles through hydrogen-bonding interactions and employed to initiate the emulsion polymerization of *n*-butyl methacrylate with a small amount of styrene under mild conditions (85 °C).

7.5 NMP in Creation of Complex Macromolecular Architectures

Data on controlled radical polymerization (CRP) of inimers or transmers (compounds chain transfer group and vinyl group in one molecule) for the synthesis of hyperbranched polymers were published [59–63]. Inimers and transmers carry numerous functional groups could be homopolymerized, or copolymerized with monovinyl monomers, using nitroxide-mediated polymerization producing hyperbranched polymers and hyperstar polymers with tunable molecular weights, compositions, and degree of branching. CRP-produced hyperbranched can be applied for encapsulation of guest molecules, nanomedicine, diagnostic imaging, and catalysis.

A typical example of production of CRP-produced hyperbranched dendric polystyrene polymers is the first bulk chain transfer polymerization of styrene with 4-vinylbenzyl dithiobenzoate [60]. Starting from diblock copolymer [poly(vinyl acetate)-b-poly(acetonitrile)] that was endcapped by a labile cobalt complex, general structures of transmers to produce hyperbranched polymers were obtained. One way to construct complex macromolecular architectures was based on employing functional nitrones, for example, (α-4-(3-(trimethylsilyl)prop-2-ynyloxy)-*N*-*tert*-butyl nitrone (Fig. 7.4) [59]. This nitron was used to introduce a secondary functionality in the midchain position of a chain, which can be an alkyne for by copper-assisted azide alkyne cycloaddition or an N-hydroxysuccinimide ester for bioconjugation.

Using nitrone spin capturing with PBN derivative, the diblock copolymer [poly(vinyl acetate)-b-poly(acetonitrile)], endcapped by a labile cobalt complex, was obtained from cobalt-mediated radical polymerization (CMRC) [61]. CMRC-derived polymers were efficiently trapped by the nitrone, leading to symmetrical alkyne-functionalized triblock copolymers. By reaction with a homotelechelic poly(ethylene glycol) azide, the synthesis of H-shaped polymers via a combination of CMRP and NMRC followed by CuAAC conjugation was performed. An approach to grow dense

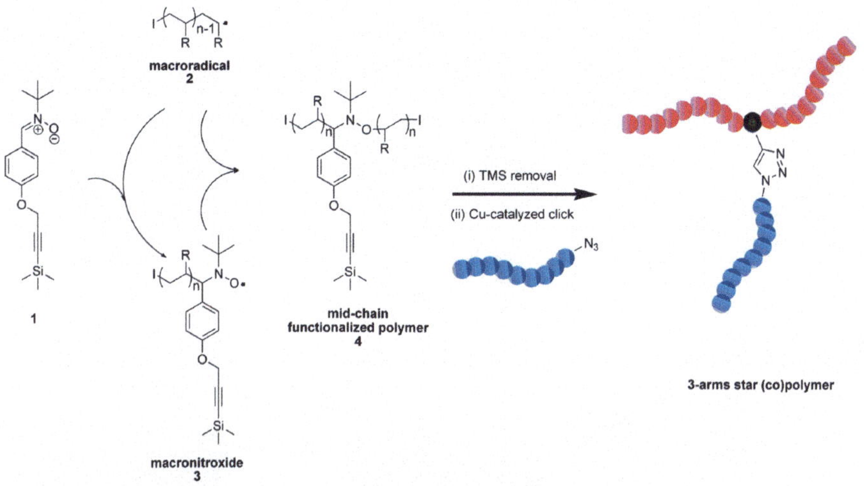

Fig. 7.4 Spin capturing of macroradicals with an alkyne derivative of PBN and the subsequent Cu-catalyzed click reaction form in three-armed star (co)polymers. I denotes the polymer end group resulting from the employed initiator, and R represents the monomer side group (COOR for iBoA or phenyl for styrene) [60]. Reprinted from [60], Copyright 2010 American Chemical Society

polymer brushes via SI-NMP from an ultrathin polymer coating was introduced (Fig. 7.5) [62]. The synthesis of NMP inimer M1 comprised three consecutive synthetic steps was described. A new NMP inimer of ethylbenzene-TEMPO derivative to grow polymer brushes via SI-NMP was designed and employed to afford polymer brushes with controlled molecular weights and low dispersity. Surface-initiated nitroxide-mediated polymerization (SI-NMP) was conducted to grow polystyrene (PS) brushes on the coating with a sacrificial layer designed to cleave the brushes

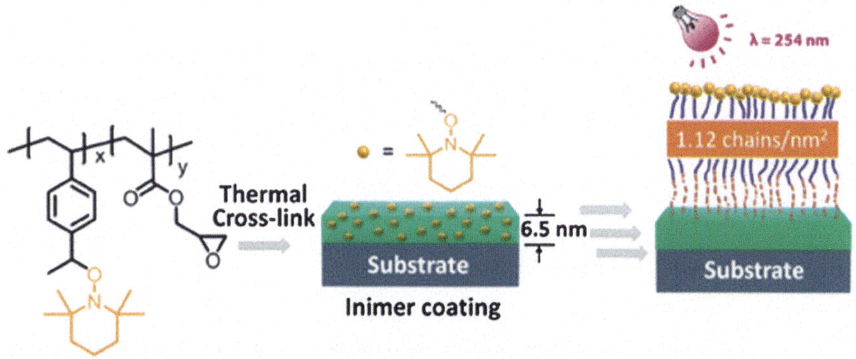

Fig. 7.5 Substrate-independent approach to dense cleavable polymer [63]. Reprinted from [63], Copyright 2018 American Chemical Society

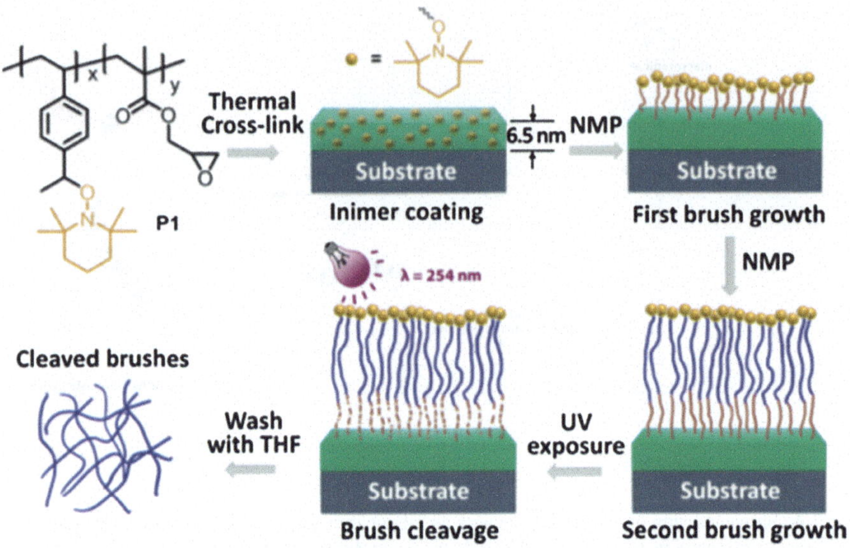

Fig. 7.6 Schematic illustration for the SI-NMP of polystyrene (PS)PS. Surface modification with polymer brushes is an attractive method for tuning of physical and chemical properties of interfaces such as wetting, adhesion, electronic, catalytic, or biological activity [64]. Reprinted from [64], Copyright 2017 American Chemical Society

The brushes are grown on an inimer coating cross-linking and bearing nitroxide-mediated polymerization (inimers and glycidyl methacrylate (Fig. 7.5). The cleaved brushes have larger molecular weights than the corresponding free polymers.

Figure 7.6 shows an overview of the SI-NMP strategy including the following steps to create bilayer polymer brushes: (1) P1 was thermally cross-linked to form an inimer coating which provided high-density NMP initiators on the substrate. (2) Then, a layer was designed to be a sacrificial layer. (3) The second brush layer is released [63].

Therefore, the regiospecificity of graft-polymerized surfaces and precision graft-chain architectures, including their surface graft-chain density, spatio segmental density, segmental or graded block chains, molecular shape or configuration, can be accomplished.

7.6 NMP and Polymer Materials

Over the past two decades, NMP is widely used for designing numerous organic electronic devices, nanoporous membranes, and other technologies [8, 64–73]. The use of NMP can result in homopolymers and copolymers that are characterized by having a narrow molecular weight distribution providing a high quality of produced materials. Chromatography, pharmaceutical purification and filtration, high surface area catalysis, microelectronics, imprint lithography, and others are fields of application of nanoporous membranes and nanoporous surfaces.

In work [64], a facile reaction method for incorporation of (TEMPO group into poly(2,6-dimethyl-1,4-phenylene oxide) (PPO) chains through the atom transfer radical coupling reaction has been elaborated. The TEMPO-functionalized PPO (PPO-TEMPO) was used as a macroinitiator for NMP of styrene to result in the corresponding PPO-g-PS graft copolymers and to carry out a thermally induced self-cross-linking reaction through radical coupling. The cross-linked PPO enhanced mechanical properties, reduced oxygen permeability coefficient, and provided low dielectric constant and low dissipation factor. Reaction route for incorporation of TEMPO groups to poly(2,6-dimethyl-1,4-phenylene oxide) chains through an atom transfer nitroxide radical coupling reaction was indicated in a scheme. The authors suggested that the obtained materials have high potential for application in dielectric layers of microelectronic. TEMPO was also used to synthesize poly(4-hydroxystyrene) homopolymers possessing preferred absorption characteristics in the UV range (220–260 nm) [66]. Using NMP, well-defined narrowly disperse glycidyl methacrylate (GMA) and pentafluorostyrene (PFS) random copolymers for application in photoresists were synthesized employing NMP. The polymer was proved to be amenable to aqueous base microlithographic fabrication technologies. Poly(styrene)-block-poly(ethylene oxide) [poly(S)-block-poly(EO)] and other block copolymers were demonstrated to be suitable materials for nanoporous membranes and templates, what is the basis for the next generation of photonics including polymer waveguides, materials for lithography, etc. [67]. Role of NMP in polymer templating was reviewed.

Poly(S) nanoporous membrane, coated with terpyridine groups, was prepared [68]. The preparation was carried out by the following stages: (1) the synthesis of functional terpyridine poly(S) homopolymers by NMP and coupling them to other terpyridine functional poly(EO) homopolymers by forming nickel–terpyridine complexes, (2) the resulting block copolymers self-assembled into a thin film with hexagonal morphology and were treated with a competing ligand, and (3) these ligands reacted with the poly(S) blocks resulting in the dissolution of free poly(EO) blocks. A formation of microporous thin films that made self-assembled poly(ethylene oxide)-block-poly(styrene) was schematically illustrated. Obtained materials can be utilized for further heterogeneous surface reactions.

Employing NMP, microporous thin films were made from acid cleaving and washing of a thin film of self-assembled poly(ethylene oxide)-block-poly(styrene) [poly(EO)-block-poly(S) [69]. The used breath figure technique is based on the principle of a solution evaporating in the presence of wet air, which results in the polymers self-assembling around the drying water droplets. A mixture of styrene and a styrenic p-conjugated oligo(fluorene) side chain was polymerized in the presence of fluoromicasilicate layers [70]. A cation exchange between the silicate layers and amine functional TEMPO molecules followed by a polymerization of the styrenic monomers in the presence of this hybrid TEMPO-containing surface was investigated. The obtained polymer exhibits in the solids a blue photoluminescence.

7.7 Nitroxide on Inorganic Templates

For high temperature and electronic applications, block copolymers can be used as a template for the nanofeatured design of inorganic materials, to make silicon oxide hierarchical nanostructures, for example, utilizing self-assembly of silicon-containing block copolymer. On this way, a series of silicon-containing styrenic polymers and block copolymers by narrow molecular weight distribution (Mw/Mno1.2) was synthesized and exposed the final materials to an oxygen plasma [71]. Poly(styrene-ran-acrylonitrile)-block-poly(EO)-block-poly(styrene-ran-acrylonitrile) [poly(S-ran-AN)-block-poly(EO)-block-poly(S-ran-AN)] block copolymers were prepared [72]. It was found that when spin coating the resulting polymers from different solvent mixtures occurred free-standing nanoparticle films were obtained.

Proton exchange membranes or polymer electrolyte membranes (PEMs) allow the transfer of protons while preventing the passage of reactive gases such as hydrogen and oxygen. These materials are an integral part of the modern fuel cell. Star polymers of sodium styrenesulfonate with regulated, molecular weight, arm length, narrow polydispersity, and with variable amounts of SO_3OH groups were prepared in a one-pot TEMPO-controlled polymerization [73]. Poly(sodium styrenesulfonate) was terminated with divinyl benzene, and star polymers were produced via stable-free radical coupling of vinylic terminal groups. The polymers based on arms of 20 and 32 repeat units possessed ~33 and ~41 arms per star, respectively. Micrographs of copolymer membranes: (poly(styrene)-graft-poly(styrene sulfonate) (poly(S)-graft-poly(S)SA) and poly(styrene)-ran-poly(styrene sulfonate) (poly(S)-ran-SSA) were obtained.

Ionic liquids are thermally stable, with negligible vapor pressure. Polymerizing an ionic liquid-containing monomer allows for the combination of the ionic liquid properties with the desirable polymer mechanical properties. These polymers are potential candidates for next generation ion conducting membranes, polymer electrolytes, and other electrochemical applications. NMP has been utilized to incorporate ionic liquid-containing monomers into a copolymer [74]. As an example, using Bloc-Builder-MA as the initiator, the controlled homopolymerization of N-vinylimidazole as well as a styrenic derivative, 1-(4-vinylbenzyl)-imidazole (ionic liquid-containing monomers), was accomplished [75]. A difunctional alkoxyamine NMP initiator based on SG173

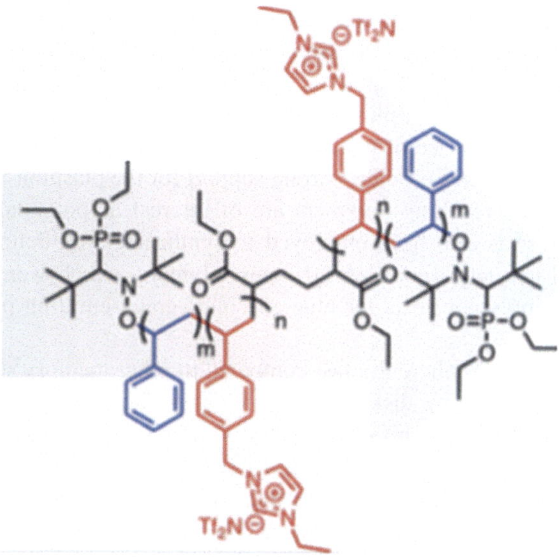

was employed for the synthesis of an imidazole-based ABA triblock copolymer.

7.8 NMP for Organic Electronics and Other Devices

Over the past two decades, significant progress has been made in the performance of relatively more complex organic optoelectronic devices, such as organic light-emitting diodes (OLEDs), thin film transistors (OTFT), and organic photovoltaic (OPV) devices [40, 76–84]. Morphology, photophysical properties, electrochemical behavior and BHJ organic photovoltaic device performance of poly(vTPA)-block-poly(PerAcr), coil–coil block copolymer synthesized by NMP, and poly(vTPA) and poly(PerAcr) homopolymers have been evaluated [76].

Donor–acceptor rod–coil block copolymers–acceptor block copolymers, composed of poly(3-hexylthiophene) homopolymers were synthesized by NMP, utilizing the poly (3-hexylthiophene) macroinitiator to homopolymerize a perylene diimide acrylate [77]. The resulting donor–acceptor block copolymers were characterized by having a narrow molecular weight distribution of 1.2–1.4. Obtained copolymers showed efficient photoluminescence quenching in the solid state, indicative of charge separation, and were employed to produce a solar cell with power conversion efficiency of 0.49%. The viability of three different synthetic pathways to achieve poly[1-(4-vinylbenzyl)-3-butylimidazolium bis(trifluoromethylsulphonyl)-imide-*co*-methyl methacrylate] (poly(VBBI$^+$TFSI$^-$-*co*-MMA)) was explored including direct NMP polymerization of the ionic liquid monomer (VBBI$^+$TFSI$^-$) and post-functionalization of precursor polymer poly(4-vinylbenzyl imidazole-*co*-methyl methacrylate) [78]. A full synthetic, kinetic, and compositional comparison between methods was described. In the first example of plasmon-assisted NMP of stimuli-responsive block copolymers poly(*N*-isopropylacrylamide)-*co*-4-vinylboronic acid, the growth of a polymer film at room temperature was achieved via plasmon-induced homolysis of alkoxyamines covalently attached to the surface of plasmon-active gold gratings at room temperature [79]. EPR and other assays provided strong support for the plasmon-initiated mechanism of NMP. The resulting polymers are of interest as potential materials for electronics. NMP has also been employed to synthesize well-defined poly(2-(N-carbazolyl)ethyl acrylate) [poly(VAK)] homopolymers, which were subsequently used as effective polymer hosts for blue and blue–green-emitting phosphor-based device [80].

Electrical memory is inherently fast compared to other memory storage devices, such as hard drives, floppy disks, DVDs, or CDs. The memory cell is an electronic circuit that stores one bit of binary information, and it must be set to store a logic 1 (high-voltage level) and reset to store a logic 0 (low-voltage level). Polymer-based memory devices comprising with memory cells can provide fast digital electrical memory. NMP can be used to synthesize polymers for applications in memory devices. An oxadiazole-containing brush polymer, poly-((5-phenyl-1,3,4-oxadiazol-2-yl-[1,10-biphenyl]carboxyloxy-(n-nonyl acrylate) was synthesized using st-TIPNO [81]. The material exhibits volatile or non-volatile memory behavior based on the packing of the oxadiazole molecules within the polymer film, which could be controlled with subsequent annealing steps. Authors of work [82] employed NMP for the synthesis of a series of random copolymers containing pendant electron-donating 9-(4-vinylphenyl)carbazole and electron-accepting 2-phenyl-5-(4-vinylphenyl)-1,3,4-oxadiazole or 2-(4-vinylbiphenyl)-5-(4-phenyl)-1,3,4-oxadiazole with various copolymer ratios. It was found that when building simple anode/polymer/cathode devices, the electrical switching behavior could be tuned through the donor–acceptor ratio. Thus, these copolymers exhibit memory behavior.

Over a decade ago, it was estimated that controlled radical polymerization would be a $20 billion market [83]. Based on the use of the Bloc-Builder alkoxyamine, the range of polymeric products was developed by the chemicals and advanced materials company (ARKEMA). Various all-acrylic poly(methyl methacrylate)-poly(n-butyl acrylate)-b-poly(methyl methacrylate) (MAM), triblock copolymers were produced at industrial scale.

To design higher capacity next generation lithium ion batteries, graphene was grafted with 4-hydroxy-2,2,6,6-tetramethylpiperidin-1-oxyl (4-hydroxy-TEMPO) via conventional esterification of its carboxylic acid groups [40]. When applied as the anode material for lithium ion batteries (LIBs), TEMPO-G exhibited a high reversible capacity with cycling stability for lithium storage in terms of 1080 mA h g^{-1} at a current density of 100 mA g^{-1} after 400 cycles. Thus, the resultant 4-hydroxy-TEMPO-functionalized graphene (TEMPO-G) was revealed to consist of an electrically conducting network of graphene sheets with abundant electrochemically active nitroxide radical functionalities. The grafting reagents and conditions were described in details, and unwanted dissolution of nitroxide radicals in the electrolyte was also discussed.

7.9 NMP and Materials for Biomedical Applications

Polymer derived from the nitroxide-mediated polymerization, glycopolymers, bioconjugates, nanoparticulate, systems, and polymer nanocomposites have been recognized as biomaterials [22, 85–91]. The main requirement for these materials is compatibility with biological and medical objects. Another avenue to use these polymers is their biologic activity.

In work [86], poly(2-(b-D-galactosyloxy)ethyl methacrylate-costyrene)-β-polystyrene (P(GalEMA-co-S)-b-PS) amphiphilic block copolymers were obtained from SG1-mediated polymerization of glycomonomer G10, using either P(AcGalEMA-co-S)-SG1 or PS-SG1 macroinitiators, followed by deacetylation of AcGalEMA moieties. The polymer was used to prepare honeycomb-structured porous films with bioactivity (Fig. 7.7). A fluorophore was conjugated with the sugar moieties on the porous film, and proteins were conjugated to glycopolymer inside the pore.

The combination of synthetic polymers with polypeptides has paved a way toward innovative materials with unprecedented properties [22, 87–90]. For example, the covalent linkage of poly(ethylene glycol) (PEG), a hydrophilic and flexible polymer, to poly(peptide)-based therapeutics resulted in the synthesis of bioconjugates

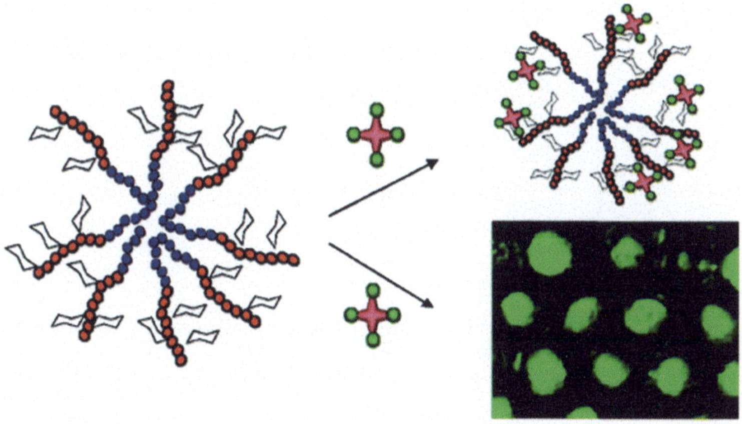

Fig. 7.7 Schematic presentation formation of porous films with a fluorophore conjugated with the sugar moieties on the porous film and proteins conjugated to glycopolymer inside the pore [86]. Reprinted from [86], Copyright 2009 American Chemical Society

exhibited improved biodistribution and pharmacokinetics, high stability and solubility, reduced immunogenicity and longer plasma half-lives due to both reduced renal filtration and proteolysis [87]. An SG1-functionalized peptide (SG1-GGGWIKVAV) was also constructed by solid-phase peptide synthesis SPPS followed by the direct attachment of the Bloc-Builder alkoxyamine to the terminal amine group of the peptide in the presence of PyBOP as the coupling agent [22].

The conjugation of preformed polymer to free peptides and proteins relied on the design of a-functional comb-shaped polymethacrylates with PEG side chains from NMP of OEGMA with a few percent of acrylonitrile initiated by SG1-based alkoxyamines bearing an N-hydroxysuccinimidyl moiety was achieved [88]. The polymers exhibited tunable reactivities toward nucleophiles. Copolymers derived from the SG1-based alkoxyamine can be quantitatively coupled with small molecules and a neuroprotective peptide; whereas, partial conjugation was obtained with lysozyme. α-functional, fluorescent comb-shaped polymethacrylates with PEG side chains for bioconjugation purposes were designed. Reversible switching of the activity in the conjugates was demonstrated. The preparation of block copolymers of PNVP and polypeptides (PNVP-b-PBLG, PNVP-b-PBLL and PNVP-b-PBLG-b-PBLL), with PBLL for poly(tert-butyloxycarbonyl-L-lysine) in a sequential fashion was performed using 4-Amino-TEMPO [89].

Carboxybetaine monomers differing in intercharge separating spacer groups were synthesized, and nitroxide-mediated free radical polymerization was conducted using alkoxyamine initiators with hydrophobic (TEMPO) and hydrophilic (β-phosphonate) terminal groups (Fig. 7.8) [90]. The effect of the charge, hydration, and end groups for corresponding films attached to silica surfaces on absorbed protein conformation was studied with three plasma proteins, lysozyme, bovine a-lactalbumin, and human serum albumin.

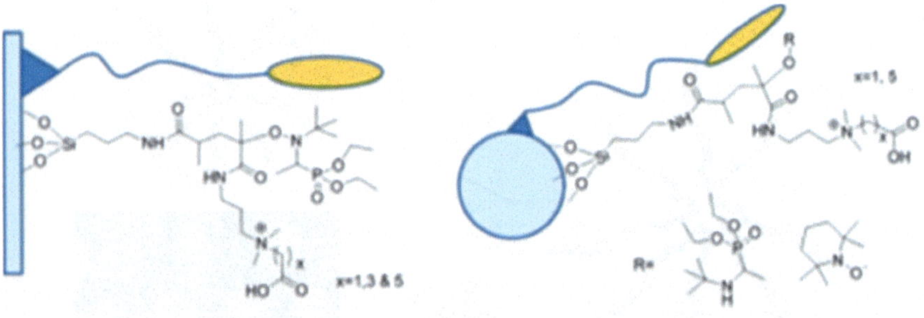

Fig. 7.8 Combination of non-fouling properties with integrated biomimetic cell adhesive end groups on silica surfaces and silica particles SiNP [91]. Reprinted from [91], Copyright 2011 American Chemical Society

7.10 NMP Miscellaneous Applications

Well-defined PMMA-rich P(MMA-*co*-AN) copolymers using the nitroxide-mediated controlled/living free radical polymerization of methyl methacrylate initiated by the SG1-based alkoxyamine Bloc-Builder were prepared [91]. It was found that a minimal amount of acrylonitrile turns the nitroxide-mediated polymerization of methyl methacrylate into the controlled/living system. The polymer structure was characterized by ^{1}H, ^{13}C, and ^{31}P NMR technics. Detrembleur et al. [92] made nitroxide-mediated polymerization of methacrylates at moderate temperature straightforward by preparing and isolating the methyl 2-methyl-3-nitro-2-nitrosopropionate (NMMA). Nitrone (a-phenyl-N-tert-butylnitrone PBN) was involved in the homopolymerization of a variety of methacrylates by an NMP process at moderate temperature (40–50 °C). The combination of a low-temperature azo-initiator with a hindered nitroso-compound produces a mixture of hindered nitroxides in the polymerization medium that act as efficient polymerization control agents. Results of ESR spectroscopy experiments combined with mass spectrometric studies support the proposed in situ NMP mechanism.

Water-soluble nitroxides based on 1,1,3,3-tetramethylisoindolin-2-oxyl and 1,1,3,3-tetraethylisoindolin-2-oxyl with an ionic group on the aromatic ring (either a quaternary ammonium or a sulfonate substituent) and TEMPO were tested as mediators in the controlled free radical polymerization of sodium 4-styrenesulfonate carried out in water at 130 °C [93]. It was found that the steric hindrance at the nitroxide site (ethyl or methyl substituent) had a much larger influence on the activation–deactivation equilibrium than the type of ionic group. For nitroxides bearing four ethyl substituents, the equilibrium constant was one order of magnitude larger than that obtained for nitroxides with four methyl substituents. Controlled radical polymerization (CRP) of inimers or transmers for the synthesis of hyperbranched polymers was summarized in review [94]. Data on nitroxide-mediated polymerization (or reversible addition–fragmentation chain-transfer (RAFT) polymerization techniques, producing hyperbranched polymers and hyperstar polymers with tunable molecular weights, compositions, and degree of branching were also reviewed.

The ability of NMP to reduce the dispersity, to increase chain end homogeneity, to synthesize block copolymers that could self-assemble, to tether the polymers to surfaces, and to incorporate novel functional monomers has been well documented. Aforementioned data showed that the use of NMP was explored in numerous applications, such as organic electronics nanoporous templating, photoresists, and membranes for fuel cells. Thus, nitroxide-mediated polymerization (NMP) is a robust and industrially relevant technique.

References

1. D.H. Solomon, Polymer group and the discovery and significance of nitroxide-mediated living radical polymerization. J. Polym. Part A: Polym. Chem. **43**(23), 5748 (2005)

2. H. Fischer, M. Souaille, The persistent radical effect in living radical polymerization – border cases and side-reactions. Macromol. Symp. **174**, 231–240 (2001)
3. D. Bertin, D. Gigmes, S.R.A. Marque, P. Tordo, Kinetic subtleties of nitroxide mediated polymerization. Chem. Soc. Rev. **40**, 2189 (2011)
4. J. Nicolas, G. Yohann, C. Lefay, D. Bertin, D. Gigmes, B. Charleux, Nitroxide-mediated polymerization. Prog. Polym. Sci. **38**(1), 63–235 (2013)
5. J. Kreutzer, Yu. Yagci, Metal free reversible-deactivation radical polymerizations: advances, challenges, and opportunities. Polymers **10**, 35 (2018)
6. D.F. Grishin, I.D. Grishin, Controlled radical polymerization: prospects for application for industrial synthesis of polymers (review). Russ. J. Appl. Chem. **84**(12), 2021 (2011)
7. E.G. Bagryanskaya, S.R.A. Marque, Kinetic aspects of nitroxide mediated polymerization. RSC Polym. Chem. Ser. **19**(Nitroxide Mediated Polymerization), 45–113 (2016)
8. K. Matyjaszewski, Macromolecular engineering: from rational design through precise macromolecular Matyjaszewski synthesis and processing to targeted macroscopic material properties. Prog. Polym. Sci. **30**, 858–875 (2005)
9. M. Maric, Application of nitroxide mediated polymerization in different monomer. Curr. Org. Chem. **22**, 1264–1284 (2018)
10. W.E. Bachmann, F.Y. Wiselogle, The relative stability of pentaarylethanes. III.[1] The reversible dissociation of pentaarylethanes. J. Org. Chem. **1**, 354 (1936)
11. G. Moad, E. Rizzard, The history of nitroxide-mediated polymerization, in *RSC Polymer Chemistry Series No. 19. Nitroxide Mediated Polymerization: From Fundamentals to Applications in Materials Science*, ed. by D. Gigmes (The Royal Society of Chemistry, 2016), pp. 1–44
12. D. Gigmes (ed.), *RSC Polymer Chemistry Series No. 19. Nitroxide Mediated Polymerization: From Fundamentals to Applications in Materials Science* (The Royal Society of Chemistry, 2016)
13. D.H. Solomon, J. Macromol. Sci. Chem. **17**, 337 (1982)
14. G. Moad, D.H. Solomon, *The Chemistry of Radical Polymerization* (Elsevier, 2006)
15. R.B. Grubbs, Nitroxide-mediated radical polymerization: limitations and versatility. Polym. Rev. **51**, 104 (2011)
16. L. Marx, P. Hemery, Synthesis and evaluation of a new polar, TIPNO type nitroxide for "living" free radical polymerization. Polymer **50**, 2752–2761 (2009)
17. L.W. Jones, R.T. Major, Substituted O-alkyl hydroxylamines chemically related to medicinally valuable amines. J. Am. Chem. Soc. **49**, 1527 (1927)
18. G.A. Kovtun, A.L. Aleksandrov, V.A. Golubev, Interaction of peroxide radicals with esters of hydroxylamines. Bull. Akad. Sci. USSR: Div. Ser. 2115 (1974); Izv. Akad. Nauk SSSR: Ser. Khim. 2197 (1974)
19. E.M. Pliss, I.V. Tikhonov, A.I. Rusakov, Kinetics and mechanism of reactions of aliphatic stable nitroxide radicals in chemical and biological chain processes, in *Nitroxides—Theory, Experiment and Applications*, ed. by A.I. Kokorin (IntechOpen). https://doi.org/10.5772/39115
20. A. Debuigne, M. Hurtgen, C. Detrembleur, C. Jerome, C. Barner-Kowollik, T. Junkers, Spin capturing with nitrones: radical coupling reactions with concurrent introduction of mid-chain functionality. Prog. Polym. Sci. **37**, 1004–1030 (2012)
21. G. Wang, J. Huang, Versatility of radical coupling in construction of topological polymers. Polym. Chem. **5**, 277–308 (2014)
22. J. Nicolas, Y. Guillaneuf, C. Lefay, D. Bertin, D. Gigmes, B. Charleu, Nitroxide-mediated polymerization. Prog. Polym. Sci. **38**, 63–235 (2013)
23. T. Junkers, L. Zang, E. H.H. Wong, N. Dingenouts, C. Barner-Kowollik, Formation of triblock copolymers via a tandem enhanced spin capturing nitroxide-mediated polymerization reaction sequence. J. Polym. Sci. Polym. Chem. **49**(22), 4841–4850 (2011)
24. J.P. Fouassier, J. Lalevée, *Photoinitiators for Polymer Synthesis—Scope, Reactivity, and Efficiency* (Wiley, Weinheim, 2012)
25. Y. Guillaneuf, D.L. Versace, D. Bertin, J. Lalevee, D. Gigmes, J.P. Fouassier, Importance of the position of the chromophore group on the dissociation process of light sensitive alkoxyamines. Macromol. Rapid Commun. **31**, 1909 (2010)

26. Y. Guillaneuf, D. Bertin, D. Gigmes, D.L. Versace, J. Lalevee, J.P. Fouassier, Toward nitroxide-mediated photopolymerization. Macromolecules **43**(5), 2204–2212 (2010)
27. J. Morris, S. Telitel, K.E. Fairfull-Smith, S.E. Bottle, J. Lalevée, J.L. Clément, Y. Guillaneuf, D. Gigmes, Novel polymer synthesis methodologies using combinations of thermally- and photochemically-induced nitroxide mediated polymerization. Polym. Chem. **6**, 754–763 (2015)
28. E. Yoshida, Nitroxide-mediated photo-living radical polymerization of vinyl acetate. Colloids Polym. Sci. **288**, 73 (2010)
29. K.T. Haraldsson, J.B. Hutchison, R.P. Sebra, B.T. Good, K.S. Anseth, C.N. Bowman, 3D polymeric microfluidic device fabrication via contact liquid photolithographic polymerization (CLiPP). Sens. Actuators B **113**, 454 (2006)
30. J. Johnson, G. Moad, D.H. Solomon, T.H. Spurling, D.J. Vearing, The application of supercomputers in modeling chemical reaction kinetics: kinetic simulation of 'quasi-living' radical polymerization. Aust. J. Chem. **43**, 1215 (1990)
31. H. Fischer, The persistent radical effect: a principle for selective radical reactions and living radical polymerizations. Chem. Rev. **101**, 3581 (2001)
32. D. Greszta, K. Matyjaszewski, Mechanism of controlled/"living" radical polymerization of styrene in the presence of nitroxyl radicals, kinetics and simulations. Macromolecules **29**, 7661 (1996)
33. T. Kothe, S. Marque, R. Martschke, M. Popov, H. Fischer, Radical reaction kinetics during homolysis of *N*-alkoxyamines: verification of the persistent radical effect. J. Chem. Soc. Perkin Trans. **2**, 155 (1998)
34. D. Gigmes, S.R.A. Marque, Nitroxide mediated polymerization and its applications, in *Encyclopedia of Radicals in Chemistry, Biology, and Materials*, ed. by C. Chatgilialoglu, A. Studer (Wiley, 2012)
35. M. Edeleva, G. Audran, S. Marque, E. Bagryanskaya, Smart control of nitroxide-mediated polymerization initiators' reactivity by pH, complexation with metals, and chemical transformations. Materials (Basel) **12**(5), 688 (2019)
36. M. Asteasuain, Deterministic approaches for simulation of nitroxide-mediated radical polymerization. Int. J. Poly. Sci. Article ID 7803702, 16 pages (2018). https://doi.org/10.1155/2018/7803702
37. H. Blas, M. Save, C. Boissière, C. Sanchez, B. Charleux, Surface-initiated nitroxide-mediated polymerization from ordered mesoporous silica. Macromolecules **44**, 2577 (2011)
38. R.-V. Ostaci, C. Celle, G. Seytre, E. Beyou, J.-P. Chapel, E. Drockenmull, J. Polym. Sci. Part A: Polym. Chem. **46**, 3367 (2008)
39. J.P. Blinco, B.A. Chalmers, A. Chou, K.E. Fairfull-Smith, S.E. Bottle, Spin-coated carbon. Chem. Sci. **4**, 3411 (2013)
40. Z. Du, W. Ai, L. Xie, W. Huang, Organic radical functionalized graphene as a superior anode material for lithium-ion batteries. J. Mater. Chem. A **2**, 9164 (2014)
41. Y. Guo, P.B. Zetterlund, Synthesis of nanosized (<20 nm) polymer particles by radical polymerization in miniemulsion employing in situ surfactant formation. Macromol. Rapid Commun. **32**, 1669–1675 (2011)
42. R. González-Blanco, E. Saldívar-Guerra, J. Herrera-Ordonez, P.B. Zetterlund, A. Cano-Valdez, TEMPO-mediated emulsion polymerization. Macromol. Symp. **325–326**, 89–95 (2013)
43. X.G. Qiao, P.Y. Dugas, B. Charleux, M. Lansalot, E. Bourgeat-Lami, Synthesis of multipod-like silica/polymer latex particles via nitroxide-mediated polymerization-induced self-assembly of amphiphilic block copolymers. Macromolecules **48**, 545–556 (2015)
44. M.E. Thomson, J.S. Ness, S.C. Schmidt, M.F. Cunningham, Nitroxide mediated polymerization: from fundamentals to applications. Macromolecules **44**, 1460–1470 (2011)
45. G. Delaittre, M. Save, M. Gaborieau, P. Castignolles, J. Rieger, B. Charleux, Synthesis by nitroxide-mediated aqueous dispersion polymerization, characterization, and physical core-crosslinking of pH- and thermoresponsive dynamic diblock copolymer micelles. Polym. Chem. **3**, 1526–1538 (2012)
46. J. Nicolas, S. Mura, D. Brambilla, N. Mackiewicz, P. Couvreur, Design, functionalization strategies and biomedical applications of targeted biodegradable/biocompatible polymer-based nanocarriers for drug delivery. Chem. Soc. Rev. **42**, 1147–1235 (2013)

47. M. Elsabahy, K.L. Wooley, Design of polymeric nanoparticles for biomedical delivery applications. Chem. Soc. Rev. **41**, 2545–2561 (2012)
48. G.J.M. Habraken, M. Peeters, P.D. Thornton, C.E. Koning, A. Heise, Selective enzymatic degradation of self-assembled particles from amphiphilic block copolymers obtained by the combination of *N*-carboxyanhydride and nitroxide-mediated polymerization. Biomacromolecules **12**, 3761–3769 (2011)
49. S. Harrisson, J. Nicolas, A. Maksimenko, D.T. Bui, J. Mougin, P. Couvreur, Nanoparticles with in vivo anticancer activity from polymer prodrug amphiphiles prepared by living radical polymerization. Angew. Chem. Int. Ed. **52**, 1678–1682 (2013)
50. B. Grignard, T. Phan, D. Bertin, D. Gigmes, C. Jerome, C. Detrembleur, Dispersion nitroxide mediated polymerization of methyl methacrylate in supercritical carbon dioxide using *in situ* formed stabilizers. Polym. Chem. **1**, 837–840 (2010)
51. R. McHale, F. Aldabbagh, P.B. Zetterlun, M. Okubo, Nitroxide SG1-mediated and conventional radical precipitation polymerizations of styrene. Macromol. Chem. Phys. **208**, 1813–1822 (2007)
52. Ch. Magee, A. Earla, J. Petraitis, J. Petraitis, Ch. Higa, R. Braslau, P. Zetterlund, F. Aldabbagh, Synthesis of fluorinated alkoxyamines and alkoxyamine-initiated nitroxide-mediated precipitation polymerizations of styrene in supercritical carbon dioxide. Polym. Chem. **5**, 5725 (2014)
53. M. Aguirre, A. Simula, J.R. Leiza, S. van Es, J.M. Asua, Nitroxide mediated suspension polymerization of methacrylic monomers. Chem. Eng. J. **316**, 655–662 (2017)
54. J. Nicolas, B. Charleux, O. Guerret, S. Magnet, Nitroxide-mediated controlled free-radical emulsion polymerization using a difunctional water-soluble alkoxyamine initiator. Toward the control of particle size, particle size distribution, and the synthesis of triblock copolymers. Macromolecules **38**, 9963–9973 (2005)
55. M.J. Monteiro, M.F. Cunningham, Polymer nanoparticles via living radical polymerization in aqueous dispersions: design and applications. Macromolecules **45**(12), 4939–4957 (2012)
56. Z.J. Wang, K. Wylie, M. Marić, Synthesis of narrow molecular weight distribution copolymers for ArF photoresist materials by nitroxide mediated polymerization. Macromol. React. Eng. **11**(3), 1600029 (2017)
57. S. Sueeng, T. Boonchuwong, P. Chaiyasat, M. Okubo, A. Chaiyasat, Preparation of stable poly(methacrylic acid)-b-polystyrene emulsion by emulsifier-free emulsion iodine transfer polymerization (emulsion ITP) with self-assembly nucleation. Polymer **110**, 124–130 (2017)
58. X.G. Qiao, P.-Y. Dugas, B. Charleux, M. Lansalot, E. Bourgeat-Lami, Synthesis of multipod-like silica/polymer latex particles via nitroxide-mediated polymerization-induced self-assembly of amphiphilic block copolymers. Macromolecules **48**, 545–556 (2015)
59. E.H.H. Wong, M.H. Stenzel, T. Junkers, C. Barner-Kowollik, Spin capturing with "clickable" nitrones: generation of miktoarmed star polymers. Macromolecules **43**, 3785–3793 (2010)
60. A.J. Heidenreich, J.E. Puskas, Synthesis of arborescent (dendritic) polystyrenes via controlled inimer-type reversible addition-fragmentation chain transfer polymerization. J. Polym. Sci. A **46**, 7621–7627 (2008)
61. C. Detrembleur, A. Debuigne, O. Altintas, M. Conradi, E.H.H. Wong, C. Jerome, C. Barner-Kowollik, T. Junkers, Synthesis of star and H-shape polymers via a combination of cobalt-mediated radical polymerization and nitrone-mediated radical coupling reactions. Polym. Chem. **3**, 135–147 (2012)
62. W. Wei, A. Balamurugan, J.H. Dwyer, P. Gopalan, Substrate-independent approach to dense cleavable polymer brushes. ACS Macro Lett. **7**(1), 100–104 (2018)
63. W.-L. Chen, R. Cordero, H. Tran, C.K. Ober, 50th anniversary perspective: polymer brushes: novel surfaces for future materials. Macromolecules **50**, 4089–4113 (2017)
64. C.-Y. Lin, C.-H. Huang, C.-C. Hu, Y.-L. Liu, Self-crosslinkable nitroxide-functionalized poly(2,6-dimethyl-1,4-phenylene oxide) through atom transfer radical coupling reaction. Polymer **154–161**, 135 (2018)
65. G.G. Barclay, C.J. Hawker, H. Ito, A. Orellana, P.R.L. Malenfant, R.F. Sinta, Macromolecules **31**, 1024–1031 (1998)

66. Z.J. Wang, M. Maric, Nitroxide mediated synthesis of low dispersity random copolymers for low-loss optical waveguides. J. Polym. Sci. Polym. Chem. **51**, 2970–2978 (2013)
67. C.G. Gamys, J.-M. Schumers, C. Mugemana, C.-A. Fustin, J.-F. Gohy, Pore-functionalized nanoporous materials derived from block copolymers. Macromol. Rapid Commun. **34**, 962–982 (2013)
68. C. Mugemana, J.-F. Gohy, C.-A. Fustin, Functionalized nanoporous thin films from metallo-supramolecular diblock copolymers. Langmuir **28**, 3018–3023 (2012)
69. K. Satoh, J.E. Poelma, L.M. Campos, B. Stahl, C.J. Hawker, A facile synthesis of clickable and acid-cleavable PEO for acid-degradable block copolymers. Polym. Chem. **3**, 1890 (2012)
70. G. Leone, U. Giovanella, F. Bertini, S. Hoseinkhani, W. Porzio, G. Ricci, C. Botta, F. Galeotti, Hierarchically structured, blue-emitting polymer hybrids through surface-initiated nitroxide-mediated polymerization and water templated assembly. J. Mater. Chem. C **1**, 6585–659 (2013)
71. K.-I. Fukukawa, L. Zhu, P. Gopalan, M. Ueda, S. Yang, Synthesis and characterization of silicon-containing block copolymers from nitroxide-mediated living free radical polymerization. Macromolecules **38**, 263–270 (2005)
72. D. Quemener, G. Bonniol, T.N.T. Phan, D. Gigmes, D. Bertin, A. Deratani, Free-standing nanomaterials from block copolymer self-assembly. Macromolecules **43**, 5060–5065 (2010)
73. J. Ding, S. Holdcroft, Polymers of sodium styrenesulfonate prepared by one-pot TEMPO-controlled. J. Chem. **65**, 1117–1123 (2012)
74. B.H. Lessard, Novel materials: from nanoporous materials to micro-electronics. Polym. Chem. Ser. **19**(Nitroxide Mediated Polymerization), 441–493 (2016)
75. M.D. Green, D. Wang, S.T. Hemp, J.-H. Choi, K.I. Winey, J.R. Heflin, T.E. Long, Synthesis of imidazolium ABA triblock copolymers for electromechanical transducers. Polymer **53**, 3677–3686 (2012)
76. S.M. Lindner, S. Huttner, A. Chiche, M. Thelakkat, G. Krausch, Charge separation at self-assembled nanostructured bulk interface in block copolymers. Angew. Chem. Int. Ed. **45**, 3364–3368 (2006)
77. Q. Zhang, A. Cirpan, T.P. Russell, T. Emrick, Donor–acceptor poly(thiophene-block-perylene diimide) copolymers: synthesis and solar cell fabrication. Macromolecules **42**, 1079–1082 (2009)
78. A.J. Peltekoff, I. Therrien, B.H. Lessard, Nitroxide mediated polymerization of 1-(4-vinylbenzyl)-3-butylimidazolium ionic liquid containing homopolymers and methyl methacrylate copolymers. Can. J. Chem. Eng. **97**(1), 5–16 (2019)
79. O. Guselnikova, S.R.A. Marque, E.V. Tretyakov, D. Mares, V. Jerabek, G. Audran, J.-P. Joly, M. Trusova, V. Svorcik, O. Lyutakov, P. Postnikov, Unprecedented plasmon-induced nitroxide-mediated polymerization (PI-NMP): a method for preparation of functional surfaces. J. Mater. Chem. A **7**, 12414–12419 (2019)
80. X. Savelyeva, B. Lessard, Amphiphilic poly(4-acryloylmorpholine)/poly[2-(N-carbazolyl)ethyl acrylate] random and block copolymers synthesized by NMP. Macromol. React. Eng. **6**, 200–212 (2012)
81. W. Kwon, B. Ahn, D.M. Kim, Y.-G. Ko, S.G. Hahm, Y. Kim, H. Kim, M. Ree, Morphology-dependent electrical memory characteristics of a well-defined brush polymer bearing oxadiazole-based mesogens. J. Phys. Chem. C **115**, 19355–19363 (2011)
82. Y.-K. Fang, C.-L. Liu, G.-Y. Yang, P.-C. Chen, W.-C. Chen, New donor–Acceptor random copolymers with pendent triphenylamine and 1,3,4-oxadiazole for high-performance memory device applications. Macromolecules **44**, 2604–2612 (2011)
83. Gerard, L. Couvreur, S. Magnet, J. Ness, S. Schmidt, in *Controlled/Living Radical Polymerization: Progress in Raft, DT, NMP & OMRP*, vol. 1024, ed. by K. Matyjaszewsik (2009), pp. 361–373
84. C. Auschra, E. Eckstein, R. Knischka, Synthetic glycopolymers: an overview. Eur. Coat. J. **156**, 162–163 (2005)
85. V. Ladmiral, E. Melia, D.M. Haddleton, Synthetic glycopolymers: an overview. Eur. Polym. J. **40**, 431 (2004)

86. S.R.S. Ting, E.-H. Min, P. Escale, M. Save, L. Billon, M.H. Stenzel, Lectin recognizable biomaterials synthesized via nitroxide-mediated polymerization of a methacryloyl galactose monomer. Macromolecules **42**, 9422–9434 (2009)
87. F.M. Veronese, Peptide and protein PEGylation. A review of problems and solutions. Biomaterials **22**, 405–417 (2001)
88. M. Chenal, C. Boursier, Y. Guillaneuf, M. Taverna, P. Couvreur, J. Nicolas, First peptide/protein PEGylation with functional polymers designed by nitroxide-mediated polymerization. Polym. Chem. **2**, 1523–1530 (2011)
89. A. Karatzas, P. Bilalis, H. Iatrou, M. Pitsikalis, N. Hadjichristidis, Synthesis of well-defined functional macromolecular *chimeras* based on poly(ethylene oxide) or poly(N-vinyl pyrrolidone). React. Funct. Polym. **69**, 435–440 (2009)
90. S. Abraham, A. So, L.D. Unsworth, Poly(carboxybetaine methacrylamide)-modified nanoparticles: a model system for studying the effect of chain chemistry on film properties, adsorbed protein conformation, and clot formation kinetics. Biomacromolecules **12**, 3567–3580 (2011)
91. J. Nicolas, S. Brusseau, B. Charleux, A minimal amount of acrylonitrile turns the nitroxide-mediated polymerization of methyl methacrylate into an almost ideal controlled/living system. J. Polym. Sci. Part A: Polym. Chem. **48**, 34–47 (2010)
92. C. Detrembleur, C. Jerome, J. De Winter, P. Gerbaux, J.L. Clement, Y. Guillaneuf, D. Gigmes, Nitroxide mediated polymerization of methacrylates at moderate temperature. Polym. Chem. **5**, 335–340 (2014)
93. W. Huang, B. Charleux, R. Chiarelli, L. Marx, A. Rassat, J.-P. Vairon, Synthesis of water-soluble nitroxides and their use as mediators in aqueous-phase controlled radical polymerization Macromol. Chem. Phys. **203**, 1715 (2002)
94. X. Wang, H. Gao, X. Wang, H. Gao, Recent progress on hyperbranched polymers synthesized via radical-based self-condensing vinyl polymerization. Polymers **9**(6), 188 (2017)

Chapter 8
Nitroxides as Materials

8.1 Introduction

Paramagnetic nitroxides show a fairly robust nature and bidentate character, and therefore, have found various practical applications [1–14]. Metal-free paramagnetic soft materials bearing nitroxides exhibit a unique dynamic, magnetic, and electrical behavior induced by applied magnetic fields at high inhomogeneous intermolecular contacts due to the molecular motion and coherent collective properties of molecules in the soft phases temperatures. Advanced chemistry techniques permit structural modifications in order to fine-tune their physical. The synthesis and development of organometallic-based magnetic materials are of great interest to current research because of their possible technological applications in recording, quantum computing, molecular spintronics, and high-density data storage.

Starting from pioneering work of Gatteschi groups [13], complexes paramagnetic metals with the nitronyl nitroxide-based ligans have received much attention after being successfully implemented as potential building blocks for molecular magnetic materials. The pure organic nitroxide liquid crystals [3–7, 10] have great advantage for directly obtaining microscopic information on dipole–dipole interactions and exchange coupling of paramagnetic centers, p–p interactions, hydrogen bonding, the molecular orientation, relative molecular position, and intermolecular magnetic interactions in the LC phases by the use of electron paramagnetic resonance (EPR) spectroscopy, X-ray diffraction, neutron scattering, and optical techniques. Metal-free paramagnetic soft materials bearing nitroxides exhibit a unique dynamic magnetic and electrical behavior induced by applied magnetic fields at high inhomogeneous intermolecular contacts due to the swift molecular motion and coherent collective properties of molecules in the soft phases temperatures. Advanced chemistry techniques permit structural modifications in order to fine-tune their physical properties.

G. I. Likhtenshtein, *Nitroxides*, Springer Series in Materials Science 292,
https://doi.org/10.1007/978-3-030-34822-9_8

8.2 Nitroxide Bearing Liquid Crystals

4-Nitrophenylnitronyl nitroxide (NPNN) was the **first** pure organic ferromagnet [10]. From the measurements of magnetization of magnetic susceptibility and heat capacity, the γ-phase crystal of 2-(4′-nitrophenyl)-4,4,5,5-tetramethyl-4 5-dihydro-1*H*–imidazol-1-oxy-3-*N*-oxide was shown to become a bulk ferromagnet below about 0.65 K. These organic nitroxide radical crystals lost ferromagnetic behavior above the Curie–Weiss temperature T_c which did not exceeded 1.46 K.

Magneto-electric or magneto-optical properties in the nitroxide liquid crystal **(LC) state** afford the following possibilities [11–14]: (i) formation of magnetic domains in applied magnetic fields, (ii) occurrence of interactions between magnetic dipole and electric dipole moments (magneto-electric effects) in the ferroelectric LC state, (iii) realization of paramagnetic susceptibility anisotropy (Δpara)-controlled molecular orientation by weak magnetic fields, and (iv) providing ferroelectricity. The latter is a characteristic of certain materials that have a spontaneous electric polarization that can be reversed by the application of an external electric field. Liquid crystalline phases combining fluidity and anisotropy are considered to be non-equilibrium dynamic states due to the molecular motion and the coherent collective properties of molecules in the LC state. Accordingly, they are sensitive to external stimuli, such as heat, light, temperature, pressure, electric, or magnetic field, and added chiral dopants, etc.

The various complexes using 3d and rare earth/4f/5f-metal ions such as cobalt(II), nickel(II), manganese(II), copper(II), and so forth and nitronyl nitroxide radicals have been synthesized and extensively studied using inelastic neutron scattering, electron paramagnetic resonance (EPR) spectroscopy, and magnetization measurements [14].

Another area of application of nitroxide as potential materials could be magnetic organic chains containing nitroxide groups. As it was predicted in [15], these polymers may be a basis for a device which acts as effective spin filter and for other spintronic constructions. The organic paramagnetic compounds nitroxides also have a great potential as magnetic resonance imaging (MRI) contrast agents.

Tamura group [11], reported in 2014 the preparation and magnetic properties of the first example of the second generation of paramagnetic all-organic rod-like liquid crystal (LC) compounds. Specifically, the LC contains a chiral cyclic nitroxil (PROXYL) unit in the mesogen core and shows various chiral and achiral phases over a wide temperature range. The molecular orientation of the bulk liquid crystals generated ferroelectric properties induced by weak magnetic fields was confirmed by magnetic susceptibility measurements and EPR spectroscopy.

The next fundamental step in this area was made by the same group in [16] when first observation of an unusual intermolecular magnetic interaction in the liquid crystalline state of an all-organic radical compound at 73 °C on water was demonstrated. The magnetic interaction allows the LC droplet to move freely on water under the influence of an ordinary permanent magnet (Fig. 8.1); whereas, the crystalline phase did not respond to the magnet. Experimental results on the magnetic field dependence of magnetization measured on a SQUID magnetometer, and the temperature

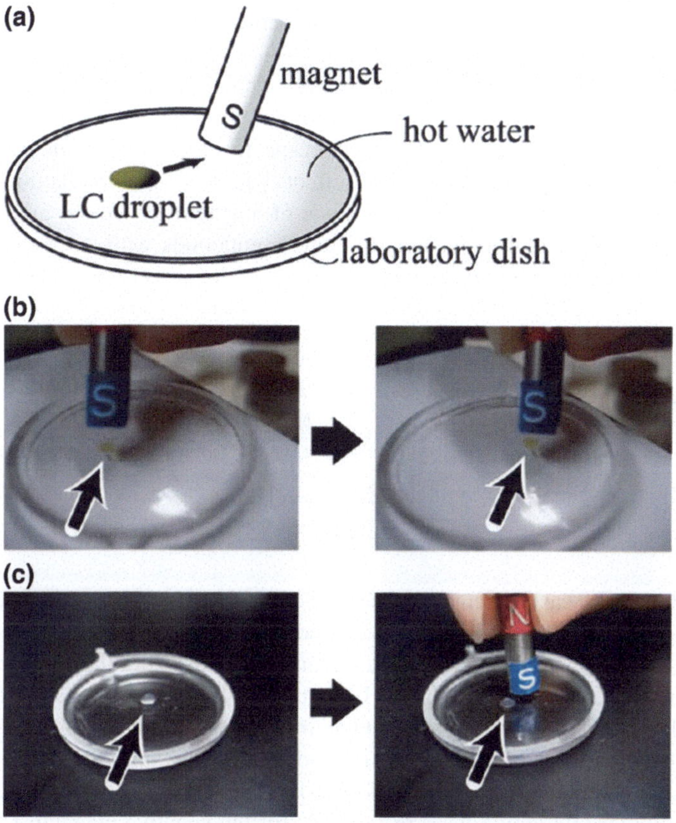

Fig. 8.1 Motion of LC droplets on water under the action of a permanent magnet (maximum 0.5 T). **a** Schematic representation of the experimental setup for observing the attraction by a permanent magnet of a paramagnetic LC droplet on water in a shallow laboratory dish. **b** Photographs showing the attraction of the yellow paramagnetic N droplet of -**1b** on water at 73 °C to the magnet. **c** Photographs showing the repulsion of the white diamagnetic N droplet of ZLI-1132 (Merck) on water at 25 °C from the magnet [16]. Reprinted from [16], Copyright 2012 American Chemical Society

dependence of g-value, and peak-to-peak line width (ΔH_{pp}) obtained by EPR spectroscopy showed the ferromagnetic effect which was discussed in terms of spin–spin dipole and exchange interactions [16].

This material exhibits of spin glass-like inhomogeneous ferromagnetic interactions (average spin–spin-exchange interaction constant $J > 0$) induced by weak magnetic fields in various chiral and achiral LC phases at high temperatures (30–150 1C). The origin of such unique magnetic interactions was referred to as positive 'magneto-LC effects. The synthesized second generation of chiral NR molecules satisfies the following four mandatory requirements having: (1) a nitroxyl group with a large electric dipole moment (ca. 3 Debye) and known principal g-values (g_{xx},

g_{yy}, g_{zz}), (2) high thermal stability, (3) the rigid core of LC molecules, and (4) both enantiomerically enriched and racemic samples.

In work [17], the generation of spin glass-like inhomogeneous magnetic interactions (the average spin–spin interaction constant: _$J > 0$ or $J < 0$) was referred to as positive or negative magneto-LC effects, respectively. To clarify the relationship between positive or negative magneto-LC effects with spin–spin interaction constant ($J > 0$ or _$J < 0$, respectively) and magnitude of magneto-LC effects and the types of rod-like LC phase and molecular structure, compounds with a terminal formyl group were synthesized and investigated.

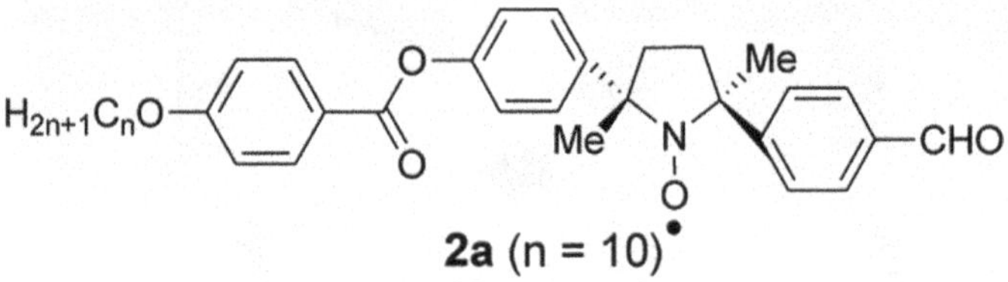

2a (n = 10)
2b (n = 18)

The following measurements were carried out: (i) the temperature dependence of molar magnetic susceptibility (χ_M) measured on a SQUID magnetometer and (ii) the temperature dependence of relative paramagnetic susceptibility (χ_{rel}), g value, and peak-to-peak line width (ΔH_{pp}) of EPR spectra. The origin of observed positive magneto-LC effects ($J > 0$) operating in the SmA* and N* phases was interpreted in terms of the generation of ferromagnetic head-to-tail spin–spin dipole interactions; whereas, antiferromagnetic interactions ($J > 0$) arising from the formation of the RS magnetic dipolar interaction were responsible for the negative magneto-LC effects in the N phase.

To elucidate the relationship between the ferroelectricity which is a characteristic of materials that have a spontaneous electric polarization and positive magneto-LC effects of (2*S*,5*S*)-1 composed of nitroxide, the electric field dependence of EPR spectra of (2*S*,5*S*)-1 confined in a surface-stabilized liquid-crystal cell was measured [18].

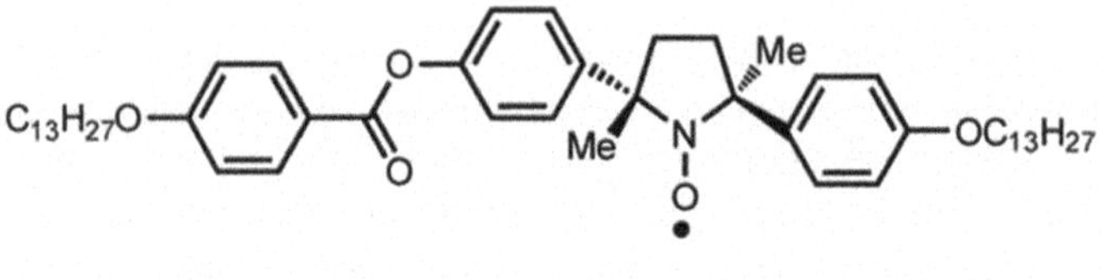

The influence of the electric fields on the positive magneto-LC effects in the FLC phase of nitroxide (2*S*,5*S*)-1 was evaluated by measuring the electric field dependence of *g*-value (*g*), paramagnetic susceptibility (χ_{para}), and the peak-to-peak line width (ΔH_{pp}) of the EPR spectra. By the application of electric field between +25 and −25 V, two magnetic bistable states resulted from the anisotropy in spin–spin dipole interactions were observed. Spin–spin dipole interactions between localized spins in the FLC phase were illustrated in a scheme.

To investigate the effects of the gelator on the positive magneto-LC effects observed in the nematic phase of (±)-**1** and the molecular alignment of (±)-**1** in the resulting nematic gels, liquid crystalline physical gels which consist of a chiral racemic nitroxide radical compound (±)-**1** indicating a nematic phase and a diamagnetic, chiral organic gelator (*R*,*R*)-**2** were prepared [4]. Experiments showed that (1) the increasing amount of the gelator decreased the domain size, (2) the molecular reorientation by a magnetic field (0.34 T) was considerably restricted in the nematic gel phase, and (3) the intermolecular magnetic interactions (positive magneto-LC effects) increased in the fine and random nematic polydomain structure formed by self-assembled fibers of organic gelator molecules.

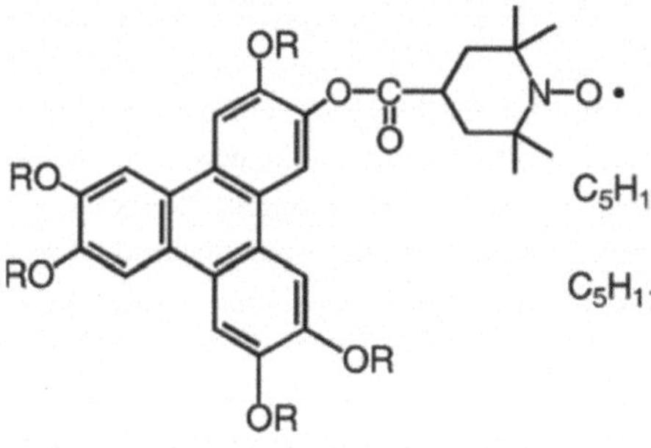

1a: R=C_5H_{11}, **1b**: R=C_6H_{13}

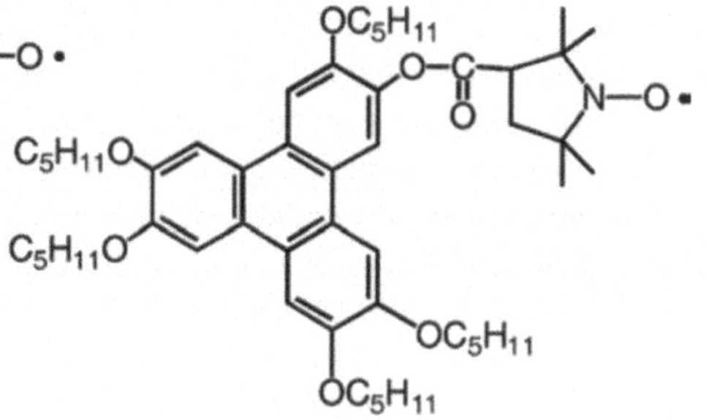

2

The magneto-responsive discotics formed by covalently linking triphenylene with TEMPO or PROXYL molecular radicals have been evaluated for the first time [19]. The phase transition behavior of the discotics was determined with the help of a polarizing optical microscope (POM), differential scanning calorimeter (DSC), and X-ray diffraction (XRD). The measurements in the fluid columnar, frozen columnar, and solid states using a SQUID susceptometer in the temperature range 2–300 K revealed the antiferromagnetic intermolecular interactions of Curie–Weiss behavior.

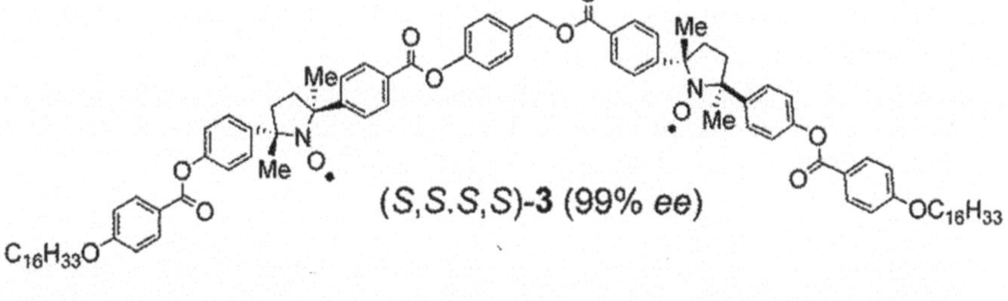

(*S*,*S*.*S*,*S*)-**3** (99% *ee*)

The preparation of liquid crystalline chiral biradical nitroxide (*S*,*S*,*S*,*S*)-3 as the first all-organic biradical LC compound which showed large magneto-LC effects in the chiral nematic (N*) phase was performed [20]. The (*S*,*S*,*S*,*S*)-3 properties and phase transition behavior of (*S*,*S*,*S*,*S*)-3 were characterized by EPR spectroscopic analysis, χ_M measurement, differential scanning calorimetry, polarized optical microscopy, and X-ray diffraction analysis. It was suggested that the occurrence of large positive magneto-LC effects originates from increased intermolecular spin–spin exchange and dipole interactions due to inhomogeneous intermolecular short contacts in the biradical N* phase. The biradical (*S*,*S*,*S*,*S*)-3 has shown much larger positive magneto-LC effects in the chiral nematic (N*) phase than the monoradical (*S*,*S*)-1. Scheme illustrated molecular structure of (*S*,*S*,*S*,*S*)-3 and LC superstructure in the N* phase.

The observation of a unique superparamagnetic-like behavior and a large positive magneto-LC effect in the nanocrystalline solid phases and the hexagonal columnar (Colh) and smectic (Sm) LC phases, respectively, of non-π-delocalized nitroxide diradical compounds **1** was reported [21]. The SQUID magnetization measurement revealed that (*R*,*S*)-**1** containing a small amount of racemic diastereomers (*R**,*R**)-**1** possessed an unusual and large temperature-independent magnetic susceptibility ($\chi_{TIM} > 0$) component in the original nanocrystalline solid **1** that was responsible for the observed superparamagnetic-like behavior under low magnetic fields.

The mesonitroxide diradical (*R*,*S*)-**1** was prepared by the condensation of the dicarboxylic acid (*R*,*S*)-**2**, which was derived from the bisnitrone, and two equivalents of 2-[3,4,5-tris(hexadecyloxy)phenyl]ethanamine (**5**) using 4-(4,6-dimethoxy-1,3,5-triazin-2-yl)-4-methylmorpholinium chloride (DMT-MM), and a catalytic amount of *N*-methylmorpholine (NMM) in dichloromethane [21]. The following three new important findings that elucidate the origin of the superparamagnetic-like behavior and positive magneto-LC effect under low magnetic fields were observed: (1) The achiral mesoorganic diradical (*R*,*S*)-1 with an intramolecular spin–spin-exchange interaction at low temperatures showed a unique phase transition behavior. (2) By SQUID magnetization measurement, it was revealed that (*R*,*S*)-1 containing small amount of racemic diastereomers (*R**,*R**)-1 possessed a large temperature-independent magnetic susceptibility ($\chi_{TIM} > 0$) component in the original nanocrystalline solid which was responsible for the observed superparamagnetic-like behavior. (3) Such unique magnetic phenomena were substantially induced by thermal processing for (*R*,*S*)-1 or by adding a small amount of racemic diastereomers (*R**,*R**)-1 to (*R*,*S*)-1 as the impurity. The origin of aforementioned magnetic phenomenon in terms of the formation and enlargement of magnetically inhomogeneous domains in the solids and LC phases was discussed.

The aim of work [22] was to obtain the detailed structural information of magnetic LC phases composed of nitroxide from analysis and the shape and angular dependence of EPR spectra of paramagnetic LC materials.

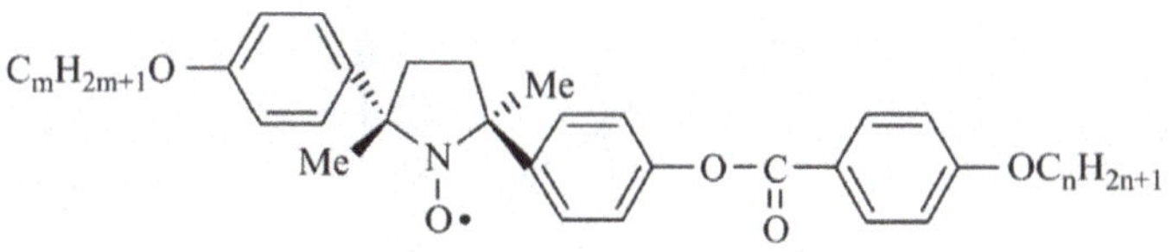

1: m = 7, n = 8 **2**: m = n = 13 **3**: m=n=4

The analysis carried out using the density functional theory calculations of spin density distribution in the interacting molecules based on the crystal structure (Fig. 8.2) and allowed to reveal the origin of the positive magneto-LC effects. The obtained structural data indicated that spin polarization interactions between neighboring molecules rather than the direct through space interactions between paramagnetic centers are responsible for the specific magnetic properties of the studied LC materials. Spin polarization is the degree to which the spin is aligned with a given direction. Figure 8.3 demonstrates the mutual location of LC molecules in layer of SmC phase along C-director neighboring molecules in two orthogonal planes.

The factors that control the orientation of guest molecules in anisotropic LC media such as the nematic 4-*n*-pentyl-4′-cyanobiphenyl (5CB) and smectic 4-*n*-octyl-4′-cyanobiphenyl (8CB) liquid crystals were evaluated [23]. For this purpose, TEMPOL and novel nitroxide spin radicals with different molecular shapes were applied. The orientation distributions of the radicals in the aligned samples in a frozen state were determined in detail by means of EPR spectroscopy. A numerical simulation of the EPR spectra recorded at 105 K was used for the determination of the anisotropic magnetic parameters. The computation of the molecular geometry

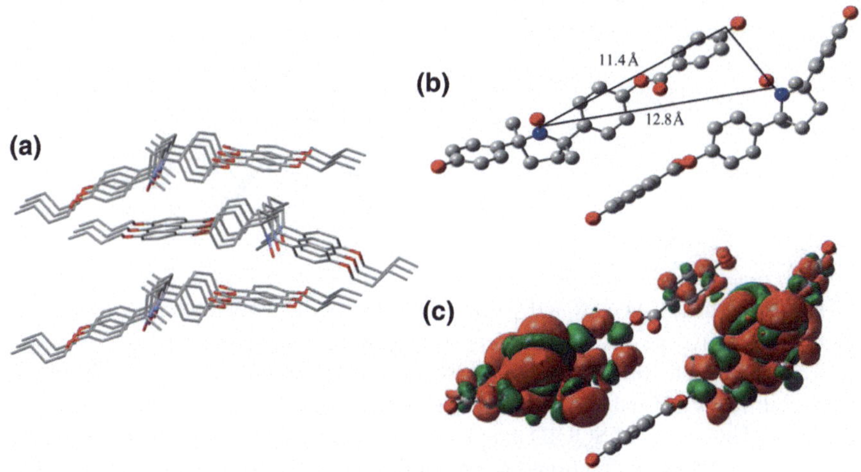

Fig. 8.2 Crystal structure of (*S*,*S*)-3 determined by X-ray crystallographic analysis (**a**), relative position of a pair of molecules in the crystal (**b**), and spin density distribution in the same pair by quantum chemical calculations (**c**). The hydrocarbon chains are omitted for clarity in panels **b** and **c** [22]. Reprinted from [22], Copyright 2014 American Chemical Society

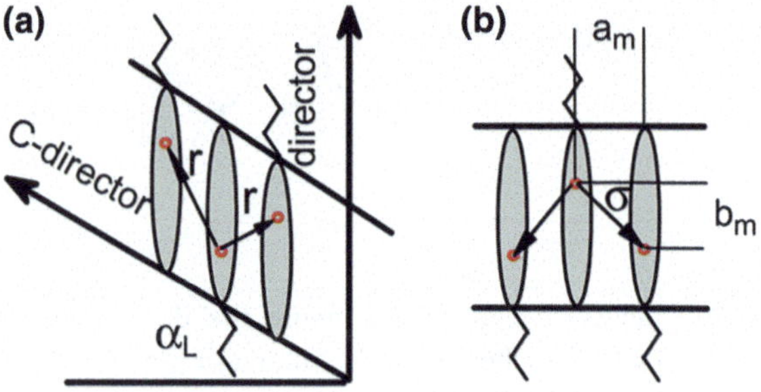

Fig. 8.3 Mutual location of LC molecules in layer of SmC phase along C-director (**a**) and in perpendicular direction (**b**). Circles denote the positions of paramagnetic nitroxide groups [22]. Reprinted from [22], Copyright 2014 American Chemical Society

of the radicals under consideration was performed with help of the ORCA software package and computational model B3LYP/N07D using optimization. Orientation distribution functions of the radicals in aligned liquid crystals were determined by the joint numerical simulation of set of 10–20 EPR spectra recorded at different angles (θ, φ) between the director of the sample and the magnetic field vector. Calculated orientation distribution functions of various radicals are shown in Fig. 8.3.

The rigid fused phenanthrene-based (A5) and 2-azaphenalene (A4) nitroxides as well as the rigid core elongated C11 and 5α-cholestane (CLS) nitroxides were found to be most sensitive to the orientation of the liquid crystal matrixes. The authors concluded that the nitroxide radicals which are the most suitable for characterization of the order of liquid crystals can be the more rigid fused systems with limited degrees of intramolecular motions and with prospects for π-stacking with the LC media (Figs. 8.4 and 8.5).

The synthesis of chiral all-organic NR compounds analogs of showing disordered Colh phases, which have a new type of mesogen core including intermolecular N–O···H–O H-bonds, was performed [3].

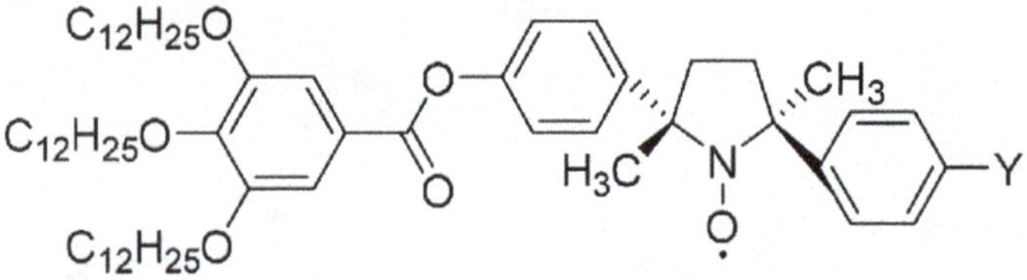

The Colh phases were found to be stable at room temperature for both (±)-1 and (2*R*,5*R*)-1. The LC compounds form hexagonal columnar phases at room temperature, which solidify as LC glasses at low temperature. Data on magnetic susceptibilities suggested that molecular mobility is one of the origins of the magneto-LC effects. The results of SQUID magnetometry and EPR spectroscopy implied that the spin–spin dipolar interactions are one of the origins of the positive magneto-LC effect.

For targeted drug delivery systems visible by magnetic resonance imaging, robust metal-free magnetic nanoemulsions (mean particle size less than 20 nm) consisting of

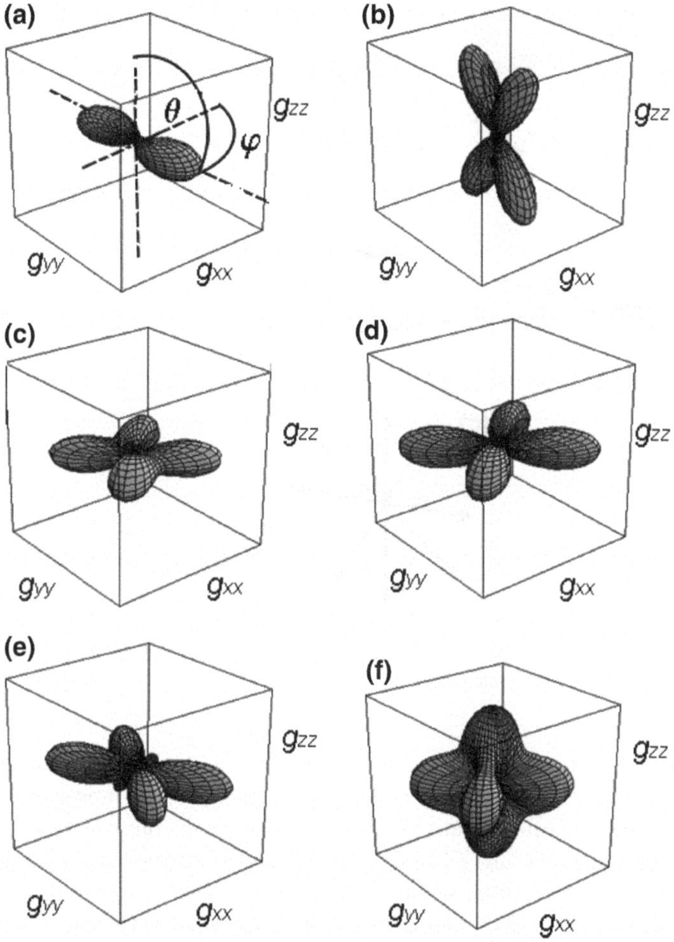

Fig. 8.4 Orientation distribution functions of radicals A4 (**a**) C11 (**b**), A3 (**c**), A6 (**d**), A2 (**e**), and A1 (**f**) in aligned liquid crystal 5CB. The distributions are presented in reference frames of the radicals [23]. Reprinted from [23], Copyright 2014 American Chemical Society

a biocompatible non-ionic surfactant polyethylene glycol hexadecyl and hydrophobic and low molecular weight 2,2,5-trimethyl-5-(4-alkoxy)phenylpyrrolidine-*N*-oxyl radicals were prepared [24]. The structure of the nanoemulsions was characterized by ESR spectroscopy, dynamic scattering, and small-angle neutron-scattering measurements. The obtained nanoemulsions possess: (1) high colloidal stability, (2) low cytotoxicity, (3) reduction resistance to excess ascorbic acid, and (4) sufficient contrast enhancement in the proton longitudinal relaxation time (T_1) in vitro and in vivo. Additional hydrophobic drugs and anticancer drug paclitaxel were incorporated into the nanoemulsions. A preliminary MRI experiment was performed with living mouse

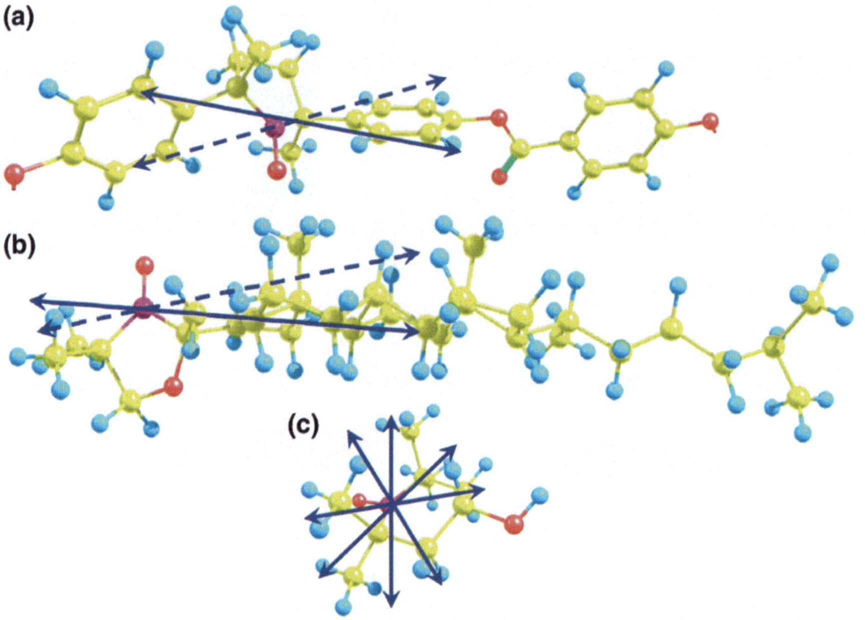

Fig. 8.5 Possible orientation axes for radicals C11 (**a**), CLS (**b**), and TEMPOL (**c**) [23]. Reprinted from [23], Copyright 2014 American Chemical Society

brain tissues. A distinct MRI contrast enhancement with high reproducibility was observed in the pituitary gland after the drugs injection to living mouse brain tissues.

Novel achiral non-p-delocalized nitroxide diradical compounds (*R*,*S*)-1 and racemic diastereomers (*R**,*R**)-1 were prepared, and its magnetic behavior was characterized by EPR, differential scanning calorimetry, polarized optical microscopy, and X-ray diffraction (XRD) and small-angle X-ray scattering analyses [25]. The SQUID magnetization measurement revealed that (*R*,*S*)-1 containing a small amount of racemic diastereomers (*R**,*R**)-1 demonstrated an unusual magnetic property in the original nanocrystalline solid in the nanocrystalline solid phases and the hexagonal columnar Colh LC phase, namely possess a large temperature-independent magnetic susceptibility component ($c_{TIM} > 0$) in the original solid I before heating. The component is responsible for the observed superparamagnetic-like M-H behavior under low magnetic fields. In addition, the experiment showed a large increase in the molar magnetic susceptibility (cM) (positive magneto-LC effect) at the solid I-to-Colh transition in the first heating run at 0.05 T. The next step of the research was molecular dynamic simulations on the mechanism of a large increase in the magnetic susceptibility in the solid and LC phases. The influence of the coherent collective molecular motion and inhomogeneous intermolecular short contacts on the magnetic susceptibility, together with the in-depth data concerning the spin glass-like properties was taken in consideration. The authors summarized that strong intermolecular spin–spin dipole as well as exchange interactions should operate in the magnetically inhomogeneous domains formed in the Colh phase to result in the χ_M and χ_{TIM} increase.

8.3 Nitroxide Crystal Structure Materials

Mixtures of new synthesized 2-(4,5,6,7-tetrafluorobenzimidazol-2-yl)-4,4,5,5-tetramethyl-4,5-dihydro-1*H*-imidazole-3-oxide-1-oxyl (F4BImNN) and 2-(benzimidazol-2-yl)-4,4,5,5-tetramethyl-4,5-dihydro-1*H*-imidazole-3-oxide-1-oxyl (BImNN) were crystallized [26]. The following results were reported: (1) Solid solutions (alloys) $(F4BImNN)_x(BImNN)_{(1-x)}$ with $x < 0.8$ gave orthorhombic unit cells, while $x \geq 0.9$ gives monoclinic unit cells. (2) In one-dimensional hydrogen-bonded chains, the dominant intermolecular packing leads to strong ferromagnetic exchange along the chains (J/k = 12–22 K). (3) For the various compositions, interchain exchange was estimated to be 33- to 150-fold weaker. An antiferromagnetic ordered phase formation (Fig. 8.6) Néel temperatures in the 0.4–1.2 K range was evaluated.

An organic biphenyl-4,4′-bis(nitronyl nitroxide) biradical with intermediately strong antiferromagnetic interactions was synthesized as a potential building block for the design of new quantum magnets [27]. In coupled $S = {}^1/_2$ dimer compounds with the antiferromagnetic intradimer coupling constant $J/k_B = -14.0$, short contacts between the oxygen atoms of the nitronyl nitroxide units and the hydrogen atoms of the benzene rings stabilize a planar geometry of the biphenyl spacer. This effect is probably responsible for a small magnetic interdimer coupling. In crystal state, the deviations from the isolated-dimer model are attributed to a small interdimer coupling J'/k_B, in the order of 1 K.

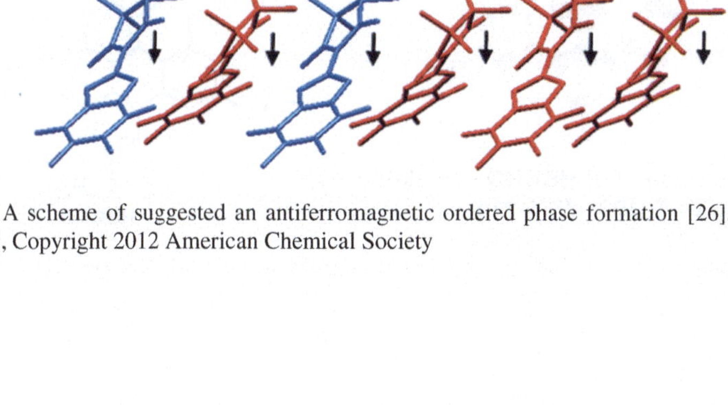

Fig. 8.6 A scheme of suggested an antiferromagnetic ordered phase formation [26]. Reprinted from [26], Copyright 2012 American Chemical Society

The first neutral paramagnetic hexabenzocoronene (HBC) derivative possessing a tert-butyl nitroxide radical moiety was prepared [28] (Fig. 8.7). As was probed by differential scanning calorimetry and ESR spectroscopy, HBC with five alkyl chains exhibits a positive magneto-LC effect in columnar hexagonal liquid crystalline phase. HBC derivative carrying a conjugated tertbutyl nitroxide radical moiety (HBCNO) was found to exist in a helical hexagonal phase at low temperature and a columnar hexagonal phase at higher temperatures. An obtained significant quenching of fluorescence indicated extensive energy or electron transfer on excitation of HBCNO. The electrochemical properties of HBCNO were investigated by cyclic voltammetric

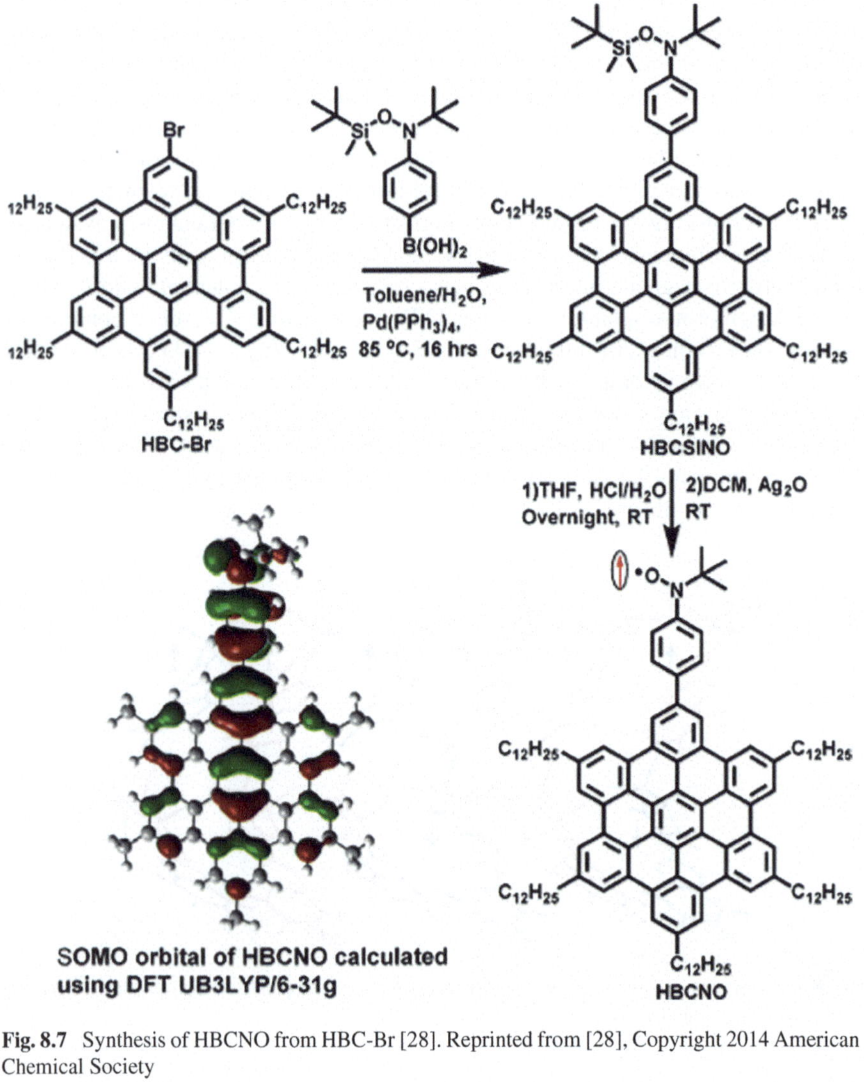

Fig. 8.7 Synthesis of HBCNO from HBC-Br [28]. Reprinted from [28], Copyright 2014 American Chemical Society

(CV) measurements. CV scans of HBCNO displayed a reversible oxidation (0.269 V vs. Fc/Fc^{+}) and a non-reversible reduction wave. The change in magnetic interaction during phase transition, variable temperature EPR spectra of a powder sample was measured during the cooling process in the temperature range 380 K (Colh phase) to 270 K (Helh phase). Closer π-stacking of disks in the Colh phase can provide stronger intracolumnar magnetic interactions.

The influence of the p-bridges on the intra- and intermolecular exchange interactions in conjugated biradicals was studied [29].

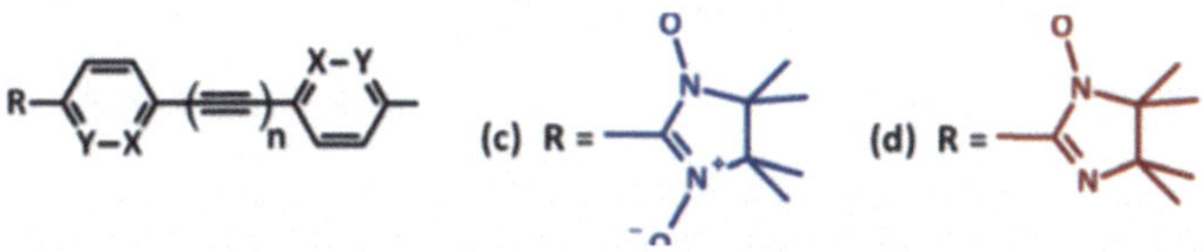

X-ray structural data and performed DFT calculations suggested that the torsion angles have crucial impacts on the overall p-conjugation in the nitroxide biradical systems, which is responsible for the efficient communication between the nitroxide fragments. Magnetic measurements, carried out on single-crystalline samples, showed that for investigated biradicals, weak antiferromagnetic intramolecular interactions are predominant and π-bridged nitroxides possess a moderate intradimer the intramolecular exchange constant J_{intra} in the range −2 to −6 K.

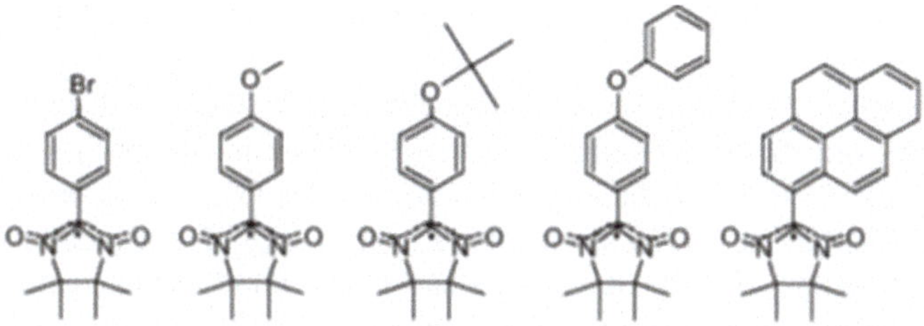

Investigation of the conduction mechanisms of nitronyl nitroxides NIT-R using structural analysis, transport measurements, low-energy, subTHz spectroscopy, and theoretical modeling was carried out [2]. Additional information on the mechanism was obtained in the analysis of data on introducing differently substituted phenyl appendages. Crystal structure of radical NIT-PhOMe, with the cylinders, highlights the presence of overlap between the electronic clouds of aromatic appendages, and the NIT moiety was established. For all substituents NIT-R attached to the NIT-R radical transport parameters, namely minimum mobility μ_{min}, surface thermal carrier density nth, sheet resistance R_S, crossover voltage V_c, Ohmic conductance and fitting exponent (G_Ω and η_Ω), charge-limited parameters G_{SCL}, η_{SCL}, and low-field mobility μ_0 for the different NIT-R radicals were measured. Below threshold voltage V_c between the Ohmic regime and the nonlinear regime, the generated free carriers inside density of the crystal were dominant with respect to the injected charge carriers. In such condition, the behavior is given by Ohm's law. Obtained findings

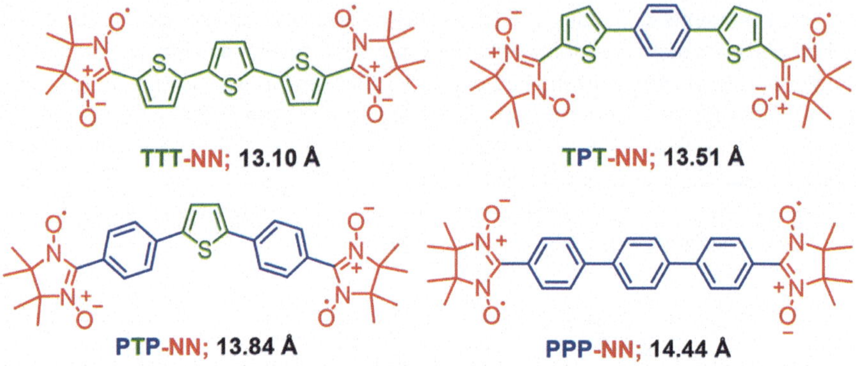

Fig. 8.8 Structures of TTT-NN, TPT-NN, PTP-NN, and PPP-NN; distance between C2–C2 of NN [30]. Reprinted from [30], Copyright 2017 American Chemical Society

show that a non-trivial surface charge-limited regime is present in addition to the standard low-voltage Ohmic conductance. Schematic representation of the different conduction regimes and the different types of traps available for NIT-R systems was presented.

Four nitronyl nitroxide (NN) biradicals (Fig. 8.8) were synthesized, and their structure, molecular packing, magnetic susceptibility, and spin state were investigated using a combined approach involving by X-ray diffraction studies, SQUID measurements, UV–Vis spectroscopy, EPR, and quantum chemical DFT calculations [30].

As it is displayed in Fig. 8.9 in the slipped π-stacked, the π–π distance between two TPT-NNs with the edge-to-edge (between thiophene and thiophene) approach of the molecules is 3.5 Å. The molar magnetic susceptibility (χ_{mol}) of the polycrystalline sample for TTT-NN and TPT-NN recorded using a SQUID magnetometer in the temperature range $K \leq T \leq 300$ K revealed that (1) The intradimer magnetic exchange coupling constant J_{intra} between two $S = 1/2$ spins are $2\,J/k_{\text{B}} = -6.2$ K for TPT-NN and $J/k_{\text{B}} = -11.9$ K for TTTNN. (2) At room temperature, the magnetic moments are close to the theoretical value 2.45 μB for magnetically uncorrelated spins of biradicals, and for TPT-NN biradical. (3) EPR studies revealed that ΔMS = 1 and ΔMS = 2 transitions at 130 K.

8.4 Nitroxides in Electric Batteries

Recent progress in the chemistry of charge transport by non-conjugated polymers with ultimate density of redox sites has demonstrated that these polymers are promising as electroactive materials for various electronic devices such as rechargeable batteries, hybrid capacitors, solar cells, electrochromic cells, sensors, and memory devices. An organic nitroxide polymers battery being notable for its flexibility, rapid

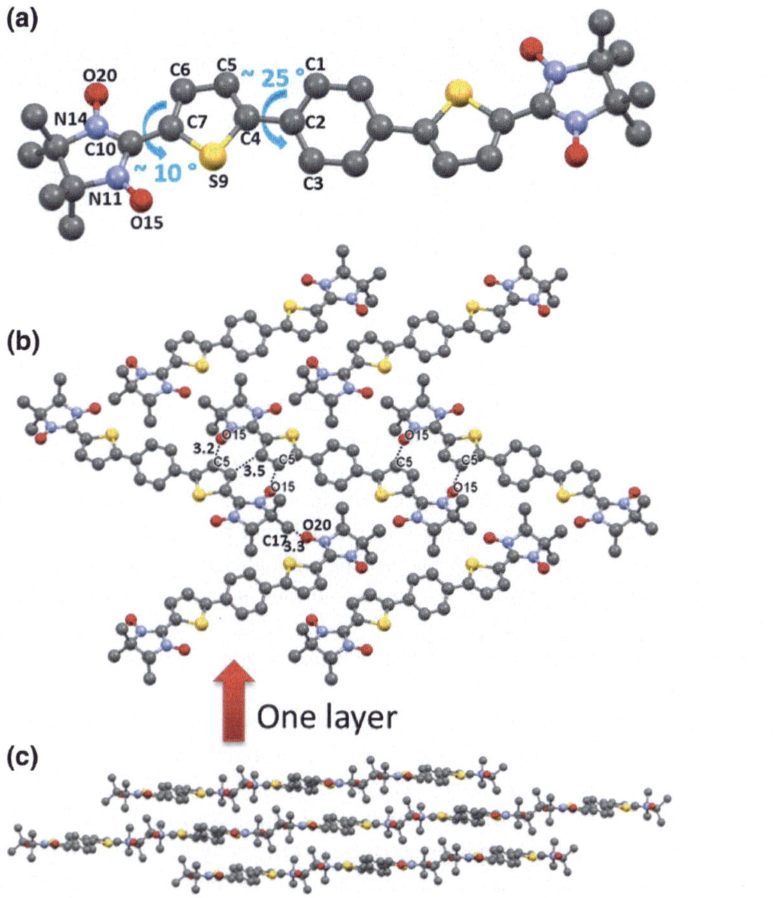

Fig. 8.9 X-ray crystal structure of the TPT-NN, **a** molecule, **b** edge-to-edge packing mode, and **c** face-to-face mode of crystal packing; hydrogen atoms are omitted for clarity [30]. Reprinted from [30], Copyright 2017 American Chemical Society

charging, and small size is candidates for replacing lithium transition metal oxide [2, 31–41]. The p-type reaction of nitroxyl radical polymers is closest to a practical alternative to lithium transition metal oxides [33]. An organic nitroxide radical battery (ORB) was first developed in 2004 [34]. The highlight review describes the performance of organic radical batteries using a nitroxyl radical polymer as the cathode-active material in publications before 2011 [35]. Below are some typical examples in this area.

Employing nitronyl nitroxide pendant, which undergoes both p- and n-type (i.e., negative) charging, potential capability of playing a dual role as cathode- and anode-active material in corresponding batteries was evaluated [35]. As an example, electrode-attached layer of poly(phenylacetylene) bearing a pendant nitronyl nitroxide group per repeating unit, obtained by the Rh-catalyzed polymerization of

2-(4-ethynylphenyl)-4,4,5,5-tetramethylimidazoline-l-oxyl 3-oxide, underwent oxidation and reduction at 0.80 and 0.84 V versus. Ag/AgCl. Galvanostatic coulometry revealed the plateau voltages for both p- and n-type charging. Galvanostatic Coulomb titration confirmed the charge-storage capability of the polymer. This finding supports suggestion that the radical survived during the course of the polymerization to allow both positive and negative charging of the pristine neutral polymer substantially per repeating unit. Electrochemical reversible bipolar charging and discharging processes in **the** dual dopable poly(phenylacetylene) with nitronyl nitroxide pendants were illustrated in a scheme.

Nitroxide polymer brushes with the poly(2,2,6,6-tetramethylpiperidin-4-yl methacrylate) (PTMPM) segment for organic radical batteries were synthesized via surface-initiated atom transfer radical polymerization (SI-ATRP) [36]. The following results were reported: (1) Patterned nitroxide polymer brush thin-film electrodes of the voltammetry and AC impedance and X-ray photoelectron spectroscopy results showed that an increase in the oxidation time could oxidize the PTMPM segment at the bottom of the brush. (2) This results in overoxidation of the brush at the top, which decreases the energy capacity of the polymer brush. (3) The energy capacity of the polymer brush electrode for organic radical batteries was determined to be approximately 94.0 mA h g^{-1} at a discharge rate of 20 C. (4) Its cycle-life performance exhibits 97.3% retention after 100 cycles. (5) Atomic force microscopy results indicated, in this condition, the polymer brush resists dissolution of polymers into electrolytes.

Poly(3-hexylthiophene)/ZnO (P3HT/ZnO) hybrid film was prepared from by in situ hydrolysis of a cross-linked block copolymer poly(3-hexylthiophene)-b-poly(zinc dimethacrylate) (P3HT-b-PZn$(MA)_2$) [37]. The device based on cross-linked P3HT/ZnO hybrid film obtained by in situ hydrolyzing P3HT-b-PZn$(MA)_2$ block copolymer yielded a power conversion efficiency of 0.45% under AM 1.5 G illumination from a calibrated solar simulator with an intensity of 100 mW/cm^2. Current density–voltage characteristics of solar cells under AM 1.5G irradiation with irradiation intensity of 100 mW/cm^2 were also determined. Normalized power conversion efficient of photovoltaic cells based on cross-linked P3HT/ZnO hybrid films obtained by hydrolyzing P3HT-b-PZn(MA)2 block copolymer and non-cross-linked P3HT/ZnO hybrid films obtained by hydrolyzing P3HT-TIPNO/Zn(MA)2 blend film during long-term exposure to air was illustrated in a figure.

Ambipolar redox-active polymers with a reversible charging and discharging capability were synthesized via ring-opening metathesis polymerization of nitronyl nitroxide radical (NN) mono- and disubstituted norbornenes [38]. The NN polymer/carbon composite electrode exhibited both p- and n-type charging/discharging with plateau potentials near the redox potentials of the polymer at 0.78 and −0.80 V versus Ag/AgCl, respectively. The spin-coated layer electrode of the NN polymer immobilized on a current collector demonstrated a fast charging/discharging performance in the range of 10–100 C rates and a cycle stability especially for the p-type reaction.

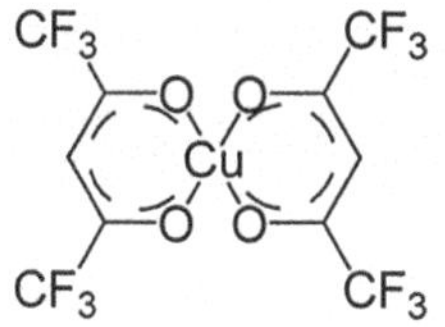

$Cu(hfac)_2$

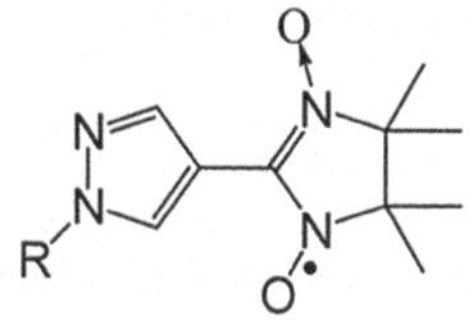

L^R

An affordable, safe, and scalable redox-flow batteries were fabricated [39]. The battery was characterized by using organic polymers as the charge-storage material in combination with dialysis membranes. The membranes separate the anode and the cathode by the retention of the non-metallic, active species, and an aqueous sodium chloride solution as the electrolyte. The water- and polymer-based RFB have an energy density of 10 W hours per liter, current densities of up to 100 milliamperes per square centimeter, and stable long-term cycling capability.

An organic molecule possessing two bipolar redox-active nitronyl nitroxide units connected via a tetraethylene glycol chain was synthesized and used as symmetric redox-flow battery (RFB) soluble in organic solvents [40]. In this battery, NN units are employed as a bipolar redox-active charge-storage material. The battery exhibits the following properties: (1) quasi-reversible redox reactions of the NN^+/NN redox couple at $E_{½} = 0.37$ V and the NN/NN^- redox couple at $E_{½} = -1.25$ V versus $AgNO_3/Ag$, (2) cell voltage of 1.62 V, (3) a stable charge/discharge performance over 75 consecutive cycles with a high-energy efficiency of 82% and an overall energy density of the electrolyte system of 0.67 W h l^{-1}, (4) an overall energy density of the electrolyte system of 4.1 W h l^{-1}, and (5) an energy efficiency of 79%.

A novel approach for the immobilization of polymer poly(2,2,6,6,-tetramethylpiperidinyl-1-oxyl methacrylate (PTMA) on reduced graphene oxide (rGO) via π–π-stacking of pyrene-functionalized PTMA-chains, forming a clayered composite material with well-dispersed PTMA among the carbon layers (Figs. 8.10 and 8.11) was developed [41]. The results in P(TMA-*co*-PyMA) random copolymers with near quantitative amounts of pyrene along the PTMA chain were suitable for greater π–π interaction with graphene oxide (rGO), while the nitroxide radicals on the polymer could simultaneously be used for energy storage. Attaching PTMA to surfaces reduced capacity loss due to dissolution of the organic polymer into the electrolyte. Faradaic reactions and a cooperative effect in the rGO-g-PTMA/rGO electrodes were evaluated.

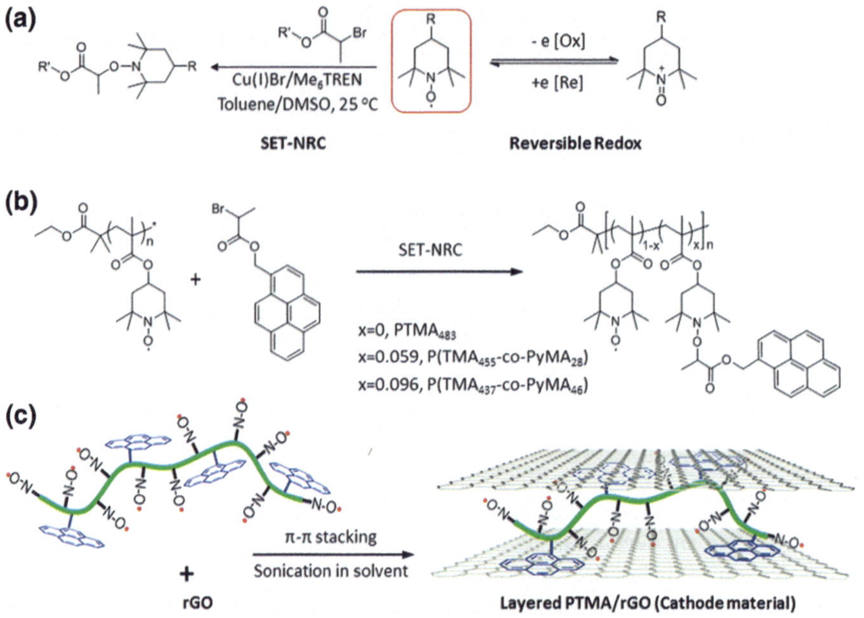

Fig. 8.10 Immobilization strategies of PTMA on rGO preventing restacking of the individual sheets. **a** π–π-Interactions between rGO and pyrenefunctionalized PTMA. **b** Surface-grafted PTMA [41]. Reprinted from [41], Copyright 2017 American Chemical Society

In study [42], ionic liquids (IL) as electrolytes with the view of increasing redox potentials of nitroxide radicals were used. A series of imidazolium, phosphonium, and pyrrolidinium-based ionic liquids (IL) coupled with widely used anions was chosen to predict redox potentials of TEMPO radical using state-of-the-art quantum chemical calculations. Some ILs showed a significant increase in the redox potential of this radical to reach as much as 5.5 eV, compared to the previously measured value of 2.2 eV in aqueous media. In particular, ILs were shown to stabilize the aminoxy anion.

Robust radical-substituted polymers PTGE with effective redox capability were used as surfaces for liquid crystal orientation [43]. The alignment of the smectic liquid crystal electrolytes with low-dimensional ion conduction pathways was reversible and switched in response to the redox states of the polymers enabling both a rapid cell response and long charge retention. PTGE cathode triggers realignment of the self-assembled liquid crystal from planar in the discharged state to homeotropic in the fully charged state (Fig. 8.12).

A novel styrenic nitroxide polymer, poly(5-vinyl-1,1,3,3-tetramethylisoindolin-2-yloxyl) (PVTMIO) was synthesized and employed as an effective organic p-dopable cathode materials [44]. Cyclic voltammetry revealed a high oxidation potential of 3.7 V versus Li. The suitability of PVTMIO for utilization in a high-voltage organic radical battery was confirmed with a discharge capacity of 104.7 mAh g^{-1}, high rate

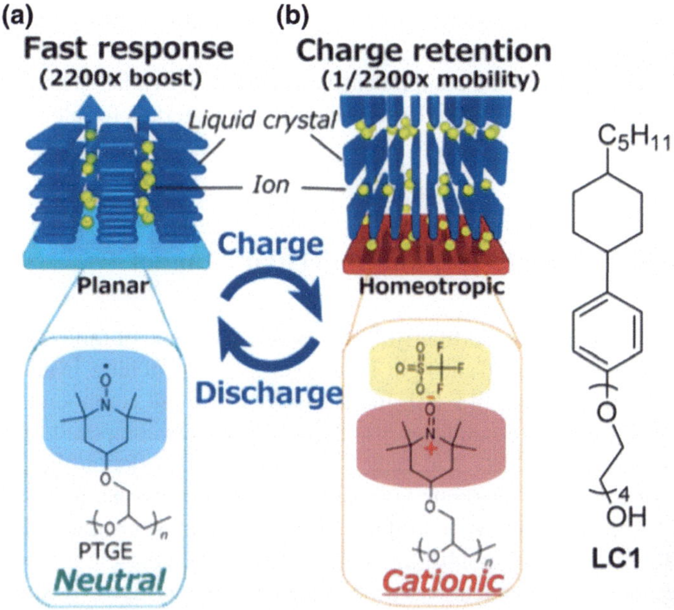

Fig. 8.11 An illustration of the radical "command surface" triggering a redox-responsive switching of the smectic liquid crystal electrolyte [43]. Reprinted from [43], Copyright 2017 American Chemical Society

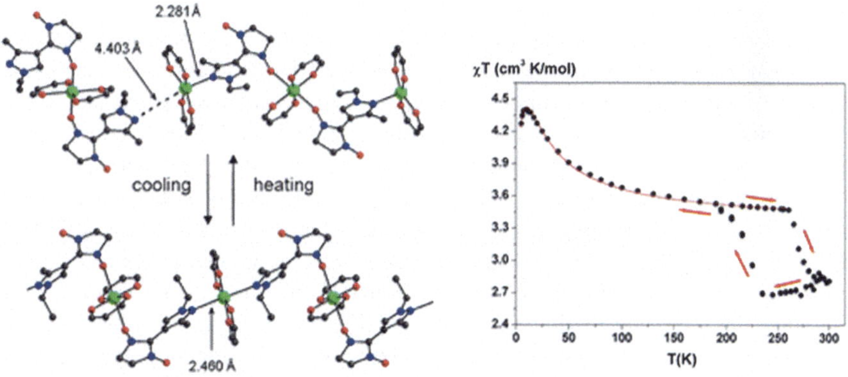

Fig. 8.12 Dependence $\chi T(T)$ for $[[Cu(hfac)_2]_3(\mu\text{-}O,N\text{-}L^a)_2][Cu(hfac)_2(O\text{-}L^a)_2]$ (data points represent experimental values; solid line denote the theoretical curve). The arrows indicate the direction of variation of χT during the cooling and subsequent heating of the sample [46]. Reprinted from [45], Copyright 2012 American Chemical Society

performance, and stability under cycling conditions (90% capacity retention after 100 cycles).

Recent review presented a comprehensive summary of synthetic strategies available for the preparation of nitroxide radical polymer materials for electric batteries [33].

8.5 Nitroxide Single-Crystal Molecular Magnet

Polymer chain complexes Cu(hfac)$_2$L^R represent a new type of molecular magnets exhibiting thermally induced and light-induced magnetic switching, in many respects similar to a spin crossover [45–53]. In the majority of these compounds, the polymer chain consists of alternating one- and three-spin units composed of copper(II) ions and nitronyl nitroxides. A review [45] highlighted works on the capabilities of EPR in investigating magnetic interactions breathing crystals published before 2017. The "breathing crystals" Cu(hfac)2LR (where hfac is hexafluoroacetylacetonate, and LR is a pyrazolyl-substituted nitroxideligand) represent a new family of switchable molecular magnets based on copper(II) ions bridged by stable nitroxide radicals (Fig. 8.1) [6–14]. In many compounds, crystals also include solvent molecules (Solv) located in the interchain space (Cu(hfac)2LR·0.5Solv).

First example of a reversible single-crystal-to-single-crystal polymerization–depolymerization accompanied by a unique magnetic anomaly was reported by Ovcharenko group [46].

A paired heterospin complex [[Cu(hfac)$_2$]$_3$(μ-*O*,*N*-La)2][Cu(hfac)$_2$(*O*La)$_2$] (Fig. 8.12) was prepared in the reaction of copper(II) hexafluoroacetylacetonate [Cu(hfac)$_2$] with the stable nitronyl nitroxide 2-(1-ethyl-3-methyl-1H-pyrazol-4-yl)-4,4,5,5-tetramethyl-4,5-dihydro-1H-imidazole-3-oxide-1-oxyl (La). Magnetic measurements were carried out on an SQUID magnetometer in the temperature range of 2–300 K and in a magnetic field of up to 5 kOe. X-ray structural analysis revealed the great shortening of intermolecular distances (from 4.403 Å at 295 K to 2.460 Å at 150 K; Δd = 1.943 Å) between the terminal Cu atoms of the trinuclear fragments. The crystals of the compound were found to be capable of a reversible single-crystal-to-single-crystal (SC–SC) transformation initiated by the variation of temperature. A reversible topotactic polymerization–depolymerization coordination reaction takes place in the solid during repeated cooling–heating cycles: [Cu(hfac)$_2$]3(μ-*O*,*N*-La)$_2$][Cu(hfac)2(*O*-La)$_2$] ⇌ Cu(hfac)$_2$(μ-*O*,*N*-La). The lengthening of distances between the paramagnetic centers on cooling below 225 K resulted in a transition from antiferromagnetic to ferromagnetic exchange. A transition from ferromagnetic exchange to antiferromagnetic exchange occurs during the heating of the heterospin polymer above 270 K (Fig. 8.12), as a result of the shortening of spin–spin distances.

The "breathing crystals" $Cu(hfac)_2LR$ where hfac is hexafluoroacetylacetonate, and LR is a pyrazolyl-substituted nitroxyl ligand also represent a family of switchable molecular magnets based on copper(II) ions bridged by stable nitroxide radicals [47].

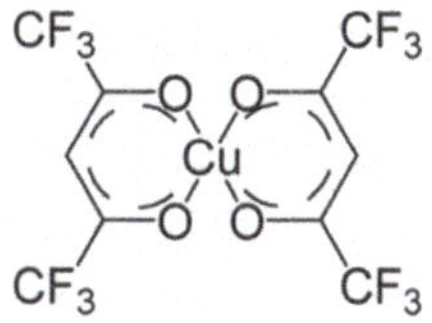

$Cu(hfac)_2$

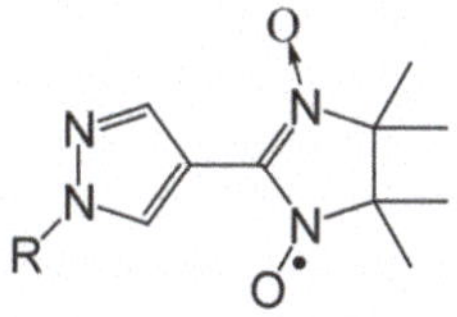

L^R

Diagram of magneto-structural transitions between the weakly coupled (SS) and strongly coupled (WS) states (top) is shown in Fig. 8.13. The energy levels of a spin triad based on spin-Hamiltonian in SS were found to be with $|J| \gg 10\,cm^{-1}$ and $J < 0$, and for WS $|J| \sim 10–20\,cm^{-1}$. A scheme of the energy levels of an exchange-coupled spin triad in SS and WS states was also presented.

Three plausible mechanisms that can cause dynamic mixing processes of mixing of the two doublets $D(S = 1/2)$ and $(S = 1/2)$ and one quartet $Q(S = 3/2)$, namely (1) modulation of exchange interaction, (2) dynamic Jahn–Teller effect, and (3)

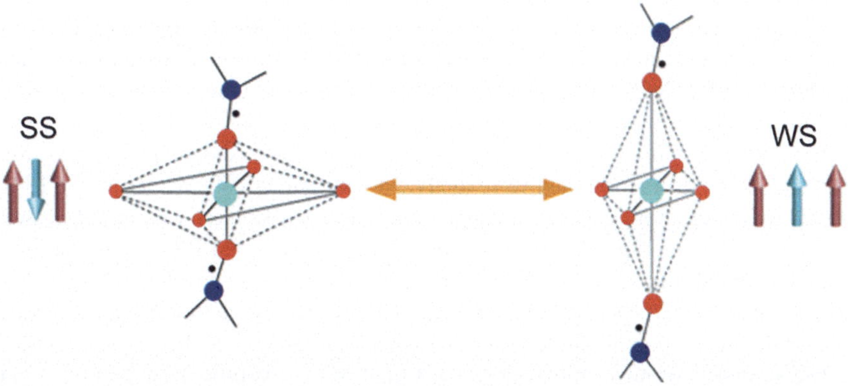

Fig. 8.13 Diagram of magnetostructural transitions between the weakly coupled and strongly coupled states [47]

intercluster exchange couplings between different spin triads were discussed [47]. It was found that for breathing crystals exhibiting abrupt spin transitions, in spin triads the jump-like change of exchange interaction from weak ferromagnetic ($J \sim 1$–$10\,cm^{-1}$) to strong antiferromagnetic ($J \sim -100$ to $-200\,cm^{-1}$) occurs. Experimental data and calculation for the SS state indicated that (1) exchange parameters ($J_{R\text{-}R}$) for the nearest nitronyl nitroxide radicals belonging to different polymer chains, (2) the magnetic chains contain only spin triads (no Cu_2 units) and owe to the nitroxide–nitroxide interchain couplings J_{inter}, (3) the week intercluster exchange interaction between Cu_2 and the spin triad is ~0.1–1.5 cm^{-1}, and (4) breathing crystals exhibits abrupt spin transitions, in spin triads the jump-like change of exchange interaction from weak ferromagnetic ($J \sim 1$–$10\,cm^{-1}$) to strong antiferromagnetic ($J \sim -100$ to $-200\,cm^{-1}$) occurs. The authors stressed on unusual properties of the copper-nitroxide-based molecular magnets in which the magnetic chains can be turned across or along the structural polymer chains by varying the type of nitroxide radicals.

Q-band (EPR) spectroscopy demonstrated that $Cu(hfac)_2L^R$ complexes are one-dimensional in the sense of the topology of their exchange channels, and the magnetic chains spread across the structural polymer chains and consist solely of spin triads of nitroxide–copper(II)–nitroxide [48]. Utilizing four selected examples of complexes $Cu(hfac)_2L^R$, it was found that the exchange coupling values between the spin triads of neighboring polymer chains range from <1 to 10 cm^{-1}. Pair exchange parameters (*JR-R*) for the nearest nitronyl nitroxide radicals belonging to different polymer chains calculated for the SS states of the corresponding breathing crystals using the spin-unrestricted broken-symmetry approach UB3LYP/6-31 were tabulated.

A comparative study of photoluminescence from new synthesized compounds substituted nitroxide derivatives of substituted 2-pyrazolylquinolines indicated that introduction of a nitronyl nitroxide radical group results in the reduction in photoluminescence from the pyrazolylquinoline moiety ($\lambda_{max} = 425$ nm, life time $\tau_1 = 0.055$ ns), and appearing a new red photoluminescence band ($\lambda_{max} = 692$ nm, $\tau_1 = 1.76$ ns) in acetonitrile from the radical fragment [49]. The experiments were performed in acetonitrile with the excitation $\lambda_{max} = 280$.

The first study on the spatial distribution of high- and low-temperature (HT/LT) magneto-structural states (HT/LT) phases of copper(II)–nitroxide-based molecular magnets Cu(hfac)2LR during gradual transitions in these compounds was carried out [50].

To explore the possibility of domain formation at intermediate temperatures, the EPR and X-ray diffraction data were analyzed, and numerical calculations of EPR spectra for different models of exchange-coupled networks were performed. Based on the models of the typical structure of breathing crystals and EPR data, the infinite chain of coupled spin triads by 16- or 32-membered rings was considered, suggesting that each triad has *g*-factor (gSS or gWS) depending on its state. A thorough analysis showed that formation of single-phase domains larger than 4–8 clusters is doubtful.

The synthesis of novel nitroxides, 4-[(3-methoxy-3-oxoprop-1-en-1-yl)oxy]-(1) and 4-[(3-ethoxy-3-oxoprop-1-en-1-yl)oxy]-TEMPO and their application as ligands for assembly of heterospin chain polymers of $[Cu(hfac)2L]_n$ type was reported

[51]. The polymers solid state was characterized by X-ray analysis, FT-IR, UV–Vis, and EPR spectroscopy. The temperature dependences of the effective magnetic moment (χ_{eff}) for heterospin complexes show the χ_{eff} values 2.56–2.67 lB (300 K) for the complexes which corresponded to the spin-only one (2.45 lB) for two non-interacting paramagnetic centers with spins $S = 1/2$ and $g = 2.00$. The χ_{eff} value increases with lowering temperature below 50 K that points to a presence of ferromagnetic exchange interactions, which are typical for Cu(II) complexes with axially coordinated nitroxides.

The electronic absorption spectra of single crystals of three thermoswitchable molecular magnets (Cu(hfac)$_2$LMe, Cu(hfac)$_2$LEt-CP, Cu(hfac)$_2$LPr), and observed thermochromism of the complexes were investigated in the visible and near-IR regions [45]. It was established that the color of the complexes is mainly determined by optical properties of the nitroxide radicals; whereas, the Cu(hfac)$_2$ fragment contributes to the near-IR range with the intensity smaller by an order of magnitude. The following trends in UV–Vis–near-IR spectroscopy of breathing crystals Cu(hfac)$_2$LR were pointed out: (1) UV/Vis–near-IR spectra are dominated by the absorption bands of nitroxides in both WS (high-temperature) and SS (low-temperature) states. (2) The spectra are changed moderately depending on the magneto-structural state. (3) The nitroxide spectrum shifts to the higher energies upon the transition from the WS to SS state. (4) Three-spin nitroxide–copper(II)–nitroxide clusters manifest a new band centered at ~500 nm in the SS state, tentatively assigned to the charge transfer transition. (5) The d–d transitions of copper(II) are relatively weak in intensity; the band positions and shapes do change upon WS ↔ SS conversion, but cover the same spectral region ~700–2000 nm.

Ferromagnetic coupling in a new bis(3,6-di-tert-butyl-catecholato) manganese complex (Fig. 8.14) was revealed by an approach combined X-ray diffraction, measurement of temperature dependence of effective magnetic moment (μ_{eff}) value in the temperature range of 100–300 K [52]. The following findings were reported:

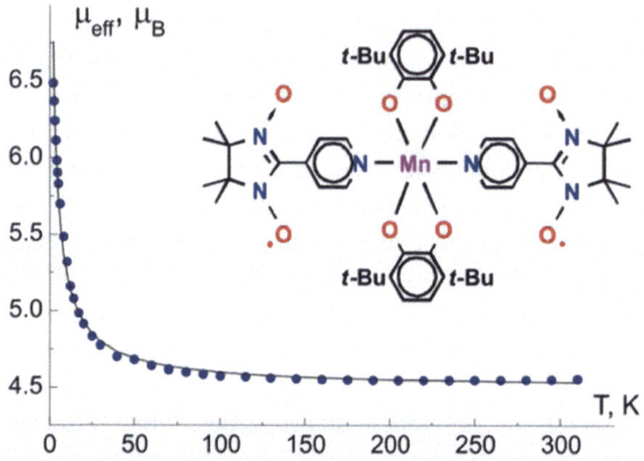

Fig. 8.14 Temperature dependence of for bis(3,6-di-tert-butyl-catecholato)manganese complex [52]. Reprinted from [52], Copyright 2017 American Chemical Society

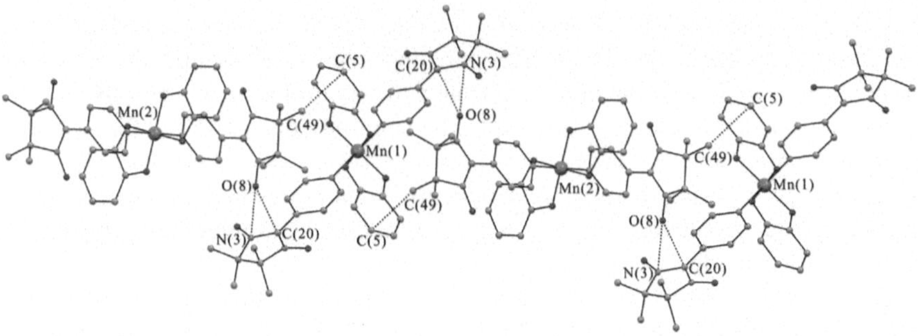

Fig. 8.15 Fragment of crystal packing of 1. The tert-butyl groups and H atoms have been omitted for the sake of clarity [52]. Reprinted from [52], Copyright 2017 American Chemical Society

(1) X-ray diffraction indicated an octahedral environment of the manganese atom with a transarrangement of ligands and the compound specific packing in solid state (Fig. 8.15). (2) Values of the bonds lengths in the inner coordination core of the metal and in the chelate cycles were found to be correspondent to the charge distribution between the metal and ligands displayed a Mn(IV)(Cat_2^{2-}). (3) Intramolecular ferromagnetic coupling between the Mn(IV) $S = 3/2$ spins and spins of nitronyl–nitroxyls and intermolecular ferromagnetic interactions of spins of adjacent nitronyl nitroxide fragments in a chain of molecules at low temperatures were revealed. (4) The μ_{eff} value in the temperature range of 100–300 K is ~4.55 μB that is close to a theoretical spin-only one (4.58 μB) for a NIT-Py ligands and the manganese(IV) ion. (5) In this condition, the Curie constant value is close to the theoretical magnitude of 1.875 K cm^{-3} mol^{-1} for a single paramagnetic center ($S = 3/2$) with ($g = 2.6$) Below 100 K, μ_{eff} increases and reaches 6.5 μB at 2 K, which points to a domination of ferromagnetic exchange interactions between spins of nitroxide ($S = 1/2$) and Mn(IV) ions ($S = 3/2$).

In work [53], a model compound $Cu(hfac)_2$Liso-Pr (Fig. 8.16) residing in the mixed SS/WS state at a room temperature was investigated using femtosecond optical spectroscopy and measurement of the temperature dependence of the effective magnetic moment (μ_{eff}) of the compound. The kinetics of photoswitching was measured in a pump-probe time-resolved experiment by pumping the system at 675 nm and probing with white light. The switching SS to WS occurs within less than 200 fs; whereas, the lifetime of the excited (WS) state is in the order of 100 ps. Suggested energy scheme of photoexcitation, switching, and relaxation processes in $Cu(hfac)_2$Liso-Pr at 300 K showing SS → WS → WS* → WS transitions was **sketched** in Figs. 8.16 and 8.17. Thus, this study is the first time evidence of the feasibility of photoswitching in nitroxide–copper(II)–nitroxide molecular magnets at room temperature.

The structural, electronic, and magnetic properties of two different models of the heterospin polymer chain complexes of Cu^{2+} hexafluoroacetylacetonate with two pyrazole-substituted nitronyl nitroxides $Cu(hfac)_2$LR have been studied by means of multiconfigurational perturbation theory. The calculation revealed an important role of the end-standing NO in the exchange interactions with Cu(II) [54].

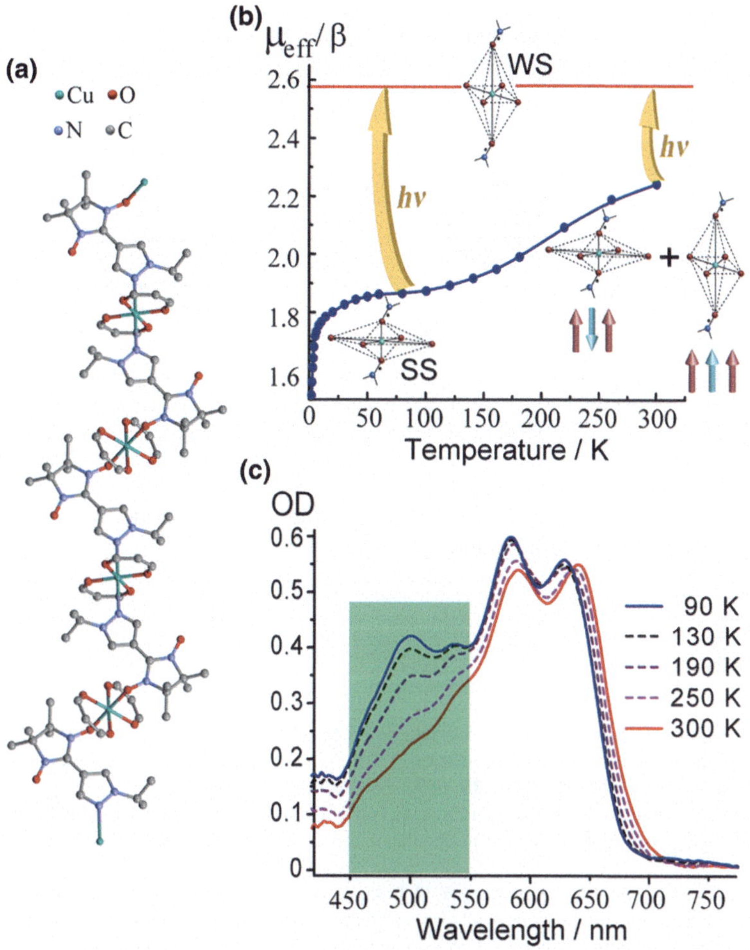

Fig. 8.16 **a** Structure of polymer-chain complex Cu(hfac)$_2$L$^{iso\text{-}Pr}$. **b** Temperature dependence of the effective magnetic moment (μeff) of Cu(hfac)$_2$L$^{iso\text{-}Pr}$. Structures of SS and WS states are sketched, and corresponding arrangements of spins are shown using arrows. Yellow arrows illustrate the photoswitching to WS state, and the horizontal red line is the magnetization level characteristic of WS state. **c** Temperature dependence of UV(vis) absorption spectrum of Cu(hfac)$_2$L$^{iso\text{-}Pr}$. Spectral region of the MLCT band is shaded with green [53]. Reprinted from [53], Copyright 2017 American Chemical Society

The calculations, employing multiconfigurational ab initio theoretical description of the structure and magnetic properties of nitroxide–Cu(II)–nitroxide spin triads *o*, revealed the presence of two minima in the electronic energy curve along the Cu–O$_L$ bond, separated by only 6 kcal/mol, and corresponding to the X-ray structures of the CuO$_6$ centers in Cu(hfac)$_2$L^{Pr} at 115 and 293 K, respectively. At low temperatures, the predicted ground state is ^2A_u, well separated from the ^2A_g, ^4A_u states and calculated

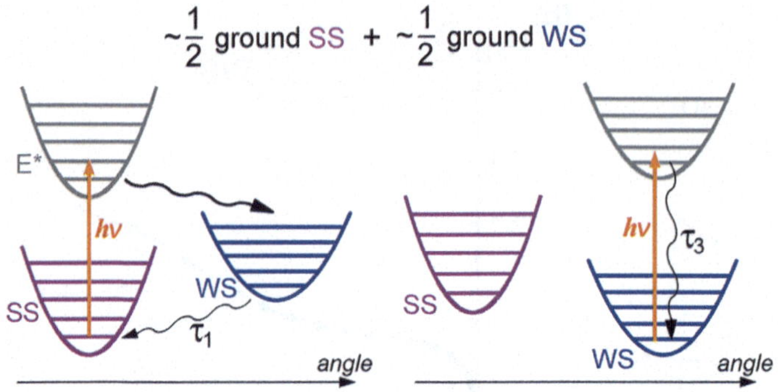

Fig. 8.17 Energy schemes of Cu(hfac)2Liso-Pr at 300 K residing in ~50% of ground SS states (left) and ~50% ground WS states (right). Corresponding photoexcitation, switching, and relaxation processes are shown [53]. Reprinted from [53], Copyright 2017 American Chemical Society

g-factors, $g_{\parallel} = 1.848$, $g_{\perp} = 1.965, 1.974$, qualitatively correspond to the observed $g < 2$ signals in the low-temperature EPR spectra.

References

1. Y. Uchida, S. Oki, R. Tamura, T. Sakaguchi, K. Suzuki, K. Ishibashi, J. Yamauchi, Electric, electrochemical and magnetic properties of novel ionic liquidnitroxides, and their use as an EPR spin probe. J. Mater. Chem. **19**, 6877–6881 (2009)
2. N. Dotti, E. Heintze, M. Slota, R. Hubner, F. Wang, J. Nuss, M. Dressel, L. Bogani, Conduction mechanism of nitronyl-nitroxide molecular magnetic compounds. Phys. Rev. B **93**, 165201 (2016)
3. S. Nakagami, T. Akita, D. Kiyohara, Y. Uchida, R. Tamura, N. Nishiyama, Molecular mobility effect on magnetic interactions in all-organic paramagnetic liquid crystal with nitroxide radical as a hydrogen-bonding acceptor. J. Phys. Chem. B **122**(29), 7409–7415 (2018)
4. Y. Takemoto, Y. Uchida, S. Shimono, J. Yamauchi, R. Tamura, Preparation and magnetic properties of nitroxide radical liquid crystalline physical gels. Mol. Cryst. Liq. Cryst. **647**(1), 279–289 (2017)
5. R. Tamura, K. Suzuki, Y. Uchida, Y. Nodac, EPR characterization of diamagnetic and magnetic organic soft materials using nitroxide spin probe techniques. Electron Paramag. Reson. **23**, 1–21 (2013)
6. Y. Uchida, N. Ikuma, R. Tamura, S. Shimono, Y. Noda, J. Yamauchi, Y. Aoki, H. Nohira, Unusual intermolecular magnetic interaction observed in an all-organic radical liquid crystal. J. Mater. Chem. **18**, 2950 (2008)
7. R. Tamura, Organic functional materials containing chiral nitroxide units, in *Nitroxides: Application in Chemistry, Biomedicine, and Materials Science*, ed. by G.I. Likhtenshtein, J. Yamauchi, S. Nakatsuji, A. Smirnov, R. Tamura (Wiley-VCH, Weinhem, 2008), pp. 303–3030
8. S. Nakatsuji, Preprations, reactions, and properties of functional nitroxide radicals, in *Nitroxides: Application in Chemistry, Biomedicine, and Materials Science*, ed. by G.I. Likhtenshtein, J. Yamauchi, S. Nakatsuji, A. Smirnov, R. Tamura (Wiley-VCH, Weinhem, 2008), pp. 161–238
9. M. Dvolaitzky, J. Billard, F. Poldy, Smectic *E*, *C* and *A* free radicals. Tetrahedron **32**, 1835–1838 (1976)

10. M. Kinoshita, P. Turek, M. Tamura, K. Nozawa, D. Shiomi, Y. Nakazawa, M. Ishikawa, M. Takahashi, K. Awaga, T. Inabe, Y. Maruyama, Chem. Lett. **20**, 1225 (1991)
11. N. Ikuma, R. Tamura, S. Shimono, N. Kawame, O. Tamada, N. Sakai, J. Yamauchi, Y. Yamamoto, Magnetic properties of all-organic liquid crystals containing a chiral five-membered cyclic nitroxide unit within the rigid core. Angew. Chem. Int. Ed. **43**, 3677–3682 (2004)
12. M. Mannini, F. Pineider, C. Danieli, F. Totti, L. Sorace, P. Sainctavit, M.-A. Arrio, E. Otero, L. Joly, J.C. Cezar, A. Cornia, R. Sessoli, Quantum tunnelling of the magnetization in a monolayer of oriented single-molecule magnets. Nature **468**, 417–421 (2010)
13. A. Caneschi, D. Gatteschi, R. Sessoli, P. Rey, Toward molecular magnets: the metal-radical approach. Acc. Chem. Res. **22**, 392–398 (1989)
14. P. Bhatt, K. Kolanji, A. Ivanova, A. Yogi, G. Jakob, M.D. Mukadam, S.M. Yusuf, M. Baumgarten, Soid state magnetic exchange interaction in nitronyl nitroxide radical-based single crystals of 3d metal complexes: a combined experimental and theoretical study. ACS Omega **3**, 2918–2933 (2018)
15. K. Zberecki, R. Swirkowicz, Transport and thermoelectric properties of magnetic organic chains. Phys. Status Solidi B: Basic Solid State Phys. **254**(7) (2017)
16. Y. Uchida, K. Suzuki, R. Tamura, N. Ikuma, S. Shimono, Y. Noda, J. Yamauchi, Anisotropic and inhomogeneous magnetic interactions observed in all-organic nitroxide radical liquid crystals. J. Am. Chem. Soc. **132**, 9746–9752 (2010)
17. K. Suzuki, Y. Uchida, R. Tamura, S. Shimono, J. Yamauchi, Observation of positive and negative magneto-LC effects in all-organic nitroxide radical liquid crystals by EPR spectroscopy. J. Mater. Chem. **22**, 6799 (2012)
18. K. Suzuki, Y. Uchida, R. Tamura, Y. Noda, N. Ikuma, S. Shimono, J. Yamauchi, Influence of applied electric fields on the positive magneto-LC effects observed in the ferroelectric liquid crystalline phase of a chiral nitroxide radical compound. Soft Matter **9**, 4687 (2013)
19. C.V. Yelamaggad, A.S. Achalkumar, D.S. Shankar Rao, M. Nobusawa, H. Akutsu, J.-I. Yamadab, S. Nakatsuji, The first examples of discotic radicals: columnar mesomorphismin spin-carrying triphenylenes. J. Mater. Chem. **18**, 3433–3437 (2008)
20. K. Suzuki, Y. Takemoto, S. Takaoka, K. Taguchi, Y. Uchida, D.G. Mazhukin, I.A. Grigor'evd, R. Tamura, Chiral all-organic nitroxide biradical liquid crystals showing remarkably large positive magneto-LC effects. Chem. Commun. **52**, 3935 (2016)
21. Y. Takemoto, E. Zaytseva, K. Suzuki, N. Yoshioka, Y. Takanishi, M. Funahashi, Y. Uchida, T. Akita, J. Park, S. Sato, S. Clevers, G. Coquerel, D.G. Mazhukin, S. Shimono, M. Sugiyama, H. Takahashi, J. Yamauchi, R. Tamura, Unique superparamagnetic-like behavior observed in non-π-delocalized nitroxide diradical compounds showing discotic liquid crystalline phase. Chem. Eur. J. **24**(65), 17293–17302 (2018)
22. A.K. Vorobiev, N.A. Chumakova, D.A. Pomogailo, Y. Uchida, K. Suzuki, Y. Noda, R. Tamura, Determination of structural characteristics of all-organic radical liquid crystals based on analysis of the dipole–dipole broadened EPR spectra. J. Phys. Chem. B **118**, 1932–1942 (2014)
23. N.A. Chumakova, T.S. Yankova, K.E. Fairfull-Smith, S.E. Bottle, A.K. Vorobiev, Molecular orientational order of nitroxide radicals in liquid crystalline media. J. Phys. Chem. B **118**, 5589–5599 (2014)
24. K. Nagura, Y. Takemoto, S. Moronaga, Y. Uchida, S. Shimono, A. Shiino, K. Tanigaki, T. Amano, F. Yoshino, Y. Noda, S. Koizumi, N. Komatsu, T. Kato, J. Yamauchi, R. Tamura, Preparation of robust metal-free magnetic nanoemulsions encapsulating low-molecular-weight nitroxide radicals and hydrophobic drugs directed toward MRI-visible targeted delivery. Chem. Eur. J. **23**, 15713–15720 (2017)
25. Y. Takemoto, E. Zaytseva, K. Suzuki, N. Yoshioka, Y. Takanishi, M. Funahashi, Y. Uchida, T. Akita, J. Park, S. Sato, C. Simon, D.G. Coquerel, S. Mazhukin, M. Shimono, H. Sugiyama, J. Takahashi, R.Tamura Yamauchi, Unique superparamagnetic-like behavior observed in non-pdelocalized nitroxide diradical compounds showing discotic liquid crystalline phase. Chem. Eur. J. **24**, 17293–17302 (2018)
26. G. Seber, R.S. Fretas, J.T. Mague, A. Paduan-Filho, X. Gratens, V. Bindilatti, N.F. Oliveira Jr., N. Yoshioka, P.M. Lahti, Magnetic tuning of all-organic binary alloys between two stable radicals. J. Am. Chem. Soc. **134**, 3825 (2012)

27. E. Mostovich, Y. Borozdina, V. Enkelmann, K. Remović-Langer, B. Wolf, M. Lang, M. Baumgarten, Planar biphenyl-bridged biradicals as building blocks for the design of quantum magnets. Cryst. Growth Des. **12**, 54 (2012)
28. P. Ravat, T. Marszalek, W. Pisula, K. Müllen, M. Baumgarten, Positive magneto-LC effect in conjugated spin-bearing hexabenzocoronene. J. Am. Chem. Soc. **136**, 12860–12863 (2014)
29. Y.B. Borozdina, E.A. Mostovich, P.T. Cong, L. Postulka, B. Wolf, M. Lang, M. Baumgarten, Spin-dimer networks: engineering tools to adjust the magnetic interactions in biradicals. J. Mater. Chem. C **5**, 9053 (2017)
30. K. Kolanji, P. Ravat, A.S. Bogomyakov, V.I. Ovcharenko, D. Schollmeyer, M. Baumgarten, Mixed phenyl and thiophene oligomers for bridging nitronyl nitroxides. J. Org. Chem. **82**, 7764–7773 (2017)
31. D. Gatteschi, R. Sessoli, J. Villain, *Molecular Nanomagnets* (Oxford University Press, Oxford, 2006)
32. S. Maekawa, *Concepts in Spin Electronics* (Oxford University Press, Oxford, 2006)
33. K.-A. Hansen, J.P. Blinco, Nitroxide radical polymers—a versatile material class for high-tech applications. Polym. Chem. **9**, 1479 (2018)
34. H. Nishide, S. Iwasa, Y.-J. Pu, T. Suga, K. Nakahara, M. Satoh, Organic radical battery: nitroxide polymers as a cathode-active material. Electrochim. Acta **50**(2–3), 827–831 (2004)
35. Kenichi Oyaizu, Takashi Sukegawa, Hiroyuki Nishide, Dual dopable poly(phenylacetylene) with nitronyl nitroxide pendants for reversible ambipolar charging and discharging. Chem. Lett. **40**, 184–185 (2011)
36. M.-K. Hung, Y.-H. Wang, C.-H. Lin, H.-C. Lin, J.-T. Lee, Synthesis and electrochemical behaviour of nitroxide polymerbrush thin-film electrodes for organic radical batteries. J. Mater. Chem. **22**, 1570–1577 (2012)
37. K. Yuan, F. Li, L. Chen, Y. Chen, Approach to a block polymer precursor from poly(3-hexylthiophene) nitroxide-mediated in situ polymerization for stabilization of poly(3-hexylthiophene)/ZnO hybrid solar cells. Thin Solid Films **520**, 6299–6306 (2012)
38. T. Sukegawa, A. Kai, K. Oyaizu, H. Nishide, Synthesis of pendant nitronyl nitroxide radical-containing poly(norbornene)s as ambipolar electrode-active materials. Macromolecules **46**(4), 1361–1367 (2013)
39. T. Janoschka, N. Martin, U. Martin, C. Friebe, S. Morgenstern, H. Hiller, M.D. Hager, U.S. Schubert, An aqueous, polymer-based redox-flow battery using non-corrosive, safe, and low-cost materials. Nature **527**, 78–81 (2015)
40. T. Hagemann, J. Winsberg, B. Häupler, T. Janoschka, J.J. Gruber, A. Wild, U.S. Schubert, A bipolar nitronyl nitroxide small molecule for an all-organic symmetric redox-flow battery. NPG Asia Mater. **9**, e340 (2017)
41. K. Zhang, Y. Hu, L. Wang, M.J. Monteiro, Z. Jia, Pyrene-functionalized PTMA by NRC for greater π–π stacking with rGO and enhanced electrochemical properties. ACS Appl. Mater. Interfaces **9**, 34900–34908 (2017)
42. L. Wylie, K. Oyaizu, A. Karton, M. Yoshizawa-Fujita, E.I. Izgorodina, Toward improved performance of all-organic nitroxide radical batteries with ionic liquids: a theoretical perspective. ACS Sustain. Chem. Eng. **75**, 5367–5375 (2019)
43. K. Sato, T. Mizuma, H. Nishide, K. Oyaizu, Command surface of self-organizing structures by radical polymers with cooperative redox reactivity. J. Am. Chem. Soc. **139**, 13600–13603 (2017)
44. K.-A. Hansen, J. Nerkar, K. Thomas, S.E. Bottle, A.P. O'Mullane, P.C. Talbot, J.P. Blinco, New spin on organic radical batteries—an isoindoline nitroxide-based high-voltage cathode material. ACS Appl. Mater. Interfaces **10**(9), 7982–7988 (2018)
45. IYu. Barskaya, S.L. Veber, E.A. Suturina, P.S. Sherin, KYu. Maryunina, N.A. Artiukhova, E.V. Tretyakov, R.Z. Sagdeev, V.I. Ovcharenko, N.P. Gritsanb, M.V. Fedin, Spin-state-correlated optical properties of copper(II)–nitroxide based molecular magnets. Dalton Trans. **46**, 13108 (2017)

46. V.I. Ovcharenko, S.V. Fokin, E.T. Kostina, G.V. Romanenko, A.S. Bogomyakov, E.V. Tretyakov, First example of a reversible single-crystal-to-single-crystal polymerization-depolymerization accompanied by a magnetic anomaly for a transition-metal complex with an organic radical. Inorg. Chem. **51**(22), 12188–12194 (2012)
47. M.V. Fedin, S.L. Veber, E.G. Bagryanskay, V.I. Ovcharenko, Electron paramagnetic resonance of switchable copper-nitroxide-based molecular magnets: an indispensable tool for intriguing systems. Coord. Chem. Rev. **289–290**, 341–356 (2015)
48. M.V. Fedin, S.L. Veber, K.Y. Maryunina, G.V. Romanenko, E.A. Suturina, N.P. Gritsan, R.Z. Sagdeev, V.I. Ovcharenko, E.G. Bagryanskaya, Intercluster exchange pathways in polymer-chain molecular magnets $Cu(hfac)_2L^R$ unveiled by electron paramagnetic resonance. J. Am. Chem. Soc. **132**(39), 13886–13891 (2010)
49. E.V. Tretyakov, V.F. Plyusnin, A.O. Suvorova, S.V. Larionov, S.A. Popov, O.V. Antonova, E.M. Zueva, D.V. Stass, A.S. Bogomyakov, G.V. Romanenko, V.I. Ovcharenko, Luminescence of the nitronyl nitroxide radical group in a spin-labelled pyrazolylquinoline. J. Lumin. 14833–14838 (2014)
50. M.V. Fedin, S.L. Veber, E.G. Bagryanskaya, G.V. Romanenkoa, V.I. Ovcharenko, Spatial distribution of phases during gradual magnetostructural transitions in copper(II)–nitroxide based molecular magnets. Dalton Trans. **44**, 18823 (2015)
51. A.V. Artemev, O.V. Vysotskaya, L.A. Oparina, A.S. Bogomyakov, S.S. Khutsishvili, I.V. Sterkhova, V.I. Ovcharenko, B.A. Trofimov, New heterospin chain-polymers based on $Cu(hfac)_2$ complex with TEMPO derivatives bearing b-(oxy)acrylate moiety: synthesis, structural and magnetic properties. Polyhedron **119**, 293–299 (2016)
52. M.P. Bubnov, I.A. Teplova, E.A. Kopylova, K.A. Kozhanov, A.S. Bogomyakov, M.V. Petrova, V.A. Morozov, V.I. Ovcharenko, V.K. Cherkasov, Ferromagnetic coupling in the heterospin bis-catecholato–manganese(IV) complex with pyridine substituted by nitronylnitroxide. Inorg. Chem. **56**, 2426–2431 (2017)
53. X. Dong, M. Lorenc, E.V. Tretyakov, V.I. Ovcharenko, M.V. Fedin, Light-induced spin state switching in copper(II)-nitroxide-based molecular magnet at room temperature. J. Phys. Chem. Lett. **8**, 5587–5592 (2017)
54. S. Vancoillie, L. Rulısek, F. Neese, K. Pierloot, Theoretical description of the structure and magnetic properties of nitroxide-Cu(II)-nitroxide spin triads by means of multiconfigurational Ab initio calculations. J. Phys. Chem. A **113**, 6149–6157 (2009)

Chapter 9
Nitroxides in Disease

9.1 Introduction

The mostly used nitroxides in biology and medicine are a group of heterocyclic nitroxide derivatives of piperidine, pyrroline, and pyrrolidine. The nitroxides are served as antioxidants and drugs for cancers, neurodegenerativity (Parkinson, Alzheimer, and others diseases), ischemia, obesity, radiation damage, age-related degeneration, immunomodulatory patology, oxidative stress, inflammation, etc.

Design and application of anticancer drugs of breast, hepatic, lung, ovarian, lymphatic, prostatic, and thyroid cancers still are challenging fundamental problem in modern medicine to be solved. Anticancer effects of nitroxides and their derivatives, involving in synergetic effects with known anticancer drugs in vivo and in vitro experiments in cancer cells and tumor tissue are as follows: (1) reducing the viability of tumor cell lines, (2) slow down ATP production, (3) damage to electron transport chain, (4) mitochondrial mitosis, (5) induction of apoptosis, (6) cell cycle arrest in G1 phase, (7) the induction of oxidative stress, and (8) protection against obesity.

Since the pioneering work of Konovalova, Bogdanov, Miller, Rozantsev, Neiman, and Emanuel in which was demonstrated on model tumor, leukemia, [1] nitroxide radicals and their biologically active derivatives have found increasing application in biomedicine and medicine [2–10].

9.2 Nitroxides in Cancer

9.2.1 Cells

Mito-CP11, a mitochondria-targeted nitroxide formed by conjugating a triphenylphosphonium cation to a five-membered nitroxide, carboxy-proxyl (CP), was used as a superoxide dismutase (SOD) mimetic [4]. The antiproliferative and cytotoxic properties of submicromolar levels of Mito-CP11 alone and in combination

G. I. Likhtenshtein, *Nitroxides*, Springer Series in Materials Science 292,
https://doi.org/10.1007/978-3-030-34822-9_9

with fluvastatin, a cholesterol lowering drug, in breast cancer cells were investigated. The nitroxide inhibited MCF-7 breast cancer cell proliferation, significantly enhanced fluvastatin-mediated cytotoxicity in MCF-7 cells. Mito-CP11 alone, and in combination with fluvastatin also inhibited nuclear factor kappa-B activity mainly in MCF-7 cells. It was concluded that Mito-CP11, being non-toxic to non-tumorigenic cells could enhance the cytostatic and cytotoxic effects of statins in breast cancer cells.

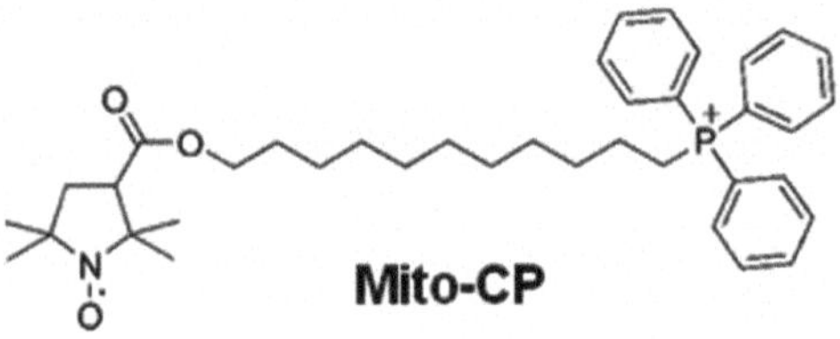

Mito-CP

Mito-carboxy-proxyl (Mito-CP), a lipophilic cationic nitroxide, accumulated in the mitochondria HepG2 cell line and primary hepatocytes was coadministered with 2-deoxyglucose (2-DG), a glycolysis disrupting agent. Mito-CP has been shown to act by disrupting the energy-producing mechanism, inducing mitochondrial-mediated apoptosis, and also enhancing the action of other chemotherapeutic agents in cancer cells [5].

Tert-butyl-2(4,5-dihydrogen-4,4,5,5-tetramethyl-3-o-1 h-imidazole-3-cationic-1-oxyl-2)-pyrrolidine-1-carboxylic ester (L-NNP) displayed cytotoxicity on human breast cancer MCF-7 and MDA-MB-231 cell lines [6]. The nitroxide was selective for liver cancer cells, while being non-toxic to normal hepatocytes. It was found that L-NNP reduced cell viability and increased ROS production and lipid peroxidation accompanied by a reduction in mitochondrial potential and GSH level in hepatoma cells. Mitochondria-targeted metabolism-interfering triphenyl-phosphonium-conjugated nitroxide can effectively suppress human B-RafV600E melanoma cell lines and their/PLX4032-resistant progenies using Mito-CP and ubiquinone [7].

Two series of novel spin-labeled combretastatin derivatives were synthesized and evaluated for cytotoxicity against tumor cell lines (K562, SGC-7901, Hela and HepG-2) [8]. The results indicated that some of the compounds showed significant cytotoxicity against four tumor cell lines in vitro with IC(50) values ranging from 0.15 to 1.05 μM, compared with values of 0.014–0.403 μM for 3-amino-deoxycombretastatin A-4.

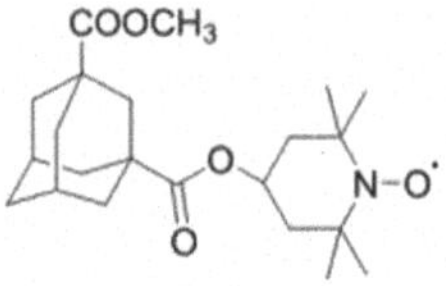

An adamantyl nitroxide derivative displayed a higher anticancer activity against all the tested human hepatoma cells (Bel-7404 cells) with IC50 of 68.1 μM in Bel-7404 cells, compared to the positive control 5-FU (IC50 = 607.7 μM) [9]. This nitroxide: (1) significantly inhibited tumor growth in a xenograft mouse model with low toxicity, (2) suppressed the cell migration and invasion, (3) induced the G2/M phase arrest, and (4) induced cell death, which was accompanied with damaging mitochondria, increasing the generation of intracellular reactive oxygen species, cleavages of caspase-9 and caspase-3, as well as activations of Bax and Bcl-2. The effect of synthetic antioxidant Pirolin (3-carbamoyl-2,2,5,5-tetramethylpyrroline-1-oxyl) on breast cancer progression in Sprague-Dawley rats was evaluated [10]. The rats were administered with 7,12-dimethylbenz anthracene (DMBA) and then treated with the nitroxide. The median number of tumors and their volume, at the end of the study, was found to be considerably smaller in antioxidant-treated groups.

A novel nitroxide derivative, 4-ferrocenecarboxyl-2,2,6,6-tetramethyl piperidine-1-oxyl (FC-TEMPO), was synthesized for evaluating the effects of spin-label compounds on tumor cells (metastatic lung cancer cell line 95-D) and its biological effects on tumor and normal cells [11]. This compound resulted in: (1) 50% reduction in the viability (390 μM, 48 h), while other nitroxide TEMPOL had no effect at the same concentration, (2) FC-TEMPO inhibiting the viability of cancer cells i, while it was less toxic to a normal human cell line, (3) suppressing the growth of tumor cells by induced apoptosis through activating caspase-3, (4) a significant release of extracellular lactatedehydrogenase (LDH), (5) enhance apoptosis and caspase-3 activity (260 μM, 48 h), and 6) increase of superoxide dismutase and catalase. It was suggested that the cytotoxicity of FC-TEMPOL may result from the presence of ferrocenecarboxyl group at position 4, which interacts with cancer cell DNA or induces genotoxicity resulting from redox reaction.

Three series of novel spin-labeled rotenone derivatives analogs

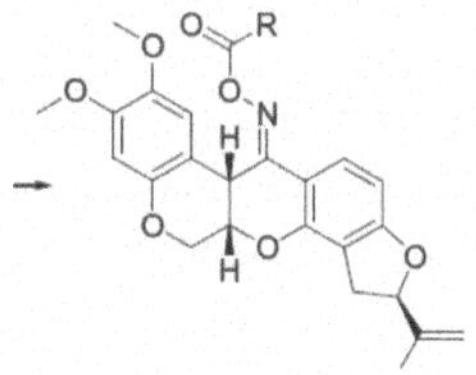

where R are nitroxides were synthesized and evaluated for cytotoxicity against four tumor cell lines, A-549, DU-145, KB and KBvin [12]. All of the derivatives exhibited promising in vitro cytotoxic activity against the tumor cell lines tested, with IC_{50} values ranging from 0.075 to 0.738 μg/mL. A series of spin-labeled fluorouracil derivatives of compound

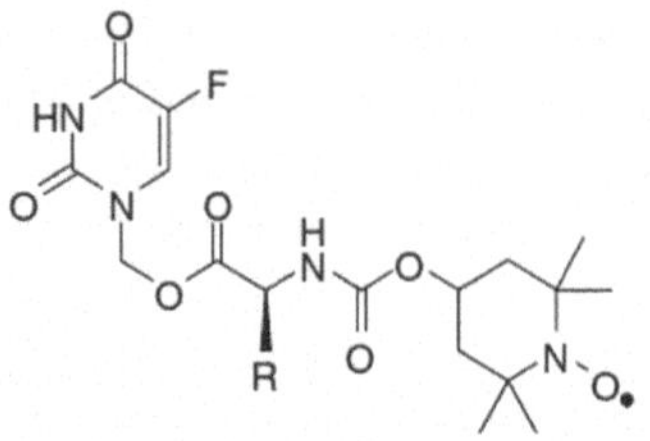

were used as a drug against lung cancer A549 cell line with anticancer activity at IC50 values of 2.76–2.38 μM [13]. These compounds were twofold more cytotoxic than 5-FU and less toxic against other tested cell lines. To evaluate the proof-of-concept anticancer-versus-antioxidant efficacy of the DAPs, a novel class of bifunctional compounds based on diarylidenyl piperidone conjugated to an N-hydroxypyrroline (NOH) and a nitroxide precursor group, HO-3867, for example, were tasted.

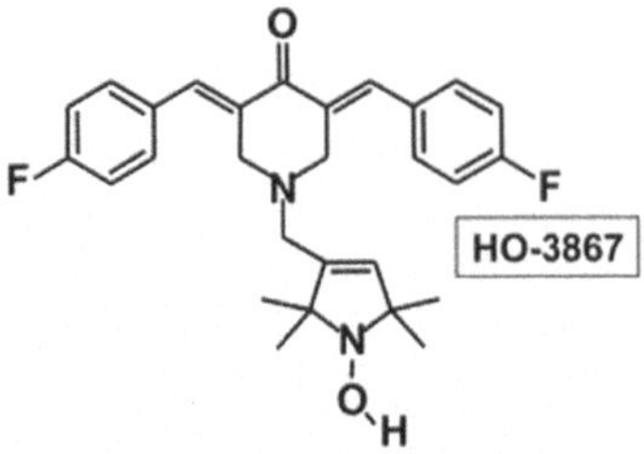

A number of cancerous (breast, colon, head and neck, liver, lung, ovarian, and prostate cancer) and non-cancerous (smooth muscle, aortic endothelial, and ovarian surface epithelial) human cell lines were subjects of interest for work [14]. The study utilized four DAPs, namely H-4073 and H-4318 without NOH and HO-3867 and HO-4200 with NOH substitution. All four compounds induced significant loss of cell viability in cancer cells; whereas, HO-3867 and HO-4200 showed significantly less cytotoxicity in non-cancerous cells. EPR and biochemical measurements indicated (1) a metabolic conversion of the N-hydroxylamine function to nitroxide, (2) significantly higher levels of the metabolite, (3) superoxide radical-scavenging activity in non-cancerous cells compared to cancer cells, (4) arrest and apoptosis in cancer cells growth mediated by inhibition of phosphorylation human protein (STAT3) at the Tyr705 and Ser727 residues, and (5) induction of apoptotic markers of cleaved caspase-3 and PARP were also evaluated.

In work [15], several principle mitochondrial effects of TEMPOL in a human promyelocytic leukemic cell line (HL-60) on 24-h exposure to TEMPOL were evaluated. Experiments revealed the following alterations: (1) decrease in both the intracellular and mitochondrial glutathione pools, (2) impairment of oxidative phosphorylation, and (3) decrease in mitochondrial membrane potential. In addition, TEMPOL specifically targeted complex I of the respiratory chain, with minor effects on complexes II and IV, suggesting that mitochondrial effects might play a role in the nitroxide-induced oxidative stress and apoptosis.

The cytotoxicity and genotoxicity of **TEMPO** in mammalian cells were investigated using the mouse lymphoma assay (MLA) and in vitro micronucleus assay [16]. Specifically, the following effects were described: (1) In the absence of metabolic activation (S9), 3 mM TEMPO produced significant cytotoxicity and marginal mutagenicity in the MLA; (2) In the presence of S9, treatment of mouse lymphoma cells with 1–2 mM TEMPO resulted in decreases of the relative total growth and increases in mutant frequency; (3) treatment of TK6 human lymphoblastoid cells with 0.9–2.3 mM TEMPO increased the frequency of both micronuclei (a marker for clastogenicity) and hypodiploid nuclei (a marker of aneugenicity), (4) TEMPO induced reactive oxygen species and decreased glutathione levels in mouse lymphoma cells; and (5) The majority of TEMPO-induced mutants had loss of heterozygosity at the *Tk*locus, with allele loss of ≤34 Mbp. The authors concluded that TEMPO is mutagenic in the MLA and induces micronuclei and hypodiploid nuclei in TK6 cells, and oxidative stress may account for part of the genotoxicity induced by TEMPO in the cell lines.

Eighteen novel spin-labeled analogs of camptothecin exhibited significant in vitro antiproliferative activity against these four tested human tumor cell lines (A-549, MDA-MB-231, KB, and KBvin) [17].

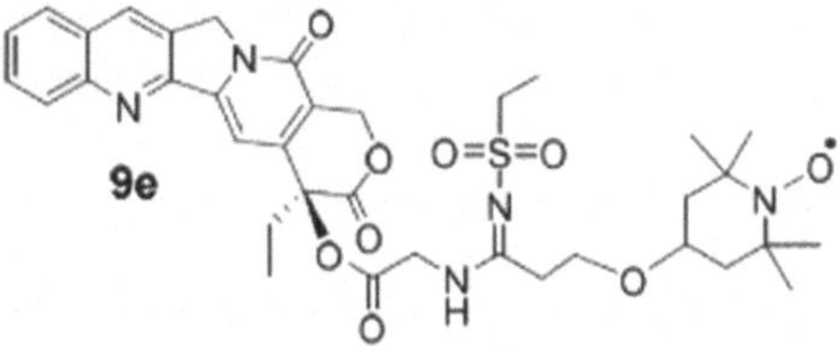

9e

For example, compounds 9e displayed the greatest cytotoxicity against the multidrug-resistant KBvin cell line.

The selective cytotoxicity of Tert-butyl-2(4,5-dihydrogen-4,4,5,5-tetramethyl-3-O-1H-imidazole-3-cationic-1-oxyl-2)-pyrrolidine-1-carboxylic ester (L-NNP on isogenetic human hepatoma HepG2 and normal L-02 cell lines was investigated [6]. Effects of L-NNP on cell growth inhibition, intracellular reactive oxygen species production, the mitochondrial membrane potential loss, malondialdehyde generation, and glutathione levels were analyzed.

Curcumin is a phytochemical agent with diverse molecular targets and well known for its antitumor potential. Curcumin-loaded pH-sensitive redox nanoparticles (RNP(N)) by self-assembling amphiphilic block copolymers conjugated with reactive oxygen species (ROS) scavenging nitroxide radicals were prepared [18]. In conditions mimicking the physiological environment, the entrapment of both curcumin and nitroxide radicals in the hydrophobic core of RNP(N) suppressed curcumin degradation. Curcumin-loaded RNP(N) induced strong cell apoptosis compared to free curcumin. Intravenous injection of curcumin-loaded RNP(N) suppressed tumor growth in vivo, which is probably due to the increased bioavailability and significant ROS scavenging at tumor sites.

The pro-oxidative effect of TEMPOL on the inhibition of cellular proliferation and its role in enhancing the effect of anticancer drug cisplatin (DDP) on the induction of apoptosis in ovarian cancer cells was studied [19]. The experiments revealed that: (1) Cell viabilities of OVCAR3 and SKOV3 cells were decreased by TEMPOL in dose-dependent manner at concentrations of 2–10 mM after 48 h incubation. (2) The cell proliferation rates of OVCAR3 and SKOV3 cells were suppressed by TEMPOL at lower toxic concentrations of 1.5 and 1 mM, respectively, compared with the control group. (3) The combination therapy significantly inhibited the cell proliferation of OVCAR3 cells compared with treatment with DDP alone. (4) The combination treatment increased the proportion of early apoptotic cells in OVCAR3 cells compared with single DDP treatment. (5) The combination treatment markedly decreased the Bcl–2:Bax expression ratio compared with treatment with DDP alone. (6) The combination therapy significantly increased cellular ROS generation compared with the DDP-only therapy.

Nitroxide 1.3. 5-carboxy-1,1,3,3-tetramethylisoindolin-2-yloxyl (CTMIO) has been investigated in the context of *Ataxia Telangiectasia* (A-T), an autosomal recessive disorder including progressive neurodegeneration, immunodeficiency, a strong predisposition to cancer [20]. This nitroxide and its derivative DCTEIO were found to regulate the expression of a collection of genes involved in oxidative stress and antioxidant defense and significantly expressed genes participation also in the inflammatory response or innate immunity. It was suggested that these nitroxides may function as immune-modulating agents. CTMIO and DCTEIO also exhibited a similar pattern of gene regulation to H_2O_2. This finding indicates that the nitroxides may have a pro-oxidant inducing oxidative stress, either by generating reactive oxygen species or by inhibiting antioxidant system rather than antioxidant role in prostate cancer cells. It is important that CTMIO demonstrated minimal toxicity in prostate cancer cells in comparison with TEMPO and TEMPOL.

Rotenone is used as a broad spectrum insecticide, piscicide, and pesticide.

Ability of twelve new derivatives of known antioxidants trolox (TroH), trolox succinate (TroS), α-tocopheryl succinate (α-TOS)-containing nitroxyl radicals to protect bacterial cells from spontaneous and peroxide-induced mutagenesis and their cytotoxicity against six different tumor cells as well as two normal cells was evaluated [21]. In contrast to TroH and TroS, all nitroxide derivatives demonstrated not only antioxidant properties, but also suppressed the growth of human tumor cells: myeloma, mammary adenocarcinoma, hepatocarcinoma, T cells leukemia, histiocytic lymphoma, and T-cellular leucosis. In addition, one group of compounds was capable of inhibiting the growth of normal mouse (LMTK) and hamster (AG17) fibroblast cells, while some nitroxide conjugates showed pronounced selectivity in suppressing the growth of several cancer cells. The authors concluded that the most perspective derivatives as antioxidants and as antitumor agents are the compounds containing pyrrolidine nitroxides.

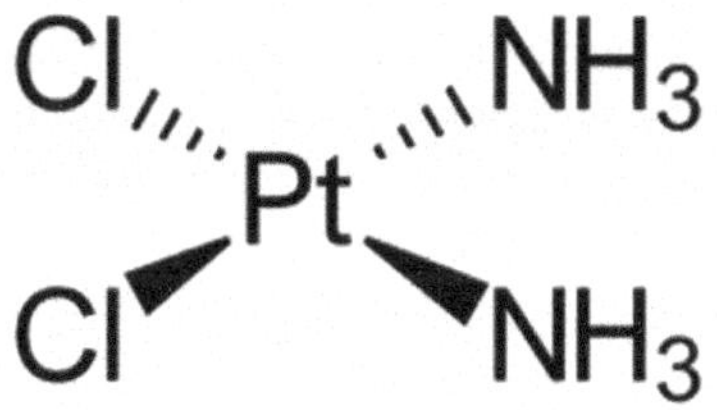

Cisplatin [Cis-Diammine-Dichloroplatinum(II)], CDDP is one of the most widely used anticancer drugs. Nevertheless, the efficacy of the anticancer drug cisplatin is restricted by tumor cell resistance and occurrence of severe side effects. It was reported that platinum (IV)-nitroxyl complexes (PNCs) are able to circumvent CDDP resistance in RT112 bladder cancer cells, thus being hybrid compounds combining platinum and biologically active nitroxyl pharmacophores [2, 22]. For further improving the therapeutic efficacy of the anticancer drug cisplatin, newly synthesized platinum-nitroxyl complexes (PNCs) (Fig. 9.1) were investigated for their potential to circumvent cisplatin resistance [23]. Intracellular platinum accumulation, DNA platination, and cytotoxicity upon treatment with the PNCs in a model system of the bladder cancer cell line RT112 and its cisplatin-resistant subline RT112-CP were evaluated. Induction of DNA strand breaks, and oxidative base modifications were assessed using the PM2 DNA relaxation assay. It was found that platinum accumulation and DNA platination were similar in RT112 and RT112-CP cells for both bivalent and four-valent PNCs. Experiments also indicated a significant decrease in the amount of Pt bound to DNA in cisplatin-treated resistant RT112-CP cells as compared to the parental RT112 line (62 vs. 152.4 nmol Pt/g DNA). A strong correlation was observed between cellular accumulation, DNA platination, and cytotoxicity in RT112 cells. Using a cell-free assay, induction of oxidative DNA damage by cisplatin but not by PNCs was observed. It was suggested that cisplatin exerts its toxic

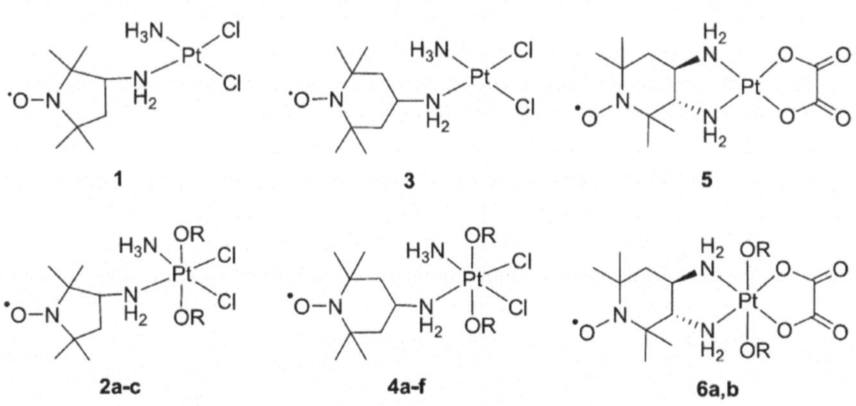

R = H(**a**), $C(O)CH_3$(**b**), $C(O)(CH_2)_2CH_3$ (**c**), $C(O)(CH_2)_3CH_3$ (**d**), $C(O)(CH_2)_4CH_3$ (**e**), $C(O)(CH_2)_6CH_3$ (**f**)

Fig. 9.1 Structure of platinumnitroxyl complexes (PNCs) [23]

action by platination and oxidative DNA damage, while cells treated with PNCs are protected against oxidatively induced lesions. The authors concluded that system PNCs were able to circumvent cisplatin resistance and may be used as improved platinating drugs in cisplatin-resistant cancer.

In work [24], the mode of action of the platinum(IV)-nitroxyl complex PN149

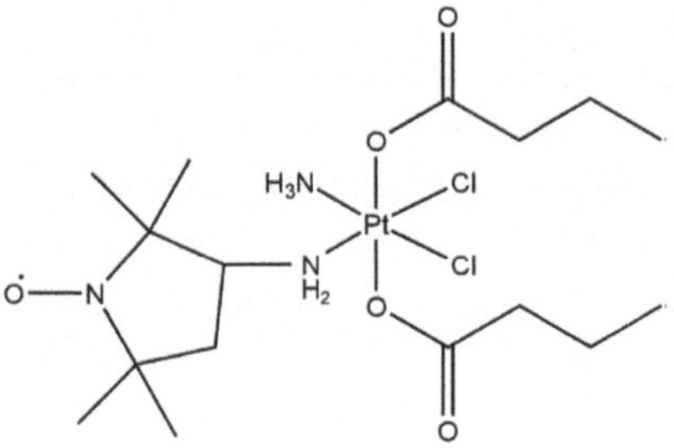

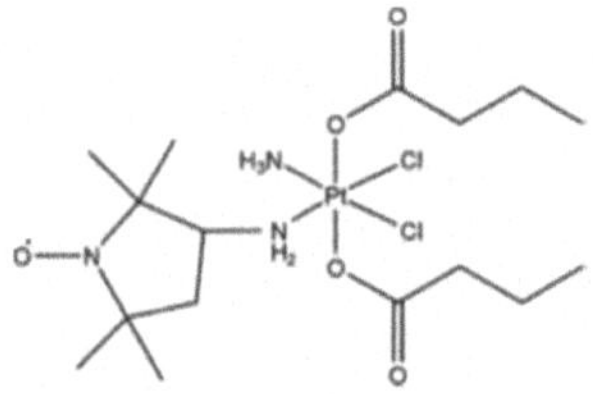

was investigated in the bladder cancer cell line RT112 and the renal cell carcinoma cell line A498 on the molecular and cellular level. For gene expression, analysis high-throughput RT-qPCR was applied. Atomic absorption spectroscopy was used to determine the intracellular platinum content. Apoptosis and cell cycle were analyzed by flow cytometry simultaneously. The p53 protein (a protein encoded by homologous genes in various organisms) level upon treatment with PN149 or CDDP was analyzed using immunoblotting. Cytotoxicity was investigated by relative cell count as well as by colony forming ability. Gene expression analysis was applied to identify the mode of action of PN149 in comparison with CDDP. The following impacts were evaluated: (1) MT gene expression and platinum uptake, (2) genes related to oxidative stress response, (3) genes related to DNA damage response (DDR) and on p53, and (4) progression and control of the cell cycle on the transcriptional and functional level. Gene expression analysis showed that PN149-induced genes related to DNA damage response (RRM2B, GADD45A), cell cycle regulation (CDKN1A, PLK3, PPM1D, and coding for the pro-apoptotic factors PUMA and Noxa.

9.2.2 Animals

The effect of synthetic antioxidant Pirolin (3-carbamoyl-2,2,5,5-tetramethylpyrroline-1-oxyl) on breast cancer progression in Sprague-Dawley rats was evaluated [10]. The rats were administered with 7,12-dimethylbenz(*a*)anthracene (DMBA) and then treated with the nitroxide. The median number of tumors and their volume, at the end of the study, were found to be considerably smaller in antioxidant-treated groups. The role of oxidative stress in carcinogenesis in the investigated animals was emphasized.

The antineoplastic properties of TEMPO in the human hormone-dependent/hormone-independent prostate carcinoma models (LNCaP, DU-145, and PC-3) were examined [25]. TEMPO treatment of LNCaP cells resulted in: (1) enhanced activities of caspase-9 and caspase-3, (2) enhanced number of cells in G_2/M phase of the cell cycle, (3) decreased BrdU incorporation indicating a decline in the number of proliferating cells, (4) a significant inhibition of tumor growth, and (5) in hormone-refractory prostate carcinoma cells, a combination of relatively low doses of TEMPO and doxorubicin or mitoxantrone caused enhanced cytotoxicity as compared with single agents.

The influence of the structure of the nitroxide on the reduction rate in normal tissue, tumor, kidney, and artery regions in mice was investigated by MRI using two cell-permeable nitroxides, 4-hydroxy-2,2,6,6,-tetramethyl-1-piperidynyloxyl (TEMPOL) and 3-carbamoyl-2,2,5,5-tetramethylpyrrolidine-1-oxyl (3CP), and one cell-impermeable nitroxide, 3-carboxy-2,2,5,5,5-tetramethylpyrrolidine-1-oxyl (3CxP) [26]. MRI measurements were done with a 4.7 T scanner. EPRI data acquisition was carried out using a home-built 300 MHz CW EPR imager. The combination of the nitroxide contrast agent and SPGR MRI studies provided pharmacokinetic information for various tissues. The decay of TEMPOL and 3CP in tumor tissue was found to be significantly faster than in normal tissue, while in the case of 3CxP (membrane impermeable), there was no difference in the reduction rates between normal and tumor tissue.

Daunorubicin (Dauno) is commonly used as a drug for acute myeloid leukemia, acute lymphocytic leukemia, chronic myelogenous leukemia, and Kaposi's sarcoma. The activity of ruboxyl (Rbx), a nitroxyl analog of daunorubicin in experimental rat models of hepatic metastases from colorectal carcinoma was compared with its parent compound and with 5-fluorouracil (5FU) [27]. Experiments indicated that in mice treated by intraperitoneal injections, Rbx and 5FU were more effective than Dauno. In addition, Rbx showed only minor and manageable toxic side effect. The regional reduction and pO_2 status in a mouse radiation-induced fibrosarcoma tumor model were evaluated using sequential three-dimensional EPR imaging after intravenous injection of a tissue-permeable nitroxyl probe, HM-PROXYL, and EPR oximetry with lithium phthalocyanine (LiPc) microcrystals implanted within the tumor [28]. The experiments indicated that the regional decay of HM-PROXYL signal obeyed first-order kinetics, and the amplitude of the reduction rate and extent of its heterogeneity in a tumor varied among six mice. The relationship between pO_2 and the reduction rate at the region of pO_2 measurement were found to be low ($r = 0.357$) in 13 tumor-bearing mice.

9.3 Miscellaneous

Aging causes decrease in the ability to cope with environmental stress resulting in increased susceptibility and vulnerability to diseases. The seminal proposal in the 1950s by Denham Harman that reactive oxygen species are a cause of aging invoked the discussion about free radical theory of aging [29]. It was concluded that certain age- and environmental-related pathophysiological changes, degenerative conditions, and diseases are driven and accelerated by radical—(i.e., $^{\bullet}OH$, H_2O_2, and O_2–) induced oxidative stress and inflammation [30].

Nitroxides mostly used in biology and medicine are a group of heterocyclic nitroxide derivatives of piperidine, pyrroline, and pyrrolidine, which may also be served as antioxidants to delay aging. Nitroxides have been proved to be preventative or therapeutic pharmaceutical drugs for diseases of aging which associated with patient genetics, smoking, diet, oxidative stress, chronic inflammation, and other environmental-induced pathophysiological conditions [31]. In addition to aging, nitroxides can prevent the destructive consequences of smoking, high sugar or high fat diet, or radiation, etc. For example, radical-induced oxidative stress and inflammation can be modulated by TEMPOL (TP)-based nitroxide drugs to prevent or treat vascular, ocular, and other pathological conditions and aging-associated disease. Effects of TP, TEMPOL hydroxylamine (TP-H), and TP-H prodrug (OT-551) on age-related degeneration (Fig. 9.2) were evaluated in groups of patients. It was found that: (1) Non-smokers versus smokers with cutaneous microvascular dysfunction, rapidly reversed by cutaneous TP. (2) Elderly cancer patients at risk for radiation-induced skin burns or hair loss were prevented by topical TP. (3) Elderly smoker or non-smoker AMD patients at risk for vision loss were prevented by daily eye drops of OT-551. (4) Both TP and TP-H topically penetrate and function in skin or mucosa, protecting and treating radiation burns and hair loss or smoking-induced cutaneous vascular dysfunction. (5) OT-551 penetrates and travels to the back of the eye, preserving visual acuity and preserving normal and low light luminance in dry AMD smokers and non-smoker patients. (6) OT-551 can activate nuclear factor E2-related factor (Nrf-2) and the antioxidant response element (ARE) and can inhibit nuclear factor kappa-light-chain-enhancer of activated B cells (NF-κB. Clinical data for TP, TEMPOL hydroxylamine (TP-H), and the more lipophilic TP-H prodrug, OT-551, were reviewed [31].

The antioxidant effects of TEMPOL on ovarian ischemia/reperfusion (I/R) injury in rats were investigated treating forty female Wistar albino rats [32]. TEMPOL (50 mg/kg) was found to be effective in the reducing ovarian I/R injury. A hypothesis that scavenging reactive oxygen species with TEMPOL would protect renal oxygenation and nitric oxide (NO) levels in the acute phase of renal ischemia/reperfusion I/R was tested [33]. Results demonstrated that scavenging ROS using TEMPOL not

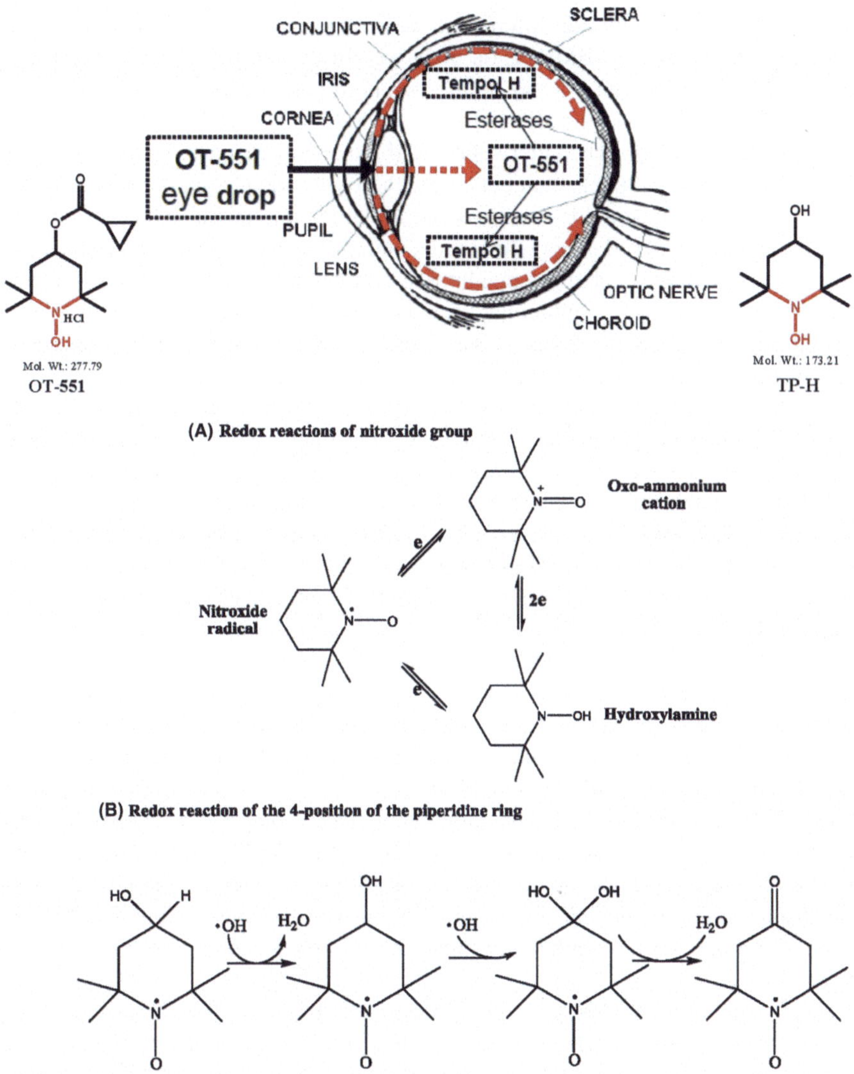

Fig. 9.2 Piperidine nitroxide OT-551 eye drop for preserving visual acuity (VA) in AMD (top), nitroxide group redox cycling (bottom, **a**), and piperidine ring redox reactions (bottom, **b**). (top) [31]. Reprinted from reference [31] with permission of Frontiers in Genetics

only reduced renal oxidative stress following I/R but also normalized renal tissue NO levels and thereby reduced rapid ventricular rate, improved drenal blood flow, renal DO_2, and renal microvascular oxygenation.

Administration of the antioxidant TEMPOL (TP) in the food of mice prevents obesity, causing significant weight loss without toxicity [34]. The results showed that the treatment of 3T3-L1 cells with TEMPOL inhibits the expression of key adipogenesis factors, adipose differentiation, lipid storage and may underlie some of the in vivo effects of TEMPOL on body weight. In addition, dietary TEMPOL decreased body mass and plasma lipid level, both parameters caused by high fat feeding of apo$E^{-/-}$ mice. Apo$E^{-/-}$ mice were fed a high fat diet without or with the nitroxide TEMPOL. The latter suppressed metabolic changes, increased leptin, plaque collagen, macrophage numbers, and atherosclerotic plaque stability [35]. TP and TP-H inhibit tissue oxidative stress and inflammation, and also have effects on the gut microbiome, food energy utilization, and metabolism. TP has been shown to prevent obesity-induced changes in adipokines in cell and animal systems. Supplementation with TB inhibits inflammation and atherosclerosis in apoE($^{-/-}$) mice fed a high fat diet (HFD). TEMPOL had a significant impact on body mass, atherosclerosis, hyperlipidaemia, and inflammation.

The immunomodulatory effects and therapeutic potential of orally delivered TEMPOL in the mouse experimental autoimmune encephalomyelitis EAE model were examined [36]. Mice receiving TEMPOL for 2 weeks prior to induction of active EAE displayed delayed onset and reduced incidence of disease compared to control-fed animals. Reduced disease severity was suggested to be associated with limited microglial activation and fewer inflammatory infiltrates. After the induction of disease, and the onset of clinical symptoms, TEMPOL reduced the severity of multiple sclerosis (MS).

Ten healthy young male smokers, who had smoked for 6 years with an average daily consumption of nine cigarettes, were tested with TEMPOL [37]. Experiments indicated the nitroxide acutely and rapidly (within 30 min) improves the thermal hyperemia response in young adult smokers, returning the response back to that typically observed in healthy non-smokers and effectively reversing their impaired endothelial function observed. The authors concluded the effect of TEMPOL was due to the removal of oxidative stress. The purpose of work [38] was to investigate the effect of the superoxide dismutase mimetic TEMPOL on decreases in retinal blood flow that are found in diabetic STZ mice. A three-week administration of TEMPOL in the STZ mice led to significantly higher blood flow rates than in the untreated mice.

Novel compound SANR was synthesized by introducing functional moiety of salicylic acid into α-nitronyl nitroxides, and the strong analgesic effects of this newly synthesized compound in a rat low back pain of a rat were examined [39]. Both acute and repeated SANR treatment exerted strong analgesic effect in low back pain mimicked by chronic compression of dorsal root ganglion in rats. Four compounds, apocynin (inhibitor of NADPH oxidase, 4-oxo-tempo, U-83836E), and tirilazad (drug for treating acute ischemic stroke), that reduce reactive oxygen species (ROS) and the downstream lipid peroxidation products, affected on nocioceptive behaviors in chronic SCI animals [40]. Specifically, apocynin and 4-oxo-tempo were found to reduce abnormal mechanical hypersensitivity measured in forelimbs and hindlimbs in a model of chronic SCI-induced neuropathic pain.

TEMPOL-H was administered in increasing doses via an intraperitoneal route to C3H mice and provided protection against the lethality of whole body radiation in the mice with a dose modification factor of 13 [41]. TEMPOL-H was associated with less hemodynamic toxicity than TEMPOL. The efficacy of TEMPOL in cell culture and animal models of the central and peripheral dysfunction associated with Parkinson's disease was examined [42]. In vivo, intraperitoneal TEMPOL protects mice from intrastriatal 6-hydroxydopamine (6-OHDA)-induced cell and dopamine metabolite loss in the striatum. In addition, TEMPOL protects mice from the ptosis, activity level decrement, and mortality induced by intraperitoneal administration of 6-OHDA, a model of autonomic dysfunction in Parkinson's disease. Adjunctive use of polynitroxylated albumin enhances the in vitro and in vivo effects of TEMPOL.

C

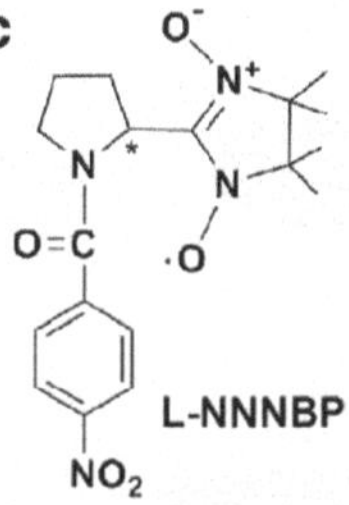

L-NNNBP

Protective effects of L-NNNBP on β-amyloid (Aβ) deposition and memory deficits in an Alzheimer disease (AD) model of APP/PS1 mice were evaluated [42]. Treatment with L-NNNBP for one month led to a marked decrease in brain Aβ deposition and tau phosphorylation in the blinded study on APP/PS1 transgenic mice, as well as decrease of astrocyte activation and improvement of spatial learning and memory compared with the vehicle-treated APP/PS1 mice. Using a rat model of paclitaxel-induced pain (intraperitoneal 2 mg/kg paclitaxel on days 0, 2, 4, 6), the effects of a non-specific ROS scavenger, N-tert-Butyl-α-phenylnitrone (PBN) and a superoxide selective scavenger, 4-hydroxy-2,2,6,6-tetramethylpiperidine-1-oxyl (TEMPOL) were compared [43, 44]. TEMPOL significantly inhibited mechanical hypersensitivity and paclitaxel-induced painful peripheral neuropathy, but in a lesser extent than PBN. Thus, such effects cannot be attributed to superoxide radicals alone.

A nitroxide radical-containing nanoparticle (RNPO) that accumulates specifically in the colon to suppress inflammation and reduce the undesirable side effects of nitroxide radicals was designed [45]. RNP(O) was synthesized by assembly of an amphiphilic block copolymer that contains stable nitroxide radicals in an ether-linked hydrophobic side chain. RNP(O), with a diameter of 40 nm and a shell of poly(ethylene glycol), had a significantly greater level of accumulation in the colonic mucosa than low-molecular-weight TEMPOL or polystyrene latex particles. The effects of RNPO that accumulates specifically in the colons of mice with colitis were determined with dextran sodium sulfate (DSS)-induced colitis and are more effective in reducing inflammation than low-molecular-weight TEMPOL or mesalamine that commonly treats inflammatory bowel disease, including ulcerative colitis.

Intravenous TEMPOL administration to hypertensive rodent models caused rapid and reversible dose-dependent reductions in blood pressure in 22 of 26 studies [46]. The administration is accompanied by vasodilation, increased nitric oxide activity, reduced sympathetic nervous system activity at central and peripheral sites, and

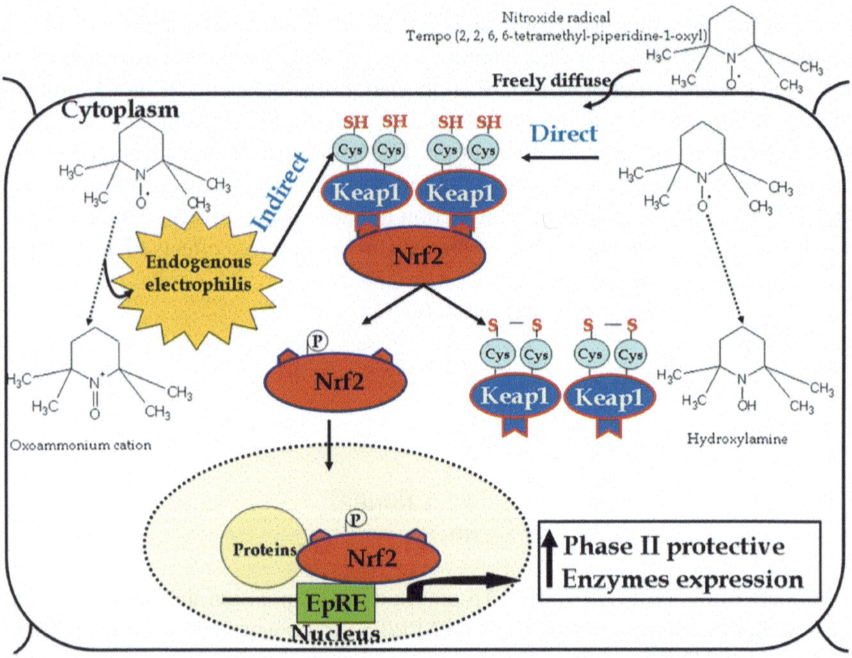

Fig. 9.3 Keap1-Nrf2-ARE pathway in human keratinocytes (HaCaT) [47]. Reprinted from [47], Copyright 2013 Elsevier

enhanced potassium channel conductance in blood vessels and neurons. Oral or by infusion administration over days or weeks to hypertensive rodent models reduced blood pressure in 59 of 68 studies and was accompanied by correction of salt sensitivity and endothelial dysfunction. In addition, the administration reduced agonist-evoked oxidative stress, contractility of blood vessels, renal vascular resistance, and increased renal tissue oxygen tension.

Study [47] showed that nitroxides may act as electrophiles, directly or indirectly, capable of activating the Keap1-Nrf2-ARE pathway in human keratinocytes (HaCaT) and in human skin (human organ culture model) (Fig. 9.3). The high potency of oxoammonium cations versus hydroxylamines in activating the system was demonstrated.

Various aspects of nitroxides as antioxidants and anticancer were recently discussed in a comprehensive review [48].

References

1. N.P. Konovalova, G.N. Bogdanov, V.B. Miller, E.G. Rozantsev, M.B. Neiman, N.M. Emanuel, Antitumor activity of stabile free radicals. Dokl. Akad. Nauk SSSR, **157**(3), 707–709 (1964).

Dokl. Chem. (Engl. Transl.)
2. Sen, A.A. Terentiev, N.P. Konovalova, Platinum complexes with bioactive nitroxyl radicals: synthesis and antitumor properties, in *Nitroxides-Theory, Experiment and Applications*, ed. by A. Kokorin (InTech, London)
3. G. Sosnovsky, S.W. Li, In the search for new anticancer drugs XII. Synthesis and biological evaluation of spin labeled nitrosoureas. Life Sci. **36**, 1479–1483 (1985)
4. G. Cheng, M. Lopez, J. Zielonka, A.D. Hauser, J. Joseph, D. Mcallister, J.J. Rowe, S.L. Sugg, C.L. Williams, B. Kalyanaraman, Mitochondria-targeted nitroxides exacerbate fluvastatin-mediated cytostatic and cytotoxiceffects in breast cancer cells. Cancer Biol. Ther. **12**, 707–717 (2011)
5. A. Dilip, G. Cheng, J. Joseph, S. Kunnimalaiyaan, B. Kalyanaraman, M. Kunnimalaiyaan, T.C. Gamblin, Mitochondria-targeted antioxidant and glycolysis inhibition: synergistic therapy in hepatocellular carcinoma. Anticancer Drugs **24**, 881–888 (2013)
6. J. Guo, Y. Zhang, J. Zhang, J. Liang, L. Zeng, G. Guo, Anticancer effect oftert-butyl-2(4,5-dihydrogen-4,4,5,5-tetramethyl-3-o-1h-imidazole-3-cationic-1-oxyl-2)-pyrrolidine-1-carboxylic ester on human hepatoma HepG2 cell line. Chem. Biol. Interact. **199**, 38–48 (2012)
7. S.K. Hong, D. Starenki, P.K. Wu, J.I. Park, Suppression of B-RafV600E melanoma cell survival by targeting mitochondria using triphenyl-phosphonium-conjugated nitroxide or ubiquinone. Cancer Biol. Ther. **18**, 106–114 (2017)
8. Y.Q. Liu, X.J. Li, C.Y. Zhao, X. Nan, J. Tian, S.L. Morris-Natschke, Z.J. Zhang, X.M. Yang, L. Yang, L.H. Li, X.W. Zhou, K.H. Lee, Synthesis and mechanistic studies of novel spin-labeled combretastatin derivatives as potential antineoplastic agents. Bioorg. Med. Chem. **21**, 1248–1256 (2013)
9. J. Sun, S. Wang, W. Bu, M.Y. Wei, W.W. Li, M.N. Yao, Z.Y. Ma, C.T. Lu, H.H. Li, N.P. Hu et al., Synthesis of a novel adamantyl nitroxide derivative with potent anti-hepatoma activity in vitro and in vivo. Am. J. Cancer Res. **6**, 1271–1286 (2016)
10. S. Tabaczar, K. Domeradzka, J. Czepas, J. Piasecka-Zelga, J. Stetkiewicz, K. Gwozdzinski, A. Koceva-Chyla, Anti-tumor potential of nitroxyl derivative Pirolin in the DMBA-induced rat mammary carcinoma model: A comparison with quercetin. Pharmacol. Rep. **67**, 527–534 (2015)
11. Y. Wu, W. Tang, C.L. Li, J.W. Liu, L.D. Miao, J. Han, M.B. Lan, Cytotoxicity of a newly synthesized nitroxide derivative of 4-ferrocenecarboxyl-2,2,6,6-tetramethylpiperidine-1-oxyl in high metastatic lung tumor cells. Pharmazie **61**, 1028–1033 (2006)
12. Y.Q. Liu, E. Ohkoshi, L.H. Li, L. Yang, K.H. Lee, Design, synthesis and cytotoxic activity of novel spin-labeled rotenone derivative. Bioorg. Med. Chem. Lett. **22**, 920–923 (2012)
13. L. Yang, M.J. Wang, Z.J. Zhang, S.L. Morris-Natschke, M. Goto, J. Tian, Y.Q. Liu, C.Y. Wang, X. Tian, X.M. Yang, K.H. Lee, Synthesis of novel spin-labeled derivatives of 5-FU as potential antineoplastic agents. Med. Chem. Res. **23**(7), 3269–3273 (2014)
14. K. Selvendiran, S. Ahmed, A. Dayton, M.L. Kuppusamy, M. Tazl, A. Bratasz, L. Tong, B.K. Rivera, K. Kalai, P.Kuppusamy Hideg, Safe and targeted anticancer efficacy of a novel class of antioxidant-conjugated difluorodiarylidenyl piperidones: Differential cytotoxicity in healthy and cancer cells. Free Radic. Biol. Med. **48**, 1228–1235 (2010)
15. E. Monti, R. Supino, M. Colleoni, B. Costa, R. Ravizza, M.B. Gariboldi, Nitroxide tempol impairs mitochondrial function and induces apoptosis in HL60 cells. J. Cell. Biochem. **82**, 271–276 (2001)
16. X. Guo, R.A. Mittelstaedt, L. Guo, J.G. Shaddock, R.H. Heflich, A.H. Bigger, M.M. Moore, N. Mei, Nitroxide TEMPO: a genotoxic and oxidative stress inducer in cultured cells. Toxicol. Vitro **27**, 1496–1502 (2013)
17. X.B. Zhao, D. Wu, M.J. Wang, M. Goto, S.L. Morris-Natschke, Y.Q. Liu, X.B. Wu, Z.L. Song, G.X. Zhu, K.H. Lee, Design and synthesis of novel spin-labeled camptothecin derivatives as potent cytotoxic agents. Bioorg. Med. Chem. **15**, 6453–6458 (2014)
18. S. Thangavel, T. Yoshitomi, M.K. Sakharkar, Y. Nagasaki, Redox nanoparticles inhibit curcumin oxidative degradation and enhance its therapeutic effect on prostate cancer. J. Controll. Release **209**, 110–119 (2015)

19. M. Wang, K. Li, Z. Zou, L. Li, L. Zhu, Q. Wang, W. Gao, Y. Wang, W. Huang, R. Liu, K. Yao, Q. Liu, Piperidine nitroxide Tempol enhances cisplatin–induced apoptosis in ovarian cancer cells. Oncol. Lett. 4847–4854 (2018)
20. G.T.A. Eng, An investigation of the effect of some stable nitroxide antioxidants in prostate cancer cells. https://eprints.qut.edu.au/82982/4/Grace_Eng_Thesis (2105)
21. O.D. Zakharova, T.S. Frolova, Y.V. Yushkova, E.I. Chernyak, A.G. Pokrovsky, M.A. Pokrovsky, S.V. Morozov, O.I. Sinitsina, I.A. Grigore'v, G.A. Nevinsky, Antioxidant and antitumor activity of trolox, trolox succinate, and α-tocopheryl succinate conjugates with nitroxides. Eur. J. Med. Chem. **122**, 127–137 (2016)
22. V.D. Sen, A. Goluvev, N.Y. Lugovskaya, T.E. Sashenkova, N.P. Konovolova, Synthesis and antitumor properties of new platinum(IV) complexes with aminonitroxyl radicals. Russ. Chem. Bull. **55**(1), 62–65 (2006)
23. M. Cetraz, V. Sen, S. Schoch, K. Streule, V. Golubev, A. Hartwig, B. Köberle, Platinum(IV)-nitroxyl complexes as possible candidates to circumvent cisplatin resistance in RT112 bladder cancer cells. Arch. Toxicol. **91**, 785–797 (2017)
24. S. Schocha, V. Senb, S. Gajewskia, V. Golubevb, B. Straucha, A. Hartwiga, B. Köberlea, Activity profile of the cisplatin analogue PN149 in different tumor cell lines. Biochem. Pharmacol. **156**, 109–119 (2018)
25. S. Suy, J.B. Mitchell, A. Samuni, S. Mueller, U. Kasid, Nitroxide tempo, a small molecule, induces apoptosis in and suppresses tumor growth in athymic mice. Cancer **103**, 1302–1313 (2005)
26. F. Hyodo, K. Matsumoto, A. Matsumoto, J.B. Mitchell, M.C. Krishna, Probing the intracellular redox status of tumors with magnetic resonance imaging and redox-sensitive contrast agents. Can. Res. **66**(20), 9921–9927 (2006)
27. I. Sirovich, N. Konovalova, G. Codacci-Pisanelli, L.M. Volkova, A. Giuliani, F. Cicconetti, P. Seminara, F. Franchi, Activity of ruboxyl, a nitroxyl derivative of daunorubicin, on experimental models of colorectal cancer metastases. Tumor Biol **20**, 270–276 (1999)
28. K. Takeshita, K. Kawaguchi, K. Fujii-Aikawa, M. Ueno, S. Okazaki, M. Ono, M.C. Krishna, P. Kuppusamy, T. Ozawa, N. Ikota, Heterogeneity of regional redox status and relation of the redox status to oxygenation in a tumor model, evaluated using electron paramagnetic resonance imaging. Cancer Res. **70**(10), 4133–4140 (2010 May 15)
29. D. Harman, Aging: a theory based on free radical and radiation chemistry. J. Gerontol. **11**, 298–300 (1956)
30. C.S. Wilcox, Effects of tempol and redox-cycling nitroxides in models of oxidative stress. Pharmacol. Ther. **126**, 119–145 (2010)
31. J.A. Zarling, V.E. Brunt, A.K. Vallerga, W. Li, A. Tao, D.A. Zarling, C.T. Minson, Nitroxide pharmaceutical development for age-related degeneration and disease. Front Genet. **6**, 325 (2015)
32. N. Pınar, K.O. Soylu, O. Özcan, E. Atik Dogan, S. Bayraktar, Protective effects of tempol in an experimental ovarian ischemia-reperfusion injury model in female Wistar albino rats. Can. J. Physiol. Pharmacol. **95**, 861–865 (2017)
33. U. Aksu, B. Ergin, R. Bezemer, A. Kandil, D.M. Milstein, C. Demirci-Tansel, C. Ince, Scavenging reactive oxygen species using tempol in the acute phase of renal ischemia/reperfusion and its effects on kidney oxygenation and nitric oxide levels. Intensive Care Med. Exp. **3**, 57 (2015)
34. Y. Samuni, J.A. Cook, R. Choudhuri, W. Degraff, A.L. Sowers, M.C. Krishna, J.B. Mitchell, Inhibition of adipogenesis by Tempol in 3T3-L1 cells. Free Radic. Biol. Med. **49**, 667–673 (2010)
35. C.H.J. Kim, J.B. Mitchell, C.A. Bursill, A.L. Sowers, A. Thetford, J. Cook, The nitroxide radical TEMPOL prevents obesity. Atherosclerosis **240**, 234–241 (2015)
36. S. Neil, J. Huh, V. Baronas, X. Li, H.F. McFarland, M. Cherukuri, J.B. Mitchell, J.A. Quandt, Oral administration of the nitroxide radical TEMPOL exhibits immunomodulatory and therapeutic properties in multiple sclerosis models. Brain Behav. Immun. **62**, 332–343 (2017)

37. N. Fujii, V. Brunt, C. Minson, Tempol improves cutaneous thermal hyperemia through increasing nitric oxide bioavailability in young smokers. Am. J. Physiol. Heart Circ. Physiol. **306**, H1507–H1511 (2014)
38. A. Yadav, N. Harris, Effect of tempolon diabetes-induced decreases in retinal blood flow in the mouse. Curr. Eye Res. **36**, 456–461 (2011)
39. W.-J. Han, L. Chen, H.-B. Wang, X.-Z. Liu, S.-J. Hu, X.-L. Sun, C. Luo, A novel nitronyl nitroxide with salicylic acid framework attenuates pain hypersensitivity and ectopic neuronal discharges in radicular low back pain. Hindawi Publishing Corporation Neural Plasticity Volume Article ID 752782, 14 pages
40. S.N. Hassler, K.M. Johnson, C.E. Hulsebosch, Reactive oxygen species and lipid peroxidation inhibitors reduce mechanical sensitivity in a chronic neuropathic pain model of spinal cord injury in rats. J. Neurochem. **131**, 413–417 (2014)
41. M. Hahn, M.C. Krishna, A.M. Deluca, D. Coffin, J.B. Mitchell, Evaluation of the hydroxylamine tempol-H as an in vivo radioprotector. Free Radic. Biol. Med. **28**(6), 953–958 (2000)
42. Q. Liang, A.D. Smith, S. Pan, V.A. Tyurin, V.A. Kagan, T.G. Hastings, N.F. Schor, Neuroprotective effects of TEMPOL in central and peripheral nervous system models of Parkinson's disease. Biochem. Pharmacol. **70**(9), 1371–1381 (2005)
43. T.-Y. Shi, D.-Q. Zhao, H.-B. Wang, S. Feng, S.B. Liu, J.H. Xing, Y. Qu, P. Gao, X.L. Sun, M.G. Zhao, A new chiral pyrrolyl-nitronyl nitroxide radical attenuates -amyloid deposition and rescues memory deficits in a mouse model of Alzheimer disease. Neurotherapeutics **10**(2), 340–353 (2013)
44. M. Fidanboylu, L.A. Griffiths, S.J.L. Flatters, Global inhibition of reactive oxygen species (ROS) inhibits paclitaxel induced painful peripheral neuropathy. PLoS ONE **6**(9), Article ID e25212, (2011)
45. L.B. Vong, T. Tomita, T. Yoshitomi, H. Matsui, Y. Nagasaki, An orally administered redox nanoparticle that accumulates in the colonic mucosa and reduces colitis in mice. Gastroenterology **143**, 1027–1036 (2012)
46. C.S. Wilcox, A. Pearlman, Chemistry and antihypertensive effects of tempol and other nitroxides. Pharmacol. Rev. **60**, 418–469 (2008)
47. M. Greenwald, S. Anzi, S. Ben Sasson, H. Bianco-Peled, R. Kohen, Can nitroxides evoke the Keap1–Nrf2–ARE pathway in skin? Free Radic. Biol. Med. **77**, 258–269 (2014)
48. M. Lewandowski, C. Gwozdzinski, Nitroxides as antioxidants and anticancer drugs. Int. J. Mol. Sci. **18**(11), 2490 (2017)

Chapter 10
Spin Labeling

10.1 Introduction

A number of problems concerning the structure and dynamics of complex molecular systems including examples from biology can be solved by a modification of a chosen portion of the object of interest by a physical label such as a stable radical, paramagnetic complex, luminophore, or Moessbauer atom followed by monitoring the label's properties by appropriate physical methods. In this series, a special place is occupied by nitroxide stable radicals and methods of electron paramagnetic resonance (EPR) and nuclear magnetic resonance (NMR), and its combination.

The birth of the spin-labeling method pioneered by McConnell group in 1965 [1] was preceded by the development of synthetic chemistry of nitroxide starting with classical synthesis of Frémy's salt [2] (Chap. 1). The work by Neiman et al. [3], in which the principal possibility of the reaction of stable radicals without affecting the center of free valence, was demonstrated, happened to be of crucial importance. Since then, synthetic chemistry of nitroxides and their use for modification of biological and non-biological molecules have received a powerful development. In the last decade, triarylmethyl (TAM) radicals and labels based on complexes of paramagnetic metal, gadolinium in particular, also found wide application. According to the generally accepted classification, a covalently bound compound is called "a label," while an adsorbing or free-moving compound is considered "a probe."

Numerous applications of labels and probes are based on four types of spin–spin interactions: (1) the dipole–dipole interaction arises from the magnetic dipole field of one paramagnetic center inducing a local magnetic field at the site of another paramagnetic group, (2) the exchange interaction is caused by overlap of the orbitals occupied by unpaired electrons as the particles approach each other, (3) the electron–nuclear dipole hyperfine interaction which can affect the powder pattern and spectra arising from rotational motion, and (4) the contact hyperfine (spin electron–spin nucleus) interaction, which arises from a finite electron spin density at the nucleus.

G. I. Likhtenshtein, *Nitroxides*, Springer Series in Materials Science 292,
https://doi.org/10.1007/978-3-030-34822-9_10

On the basis of these fundamental phenomena, the following methods were invented and developed to determine: (1) the electron spin–electron spin and electron spin–nuclear spin distances (spin ruler), (2) local viscosity and molecular motion (spin viscose meter), (3) local acidity (spin pH meter), (4) oxygen dynamics (spin oximeter), (5) functional group analysis (spin assay), and (6) spatial distribution of labeled molecular objects (spin imaging). This fundamental approach can provide a wealth of information for establishing the system's structural and dynamic properties. Since pioneered McConnell group work, application of spin labeling in chemistry, biology, and physics has received increasing attention. At present, the SciFinder program offers 4100 references with the keywords "nitroxide spin label and spin probe." The progress of the spin-labeling approach is largely associated with the use of modern methods of magnetic resonance, with their pulsed versions in particular (Chap. 4).

10.2 Nitroxide Labeling

10.2.1 Proteins and Peptides

The method of double spin labeling (DSL), pioneered by Likhtenshtein in 1968 [4], is based on specific modification of chosen groups in the object of interest by two or several spin labels, nitroxides or complexes of paramagnetic metal followed by analysis of the effects of the spin–spin interactions on the label ESR spectra or spin relaxation times. The principle feasibility of the method was demonstrated, for example, albumin covalently labeled with copper complexes. Later on, at the beginning of the 1970, a method using spin dipole–dipole interaction between nitroxides was developed on a quantitative level and then widely employed [5–21]. The method for determining the distance between nitroxides based on the phenomenon of spin exchange was first used by Likhtenshtein group [16, 17] on the example of non-heme iron proteins and nitrogenase.

The first applications of dipole–dipole spin–spin interactions for the investigation of protein surface topography were based on changes in the lineshape of nitroxide radical ESR spectra, arising from interactions with a second nitroxide radical or a paramagnetic ion, allowed one to estimate distances between the paramagnetic centers up to 2.5 nm. Later, the higher sensitivity of power saturation curves of a radical to interactions with paramagnetic species and development advance pulse EPR spectroscopy allowed to expand the distance determination maximum in specific cases up to 15 nm [15–21].

As a first example of quantifying the distance between selected groups in proteins, estimation of distances between the heme group of human hemoglobin and the nitroxide fragment of three spin labels attached covalently to the β-93 SH-group was carried out using analysis of CW power saturation curves in solids based on static dipolar interaction [10, 13, 22]. The values of r (in Å with an accuracy 5–6%)

Table 10.1 Comparison of distances (r in Å) between heme iron and spin labels on hemoglobin [22]

Labels	$\Delta(1/T_{1s})$	$\Delta(1/T_{2s})$	$\Delta(1/T_{1s})\Delta(1/T_{2s})$	r
IA-TEMPO[a]	17.0	15.1	15.9	18.6
M-TEMPO[a]	15.2	15.2	15.2	16.3
PCMB-TEMPO[a]	16.5	15.1	15.8	17.3

[a]*TEMPO* 2,2,6,6-tetramethylpiperidine-N-oxyl; *IA* 4-iodoacetamido; *M* 4-maleimido; *PCMB* p-chloromercuribenzoate

between heme and the nitroxide fragment were found to be in reasonable agreement with the X-ray diffraction structure of hemoglobin (Table 10.1).

Two heme-containing proteins, human neuroglobin (NGB) and sperm whale myoglobin (Mb), were used to test a method exploiting wideband excitation techniques based on composite pulses and relaxation properties of low-spin Fe(III) in high magnetic fields [23]. The proteins were both prepared to have low-spin Fe(III) heme centers. Four different samples (NGB-C120R1, Mb-S3R1, Mb-S117R1, and the doubly spin-labeled Mb-S3R1-S117R1) were obtained using mutagenesis to provide free cysteine amino acids for attachment of the (1-oxyl-2,2,5,5-tetramethyl-Δ3-pyrroline-3-methyl) methanethiosulfonate (MTS) spin label. The results on the label—heme distance measurement (3.2 nm) and schematical presentation of the DEER finding—are shown in Fig. 10.1.

To investigate the difference between native and melted global (MG) state in open- and closed-form mutants, maltose-binding protein (MBP) 09-11, continuous-wave

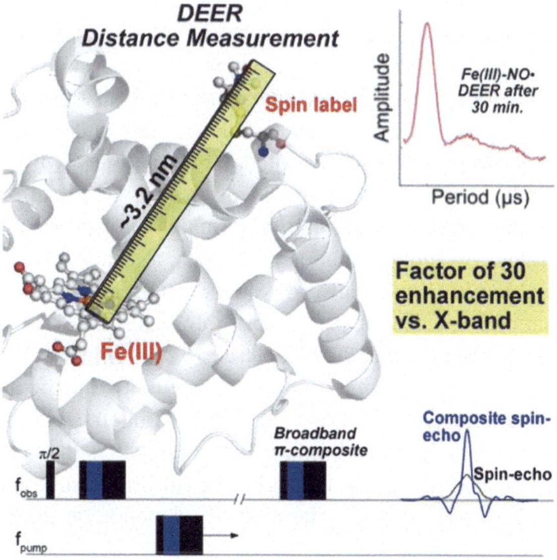

Fig. 10.1 Schematical presentation spin labeled heme protein and DEER traces [23]. Reprinted from [23], Copyright 2016 American Chemistry Society

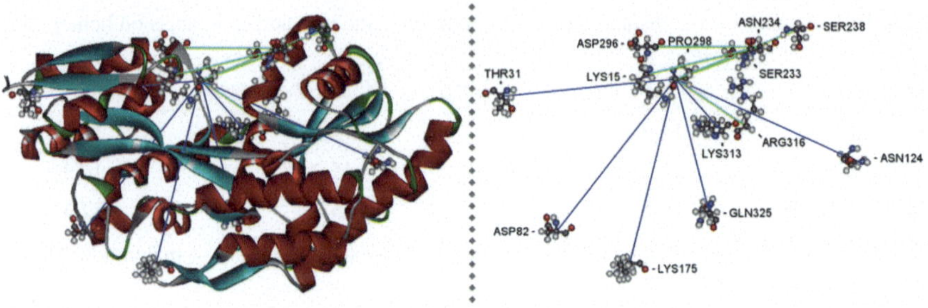

Fig. 10.2 Overview of the complete set of MBP mutants in open-state MBP (Protein Data Bank entry: 1OMP). Blue lines indicate Cα amino acid distances of mutants MBP 01–06. Green lines indicate Cα amino acid distances of mutants MBP 07–11 [24]. Reprinted from [24], Copyright 2018 American Chemical Society

and double electron–electron resonance (DEER) EPR spectroscopy in combination in spin labeling were employed [24]. The spin label (*S*-(1-oxyl-2,2,5,5-tetramethyl-2,5-dihydro-1H-pyrrol-3-yl)methyl methanesulfonothioate, MTSL) positions were placed near the active site (Fig. 10.2). Binding of the protein ligands leads to a conformational change from open to closed state, where the two domains are more closely together. All distances in MBP samples were measured by double-quantum coherence (DQC) EPR using the six-pulse sequence $\pi/2–t_p–\pi–t_p–\pi/2–t_d–\pi–t_d–\pi/2–(t_m–t_p)–\pi–(t_m–t_p)$–echo. The experimental complete set of MBP mutants obtained at pH 3.2 (MG) and pH 7.4 (native state) was in agreement with theoretic predictions of distances between the labels in biradicals constructed by molecular modeling from the crystal structures of MBP in open and closed forms. Measurements showed a defined structure around the binding pocket of MBP in MG, which explains maltose binding. In both states, ligand-free MBP can be found in open and closed forms, while ligand-bound MBP only appears in closed form due to maltose binding. In combination with MD simulation, analysis of MTS distances in mutants MBP 01–11 reveals the changes between open and closed state and ligand binding.

Hsp70s are allosteric molecular chaperones with conformational landscapes that involve large rearrangements of their two domains (viz. the nucleotide-binding domain and substrate-binding domain) in response to adenine nucleotides and substrates. Two complementary methods, ion-mobility mass spectrometry (IMMS) and double electron–electron resonance (DEER), were employed to map the conformational species present under biologically relevant ligand-bound conditions (ATP, ADP/substrate, and ATP/substrate) [25]. Spins are introduced by attaching MTSL to Cys pairs engineered into the protein at strategic and tolerant locations. A DnaK mutant carrying the mutation T199A impairs ATP hydrolysis and thus allows the study of the protein-binding ATP and ATP/substrate. The mobility of the protein ions in the gas phase through a helium-filled tube under a weak electric field was studied using IMMS. The mobility depends on mass and conformation on protein under investigation. DEER yielded the distance distributions between two spin systems for a frozen sample and is sensitive to distances between 20 and 80 Å. Energetic profile and allosteric conformational rearrangement, the interconversion between the domain-docked and domain-undocked states of DnaK in of DnaK were also established.

The architecture of Aer2, a soluble gas-sensing receptor that has emerged as a model for PAS (Per-Arnt-Sim) and poly-HAMP (histidine kinase-adenylyl cyclase-methyl-accepting chemotaxis protein-phosphatase) domain signaling, was investigated by X-ray analysis and pulsed dipolar ESR measurements [26]. Aer2 was labeled by MTSL at position S183, and the interdomain PAS–PAS distance was assessed by monitoring the magnetic dipolar interactions between spins (Fig. 10.3). PDS distance measurements of MTSSL spin-labeled Aer2 1–402 S183C gave sharp $P(r)$ distribution centered at 26.7 Å and were in good agreement with the expected distance based on the crystal structure. It was found that the helical lid of DnaK is a highly dynamic unit of the structure in all ligand-bound states, and DnaK populates

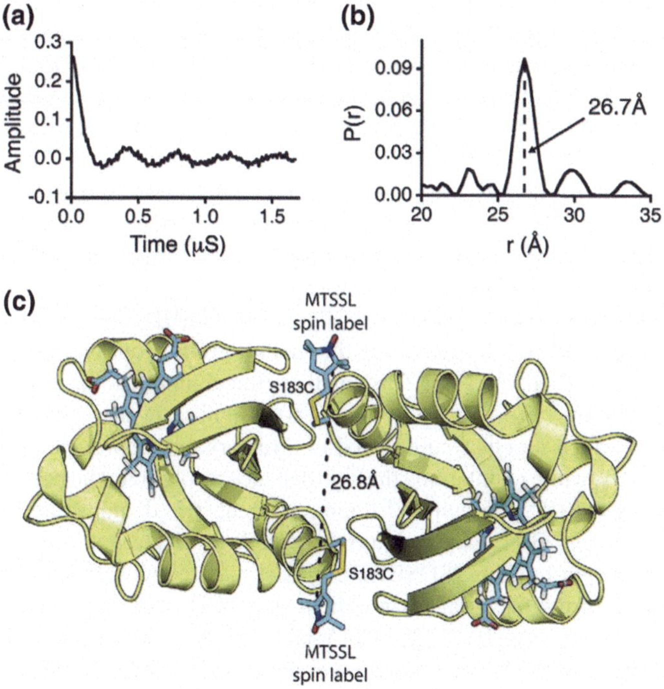

Fig. 10.3 Interdomain distance is only consistent with a PAS dimer. **a** Time-domain signal and **b** corresponding probability distance distribution [$P(r)$] for spin-labeled Aer2 1–402 S183C. A sharp $P(r)$ centered at 26.8 Å is smaller than the diameter of a HAMP domain and fully consistent with the ferric Aer2 PAS dimer. **c** Model of MTSSL spin-labeled Aer2 S183C PAS dimer. The dimer interface directs the two spin labels (blue) away from each other to add approximately 15 Å to the S183 Cα–Cα atom distance [26]. Reprinted from [26], Copyright 2013 Elsevier

a partially docked state in the presence of ATP and substrate. This state represents an energy minimum on the DnaK allosteric landscape.

An overview of available spin labels, the strategies for their introduction into proteins, and the associated potentials for protein structural studies in vitro and in the context of living cells was published.

Figures 10.4 and 10.5 describe commonly employed nitroxide spin labels and protein SDSL strategies.

Site-specific incorporation of Cu^{2+} into non-metalloproteins was accomplished through the use of a genetically encodable non-natural amino acid, 3-pyrazolyltyrosine (PyTyr), in cyan fluorescent protein CFP) (Fig. 10.6) [27]. Then, distances between Cu^{2+}- and nitroxide spin label were measured by the DEER spectroscopy, and the effects of solvent conditions on Cu^{2+} binding and protein aggregation were examined. This method was also applied to characterize the complex formed by the histidine kinase CheA and its target response regulator CheY.

N-terminal-labeled gramicidin A (GALN) and double-labeled gramicidin C (GCDL) modified by 2,2,5,5-tetramethyl-3-pyrrolin-1-oxyl-3-carboxylic acid radicals (Fig. 10.7) were reconstituted into phospholipid bilayers proteoliposomes with consequence investigation by ESR and dynamic nuclear polarization-cryogenic solid-state nuclear magnetic resonance DNP-SSNMR [28]. It was found that the doubly spin-label tagged protein can be used as enhancement agents. Analysis of the labeled GALN and GCDL incorporated into the membrane allowed to establish distances between the labels and the protein mutual disposition (Fig. 10.7). It was also demonstrated that the doubly spin-label tagged protein can be used as enhancement agents. Specifically, $^{1}H–^{13}C$ cross-polarization ^{1}D ^{13}C NMR spectra of GALN and GCDL showed obtained signal enhancements of sixfold for the dimeric protein and the overall impact of spin label to the sample T_1.

IA new Gd(III)-based spin label, BrPSPy-DO3A-Gd(III), derived from a label, BrPSPy-DO3MA-Gd(III), by removing the methyl group from the methyl acetate pending arms, was prepared [29]. This small chemical modification (Fig. 10.8) led to a reduction in the zero-field splitting and to a significant increase in the phase memory time, which together culminated in a remarkable improvement of in-cell DEER sensitivity, while maintaining the high distance resolution. The performance of BrPSPy-DO3A-Gd(III) in in-cell DEER measurements was demonstrated on doubly labeled ubiquitin and GB1 delivered into HeLa cells by electroporation.

Site-directed spin-labeling (SDSL) and EPR spectroscopy were applied to study oligomerization of the bacterial toxin colicin A (ColA) upon membrane insertion in vitro and in vivo (Fig. 10.9) [30] ColA labeling was carried out with purified proteins in vitro, and the labeled protein was incubated with *Escherichia coli* cells. Labeling ColA with MTSSL at position 192, located on helix H10. DEER interspin distance measurements were performed on proteoliposomes containing ColA-A192R1. Reconstitution of spin-labeled ColA into liposomes for the EPR experiments was carried out with a high-protein lipid ratio of (1:500) to maximize the EPR signal strength. Analysis of the data in terms of a distance distribution yields a mean distance of 2.6 nm, comparable to the results obtained for proteoliposomes. Applying SDSL-EPR led to conclusion that the pore-forming domain of colicin A

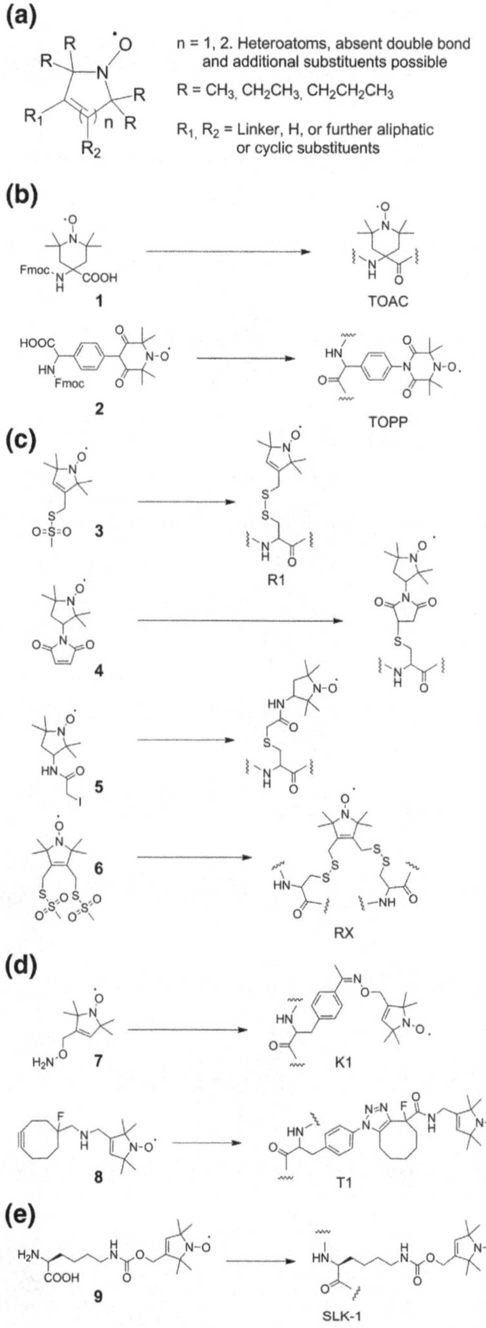

Fig. 10.4 Nitroxide spin labels. **a** General structure of commonly employed nitroxide spin labels. **b** Nitroxides for peptide labeling by solid-phase synthesis. **c** Sulfhydryl-reactive nitroxides for protein labeling by cysteine conjugation reactions. **d** Nitroxides for protein labeling by conjugation with genetically encoded ncAA. **e** Nitroxide ncAA for protein labeling by direct genetic encoding [118]. Reprinted from [118], Copyright 2016 RCSr

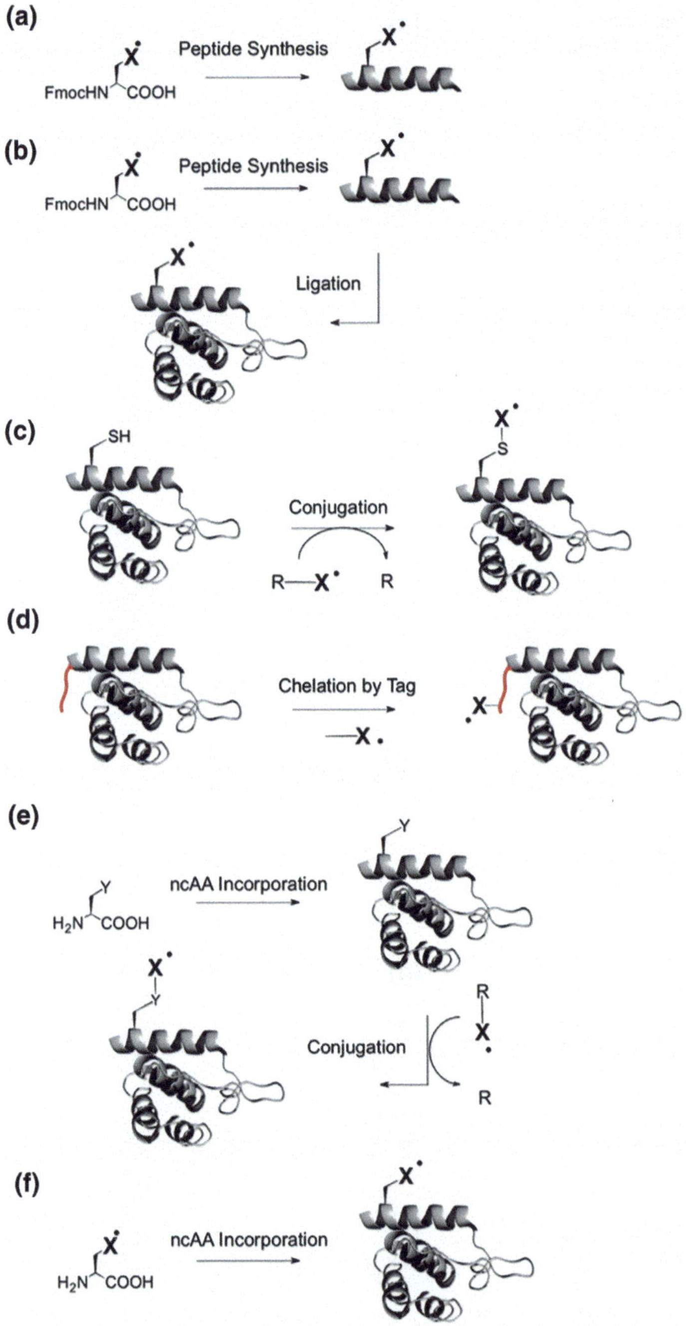

Fig. 10.5 Protein SDSL strategies. **a** Solid-phase peptide synthesis. **b** Peptide–protein ligation. **c** Conjugation reactions with canonical amino acids. **d** Chelation by genetically encoded peptide tags. **e** Conjugation reactions with genetically encoded ncAA. **f** Direct genetic encoding of spin-labeled ncAA [118]. Reprinted from [118], Copyright 2016 RCS

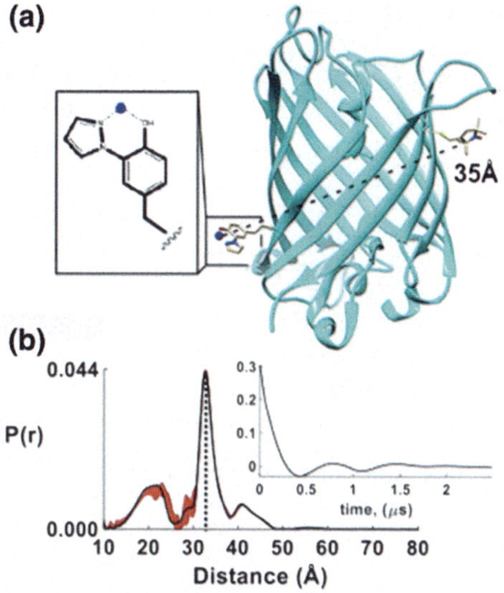

Fig. 10.6 Targeted Cu^{2+} incorporation into CFP. **a** Cu^{2+}-binding amino acid PyTyr shown with Cu^{2+}–NO separation in CFP expected from the crystal structure of CFP (PDB: 3ZTF18) modeled with a nitroxide on residue 208. **b** Distance distribution and denoised time-domain spectrum (inset) obtained by Cu^{2+}–NO DEER spectroscopy in H_2O-based buffer The red regions indicate the small uncertainty arising from the SVD of the denoised signal [27]. Reprinted from [27], Copyright 2018 American Chemical Society

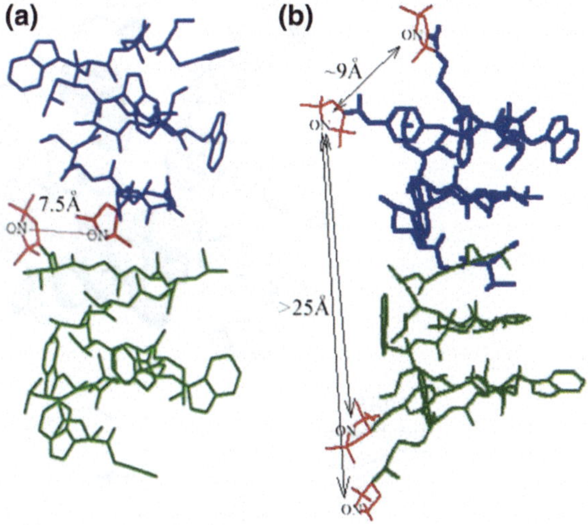

Fig. 10.7 Spin-labeled gramicidin compounds used in this study; **a** N-terminally labeled gramicidin A (GALN). Two spin labels, each from a different monomer, form a membrane embedded spin pair in the dimer, with an experimentally characterized interspin distance of 7.5 Å. **b** Doubly labeled gramicidin C (GCDL) has spin pairs on each gramicidin monomer, located close to the membrane surface and separated by an average interspin distance of 9 Å. The intermolecular interspin distance between spin labels on neighboring GCDL molecules is 25 Å. The right-handed b6.3-helical structure (PDB: 1MAG) is depicted for both compounds, as confirmed by ESR [28]

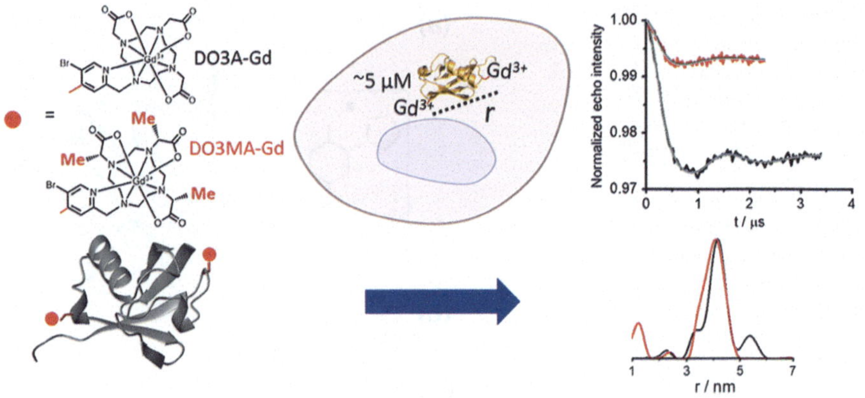

Fig. 10.8 Structure of BrPSPy-DO3A-Gd(III), BrPSPy-DO3MA-Gd(III) label protein and spin the results of DEER measurements [29]. Copyright 2018 American Chemistry Society

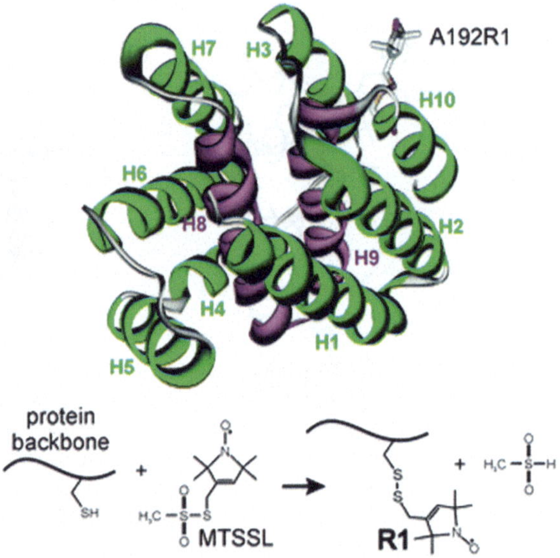

Fig. 10.9 Top panel: crystal structure of the ColA pfd (PDB: 1COL7), with a spin label side chain attached at position A192. Amphipathic helices are colored in green, and the hydrophobic hairpin is shown in magenta. Bottom panel: site-directed spin labeling. After site-directed mutagenesis to replace the residue of interest by cysteine, reaction of the methanethiosulfonate spin label (MTSSL) with the thiol of the cysteine yields the spin-label side R1 [30]. Reprinted from [30], Copyright 2015 RCS

forms oligomers in artificial membranes as well as in the membrane of live *E. coli* cells, most likely in the form of dimers.

ESR was used to investigate the direct interaction between a protein and a particular lipid, illustrating the case of lipid binding into a hydrophobic pocket of chlorocatechol 1,2-dioxygenase, a non-heme iron enzyme responsible for catabolism of aromatic compounds [31]. The effects of GPI-anchored tissue-non-specific alkaline phosphatase, a protein that plays a crucial role in skeletal mineralization, on the ordering and dynamics of lipid acyl chains were revealed. The interaction saturated, or unsaturated fatty acids with model membranes of the brain fatty acid binding

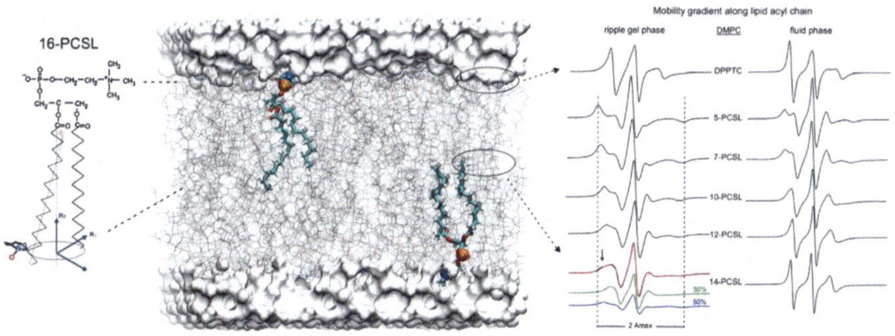

Fig. 10.10 Spin-labeling ESR from the membrane perspective and structural dynamics-line-shape correlations. The lipid bilayer was built with CHARMM-GUI Membrane Builder and rendered with Visual Molecular Dynamics. Reprinted from [31], Copyright 2016 Elsevier

protein, was evaluated. Different *n*-PCSL ($n = 5, 7, 10, 12, 14$, and 16) spin labels reported on specific regions of the lipid bilayer (Fig. 10.10).

An elegant method for determining the position of proteins in membranes was proposed in [32]. The method utilized an oxygen permeability gradient, which is a smooth continuous function of the distance from the center of the membrane. The relaxation gradient of oxygen in zwitterionic and anionic phospholipid membranes was detected by attaching a single nitroxide probe to the transmembrane α-helical polypeptide followed by studies of the oxygen effect on spin relaxation parameters of the spin labels. The peptide was used to determine the depth of penetration of the calcium-binding loops of the C2 domain of cytosolic phospholipase A2. Experiments showed that the relaxation gradient constrains the possible values of the membrane-dependent oxygen concentration and the oxygen diffusion gradients. The average oxygen diffusion coefficient is estimated to be at least twofold smaller in the membrane than that in water.

Complexin is a small soluble presynaptic protein that interacts with neuronal SNARE proteins in order to regulate synaptic vesicle exocytosis. Optical and magnetic resonance spectroscopy were employed to precisely define the boundaries of the two disordered C-terminal domain (CTD) membrane-binding motifs of complexin and to characterize their conformations [33]. The structure of the CTD in the presence of dodecylphosphocholine (DPC) micelles was characterized by standard solution-state NMR experiments. The AH and CT motifs of phospholipid vesicle bound CPX-1 CTD were investigated by electron spin resonance (ESR) spectroscopy with site-directed spin labeling. Together, these experiments provided a model of the structure and boundaries of the membrane-binding motifs of the CPX-1 CTD with MTSL. The crystal structure of full-length glycoprotein B (gB), the fusogen from herpes simplex virus, complemented by electron spin resonance measurements was examined [34]. To determine the structure of the CTD C terminus and its orientation within the membrane, the continuous-wave electron spin resonance spectroscopy on isolated CTD, residues 801–904, bound to synthetic membrane vesicles composed of 1-palmitoyl-2-oleoyl-sn-glycero-3-phosphocholine (POPC) and 1-palmitoyl-2-oleoyl-sn-glycero-3-phosphate (POPA) was employed. Single-cysteine CTD mutants from H861C to K885C were expressed in *E. coli*, purified, and spin labeled with 3-(2-iodoacetamido)-PROXYL. Four-pulse DEER measurements at 17.3 GHz and 60 K were performed using the $\pi/2$–π–π pulse widths were 16 ns, 32 ns, and 32 ns, respectively, and the π pump pulse was 32 ns. Distance distributions for all mutants in solution were found to be broad and centered at ~40 Å but narrowed in the presence of anionic liposomes, indicating that highly dynamic regions become more ordered upon membrane binding. This concurs with the increase in helicity and decrease in proteolytic susceptibility of the CTD in the presence of anionic liposomes.

Phosphocholine spin labels on the lipid headgroup and different positions on the acyl chain (Fig. 10.11) were employed to detect the HIV gp41 fusion peptide (FP) perturbation on lipid bilayers containing different cholesterol concentrations by DEER [35]. The work findings showed that: (1) gp41 FP affects the lipid order via cooperative effect versus the peptide/lipid ratio, (2), gp41 FP induces membrane ordering in all lipid compositions, and (3) in high-cholesterol-containing lipid bilayers, gp41 FP is in the b-aggregation conformation. Schematic representation of the modes of HIV gp41 FP insertion into lipid bilayers is presented in Fig. 10.12.

A distance r between two 2,2,6,6-tetramethylpiperidine-1-oxyl-4-amino-4-carboxylic acid spin labels synthetically incorporated at positions 1 and 16 of this 19-mer peptide was measured at the nanometer scale via the electron dipole–dipole interaction between two labels [18]. Data on distance distribution functions $F(r)$ between the two spin labels for frozen Alm at 77 K are displayed in Fig. 10.13. Two data sets were collected at 77 K: (i) from aggregates of Alm in hydrated egg-yolk phosphocholine (ePC) vesicles (at peptide-to-lipid ratios of 1:200 and 1:75) and (ii) from non-aggregated Alm in pure (non-hydrated) ePC and in solvents of different polarity. The intramolecular distance between the two labels obtained in this manner was found be independent of the molecular aggregated state and the environment polarity as well.

DEER distance measurements between spectroscopically orthogonal Gd(III) and nitroxide spin labels in the nanometer range were performed on a series of membrane-incorporated orthogonally labeled WALP23 polypeptides [36]. Four WALP23 constructs were synthesized by solid-phase synthesis with cysteine at position 7, 11, 15, or 19 for labeling with MTSSL and with a DOTA–Gd(III)–lysine derivative at the N-terminus. The distance distributions (r = 3–3.5 nm) obtained in 10 GHz (X-band) and 35 GHz (Q-band) experiments exhibit a shift of the mean distance by the α-helical pitch when the nitroxide labeling site is shifted by four residues. The developed orthogonal labeling approach allowed for site-specific studies of the nitroxide environment by pulse and CW-EPR in the presence of the lanthanide label and for an independent check of the polypeptide aggregation within the lipid bilayer by nitroxide–nitroxide DEER measurements. The local environment of the nitroxide spin labels was also characterized.

Relaxation time measurements and electron spin echo envelope modulation (ESEEM) experiments using pulse EPR in combination Overhauser dynamic nuclear polarization (DNP) at X-band frequency and a magnetic field of 0.33 T were

(a)

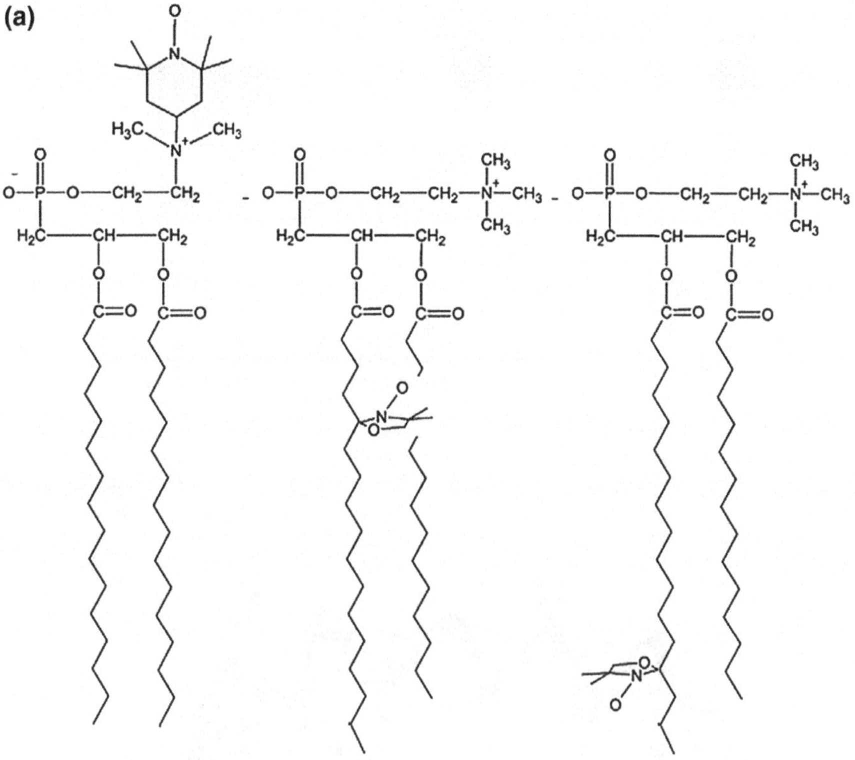

(b) DPPTC 5PC 14PC

WT: AVGIGALFLGFLGAAGSTVGAAS

V2E: A**E**GIGALFLGFLGAAGSTVGAAS

I4C: AVG**C**GALFLGFLGAAGSTVGAASGGGKKKK

A15C: AVGIGALFLGFLGA**C**GSTVGAASGGGKKKK

A21C: AVGIGALFLGFLGAAGSTVG**C**ASGGGKKKK

Fig. 10.11 **a** Structures of spin-labeled lipids DPPTC, 5PC, and 14PC. **b** Peptides used in this study, with mutations shown in bold letters [35]. Reprinted from [35], Copyright 2014 Elsevier

employed to investigate water accessibility in the hydrophobic α-helical peptide WALP23 inserted in unilamellar liposomes of DOPC, a membrane-inserting peptide [37]. Information on the water accessibility of the labeled site was obtained by the measurement of the change in the intensity of the ^{1}H NMR spectrum of H_2O at a Larmor frequency of 14 MHz under a continuous-wave microwave irradiation

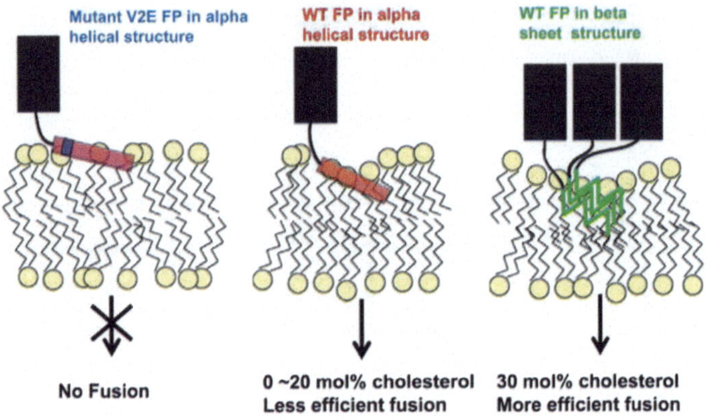

Fig. 10.12 Schematic representation of the modes of HIV gp41 FP insertion into lipid bilayers, and its effects on membrane order and fusion activities. The FP in b-sheet conformation is represented as a trimer simply because gp41 is a trimer. Its real stoichiometry during the membrane fusion process has not been confirmed. FP conformations shown are a-helical (red), b-sheet (green); the gp41 ectodomain is black [35]. Reprinted from [35], Copyright 2014 Elsevier

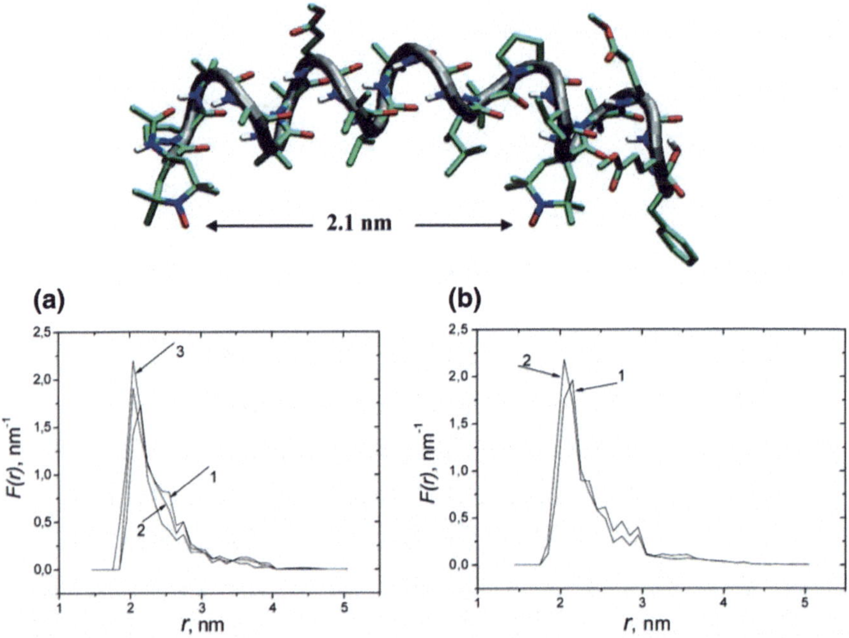

Fig. 10.13 Distance distribution functions $F(r)$ between the two spin labels for frozen Almat 77 K. (left) Curves 1, 2, and 3 are related to the PELDOR decays of Almin frozen methanol, methanol/toluene (1:4), and non-hydrated ePC. (right) Curves 1 and 2 are related to the PELDOR decays of Alm + unlabeled alamethicin (1:10) in hydrated ePC vesicles (P/L molar ratios of 1:75 and P/L = 1:200)). In the upper part of the figure, a model of Alm [1, 16], built from the crystal structure of Alm (molecule A) (27), is shown [18]. Reprinted from [18], Copyright 2008 American Chemical Society

of the nitroxide spin label. Water accessibility measurements with all techniques are conducted for eight peptides with different spin-label positions and low radical concentrations (10–20 μM). Consistently in all experiments, the water accessibility appears to be very low, even for labels positioned near the end of the helix.

10.2.2 Lipid Membranes

Spin-labeling method provides wide facilities for quantitative measurements of chemical, biochemical, and physical processes in biological and model membranes. Principles of the spin-labeling application to investigating biological membranes and recent results have been reviewed [38–45]. Location of the labels, water, and oxygen in membranes, membrane microstructure, and dynamics has been discussed in detailed. Here the recent current approaches and results in this area are illustrated by several examples.

The effectiveness of nitroxide-labeled lipids (TOTAPOL3) as polarization agents for lipids and a membrane-embedded peptide was demonstrated [43]. The chemical structure of the nitroxide lipid labels used in this work is shown in Fig. 10.14.

Rotational motion of the label was characterized by values of the ($2A_{max}$). DNP enhancements for SL-lipids in MLVs versus 20 mM TOTAPOL3 suspended with MLVs, measured via ^{13}C CP NMR spectra, are indicated in Fig. 10.15.

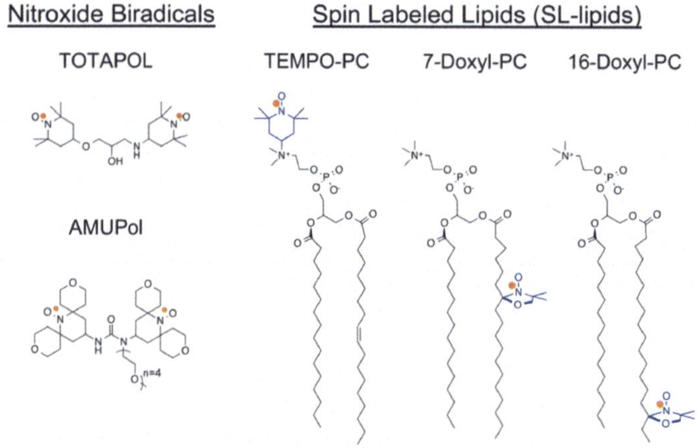

Fig. 10.14 Molecular structures of (left) common water-soluble nitroxide biradicals and (right) SL-lipids used in this study. The unpaired electron, indicated by an orange dot, is distributed over the N–O bond [43]. Reprinted from [43], Copyright 2016 American Chemical Society

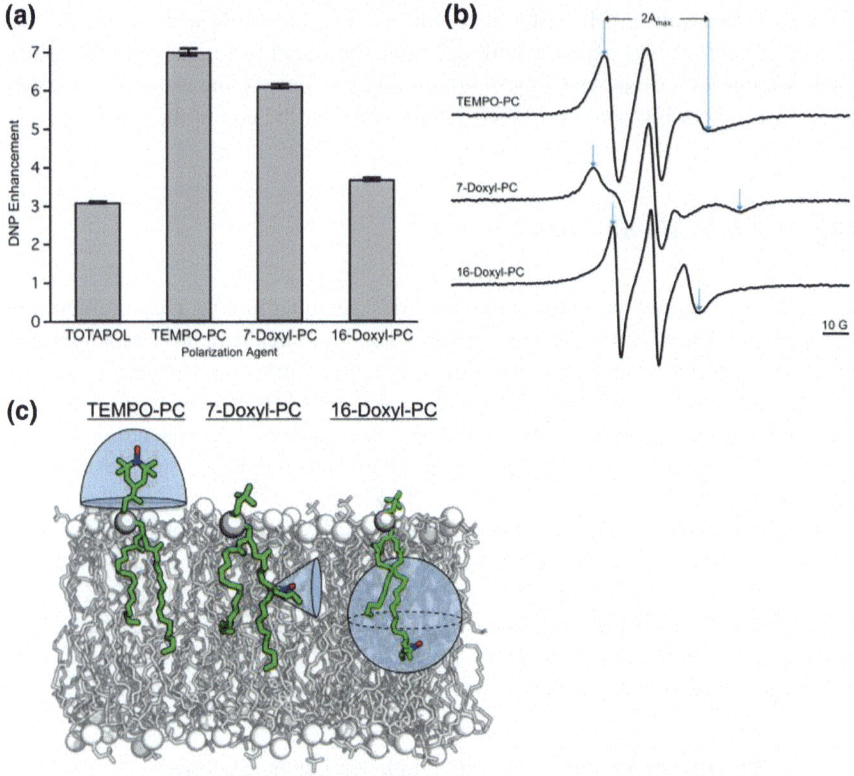

Fig. 10.15 **a** DNP enhancements for 3 mol% of indicated SL-lipids in MLVs versus 20 mM TOTAPOL3 suspended with MLVs, measured via ^{13}C CP NMR spectra. Error bars were calculated by measuring the RMS noise in the baseline of each ^{13}C CP spectrum and propagating through the DNP enhancement calculation. **b** Area normalized X-band CW EPR spectra of 3 mol% of the indicated SL-lipids reconstituted in MLVs at 282 K. Values of the super fine splitting of the label ESR spectra ($2A_{max}$) are indicated with blue arrows for each spectrum. **c** The distribution of available conformations for nitroxides in TEMPOPC, 7-Doxyl-PC, and 16-Doxyl-PC relative to the lipid bilayer normal is depicted by the shaded blue region around the respective nitroxide moiety [43]. Reprinted from [43], Copyright 2016 American Chemical Society

The relative DNP NMR sensitivity enhancement between 15-{[(7-oxyl-3,11-dioxa-7-azadispiro[5.1.5.3]hexadec-15-yl)carbamoyl][2-(2,5,8,11-tetraoxatridecan-13-ylamino)}-[3,11-dioxa-7-azadispiro[5.1.5.3]hexadec-7-yl])oxidanyl (AMUPol) and 1-[(2,2,6,6-tetramethyl-1-oxidopiperidin-4-yl)amino]-3-(2,2,6,6-tetramethyl-1-oxopiperidin-1-ium-4-yl)oxypropan-2-ol (TOTAPOL) and between deuterated and protonated lipid membranes was quantified [44]. Several lipid membranes were used in this study, namely 1,2-dimyristoyl-sn-glycero-3-phosphocholine (DMPC), d54-DMPC, 1,2-dilauroyl-sn-glycero-3-phosphoethanolamine (DLPE), 1-palmitoyl-2-oleoyl-sn-glycero-3-phosphoethanolamine (POPE), and a eukaryotic membrane mixture which contains 1-palmitoyl-2-oleoyl-sn-glycero-3-phophocholine (POPC), POPE, egg sphingomyelin (SM), and cholesterol (Chol) at molar ratios of 25.6:25.6:25.6:23%. ^{1}H T_1 relaxation times were measured using the inversion recovery experiment. ^{1}H/^{13}C/^{15}N MAS probe 2D ^{13}C spectra and 2D ^{15}N ^{13}C correlation spectra were measured without DNP. AMUPol showed fourfold higher sensitivity enhancement than TOTAPOL. Deuterated lipid membrane does not give net higher sensitivity for the membrane peptides than protonated membrane. Overall, a 100-fold enhancement between the microwave-on and microwave-off spectra can be achieved on lipid-rich membranes containing conformationally disordered peptides. Data on measurement of the paramagnetic relaxation enhancement of lipid signals by TOTAPOL and AMUPol indicated a bimodal distribution of both radicals, a surface-bound fraction and a membrane-bound fraction where the nitroxides lie at 10 Å from the membrane surface. TOTAPOL appears to have a higher membrane-embedded fraction than AMUPol and TOTAPOL preferentially binds to the middle of the membrane, 10 from the membrane surface [46].

10.2.3 Nucleic Acids

The spin labeling in combination with classical and advance EPR and NMR methods and its pulse modification, in particular, have proved to a powerful approach to study of structure and motion/rearrangements of the nucleic acids under biologically similar conditions [47–53].

Modern synthetic chemistry developed elegant methods of tethering nitroxide and other labels to all fragments of nucleotide including the nucleobase, the sugar, and the phosphate backbone of nucleic acids (Figs. 10.16 and 10.17, as example). The spin labels can be incorporated during chemical synthesis of the oligomer (phosphoramidite approach) or post-synthetically, by reaction of a spin-labeling reagent with a reactive functional group on the oligonucleotide [47].

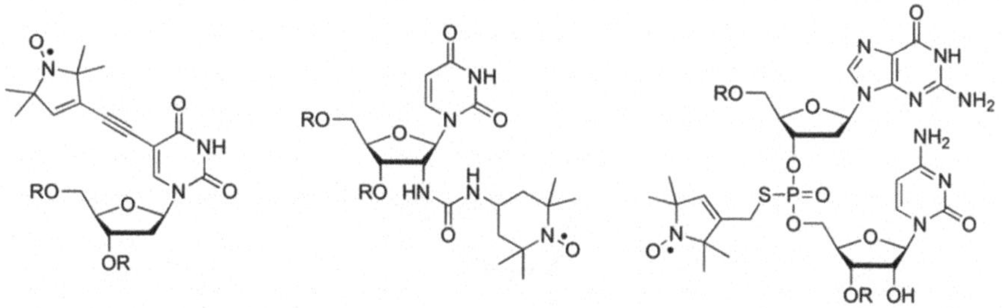

Fig. 10.16 Examples of nitroxides that have been conjugated to the nucleobase (left), the sugar (middle), and the phosphate backbone of nucleic acids (right) [47]. Reprinted from [47], Copyright 2011 De Gruyter

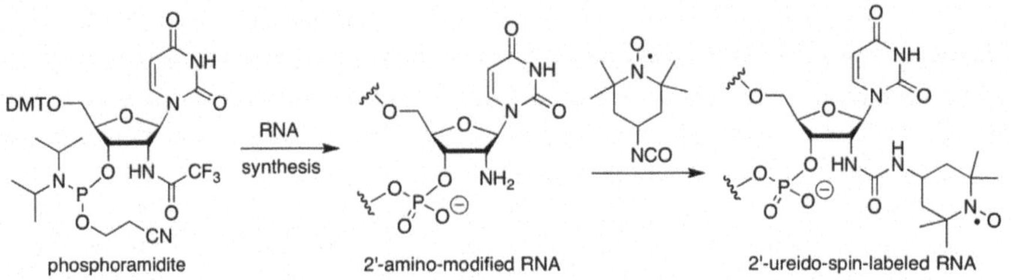

Fig. 10.17 Incorporation of 2′-amino-modification into RNA using the phosphoramidite approach (left) and post-synthetic spin labeling of 2′-amino-modified RNA with a spin-labeled isocyanate (right). DMT = 4,4′-dimethoxytrity [47]. Reprinted from [47], Copyright 2011 De Gruyter

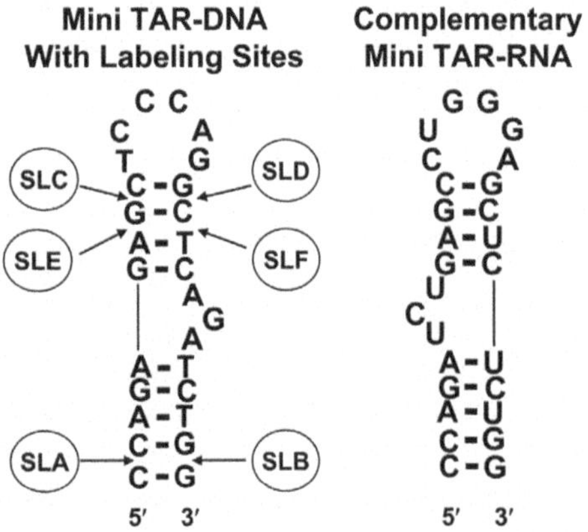

Fig. 10.18 Secondary structure of mini TAR DNA together with positions of spin labels attached via phosphorothioate linkages. The complementary mini TAR RNA is also shown. On the mini TAR DNA, the spin labels were attached in pairs, SLA-SLB, SLC-SLD, SLE-SLF, and the constructs were called SLAB, SLCD, and SLEF, respectively [48]. Reprinted from [48], Copyright Elsevier

The post-synthetical covalent modification RNA with labels was illustrated in Fig. 10.17,

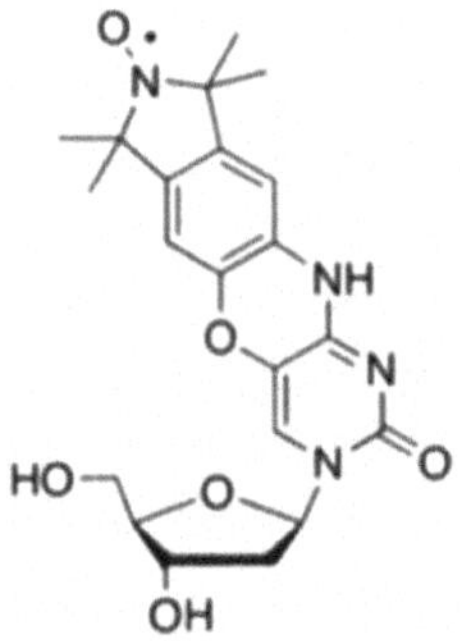

and their derivatives used for non-covalent modification were also performed.

DEER was applied to strategically placed, site-specifically attached pairs of nitroxide spin labels to monitor changes in the mini TAR DNA stem-loop structure brought on by the HIV-1 nucleocapsid protein NCp7 [48]. Spin labels (3-(2-miodoacetamide) proxyl (IPSL) for thio-amido attachment and 3-iodomethyl-(1-oxy-2,2,5,5-tetramethylpyrroline) for thio-methyl attachment) were attached at the positions SLA, SLB, SLC, SLD, SLE, and SLF of mini TAR DNA (Fig. 10.18). Measurement of distance between from several pairs of double labels (SLAB, SLCD, SLEF) within mini TAR DNA in the presence of NCp7 and complementary mini TAR RNA revealed marked changes both from the initial stem-loop structures in the absence of NCp7 to the case of duplex DNA–RNA. Specifically, when duplex TAR DNA-TAR RNA formed, double labels initially located 27.5 Å apart at the 30- and 50-termini of the 27-base mini TAR DNA and then relocated to opposite ends of a 27 bp RNA–DNA duplex with 76.5 Å between labels. Different sets of double labels initially located 26–27 Å apart in the mini TARm DNA upper stem and altered their interlabel distance to ~35 Å when 27 bp TAR DNA-TAR RNA duplex formed.

Structure of the two N-terminal dsRNA-binding domains (dsRBDs) of Dicer-binding partner TRBP in complex with a functionally asymmetric siRNA using *NOESY* NMR, DEER, and single-molecular FRET spectroscopy was solved [50]. Using suggested model a Dicer-TRBP-siRNA ternary complex, it was shown that

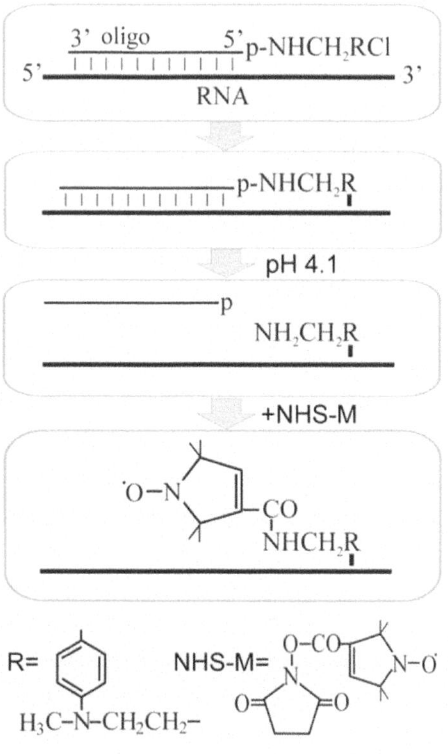

Fig. 10.19 Scheme of site-specific introduction of spin labels into definite RNA sites based on the complementary-addressed alkylation of the RNA with [4-(N-2-chloroethyl-N-methylamino)benzyl]-phosphoramides of oligodeoxyribonucleo-tides, hydrolysis of phosphoramide bond in the covalent adduct formed and selective acylation of the released aliphatic amino group by N-hydroxysuccinimidyl derivative of the spin label [51]. Reprinted from [51], Copyright 2016 Oxford International Press

TRBP's dsRBDs and Dicer's RNase III domains bind a canonical 19 base pair siRNA on opposite sides. This finding supports a mechanism whereby TRBP influences Dicer-mediated cleavage accuracy by binding the dsRNA region of the pre-miRNA during Dicer cleavage.

Hepatitis C Virus RNA internal ribosome entry site consisting of ≈330 nucleotides and having a complicated spatial structure was modified by spin labels (Fig. 10.19).

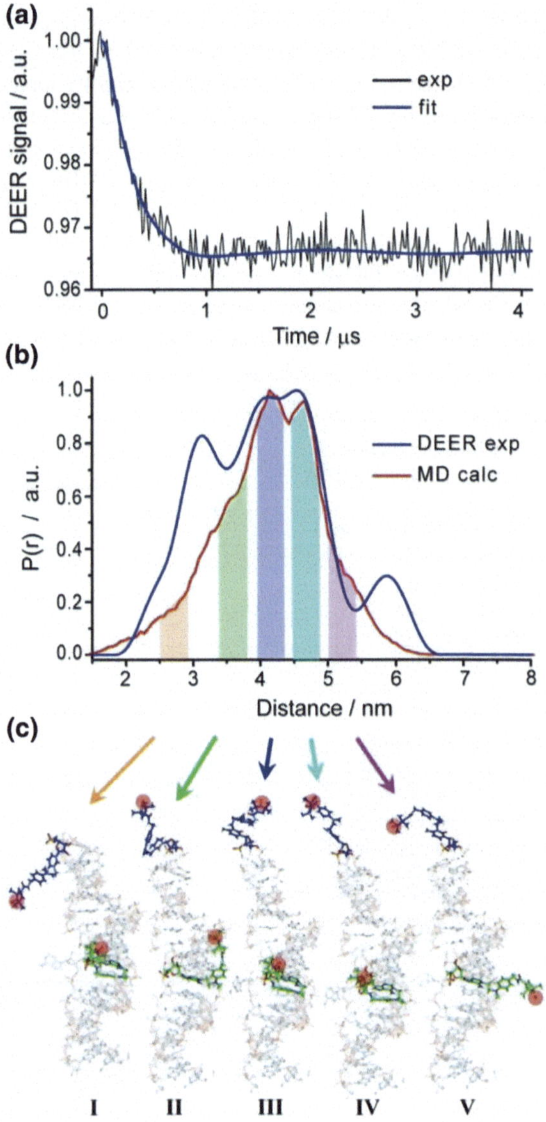

Fig. 10.20 Distance measurements on dslRNA. **a** Background-corrected Q-band DEER/PELDOR time trace (exp) and Deer analysis fitting (fit); **b** obtained distance distribution using Tikhonov regularization parameter 1000 (DEER exp) and calculated MD distribution (MD calc). **c** Typical conformations of spin labels corresponding to the selected ranges of distances (highlighted by colored bars in (**b**) and pointed out by corresponding arrows). Red circles indicate the NO group of the label, for clarity. Spin-labeled C83 (top) is shown in blue, and spin-labeled A73 (middle) is shown in green [51]. Reprinted from [51], Copyright 2016 Oxford International Press

Application of pulsed double electron–electron resonance provided spin–spin distance distribution, which agrees well with the results of molecular dynamics (MD) calculations (Fig. 10.20).

An approach that combines high-resolution NMR restraints with low-resolution long-range constraints, based on site-directed spin labeling and measurements of distance distribution restraints in the range between 15 and 80 Å by the four-pulse double electron–electron resonance (DEER), EPR technique, was developed and discussed on example of determination of structures of large RNAs and protein–RNA complexes in solution [52].

The utility and capability of EPR line shape analysis and distance measurements to monitor and describe site-specific changes in the conformational dynamics of internal loop nucleobases as well as helix–helix interactions of the kink-turn motif in the Vibrio cholerae (VC) glycine riboswitch that occur upon sequential K^+-, Mg^{2+}-, and glycine-induced folding were explored by spin labels incorporated into the 232-nucleotide sequence via splinted ligation strategy [53]. Results demonstrate the ability of SDSL to interrogate site-specific base dynamics and packing of helices in large RNAs and demonstrate ion-induced stability of the kink-turn fold of the VC riboswitch.

Data on spin labeling of nuclear acids published by 2016 were briefly reviewed in [54].

10.3 Spin Tools

10.3.1 Background

In the last decade, spin ruler method which is long-distance measurements (2–8 nm) in biomolecules obtained from pulse dipolar electron paramagnetic resonance (PD-EPR) spectroscopy in general, and double electron–electron resonance (DEER) technique in particular have been developed and widely used for solving various structural problems such as investigation of structural changes upon ligand binding, protein–protein, or protein–DNA interactions, the assembly of subunits of proteins, and for accessing the flexibility of particular regions in the biomolecule.

The recent advances of in vivo molecular EPR-based spectroscopy and imaging of tumor microenvironment (TME) and redox using functional paramagnetic probes, and applications of these approaches in various animal models of cancer magnetic resonance imaging (MRI) and low-field electron paramagnetic resonance (EPR)-based techniques were demonstrated in a series excellent works by Khramtsov group [55] and references therein. A principle key role of the important physiological components, such as acidity (pH), oxygen concentration (pO_2), intracellular glutathione (GSH) and other sulfide, redox status, and interstitial inorganic phosphate (Pi) in normal and pathological biological processes (cancer progression in tumor for example) have been established. Progress in development of functional paramagnetic probes together with recent advances of low-field electron paramagnetic resonance (EPR)-based spectroscopy, which provided a reasonable radiofrequency penetration depth in living tissues and allows larger samples to be used, opens the way for quantitative investigation of imaging in vitro and vivo of tumor macroenvironment and redox properties.

The following chemical structures of the EPR multifunctional probes were used in [55]:

pH and redox probes, disulfide biradical probe for GSH detection, pO_2 probe.

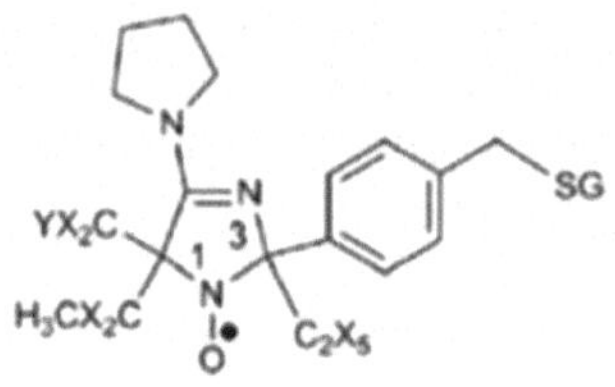

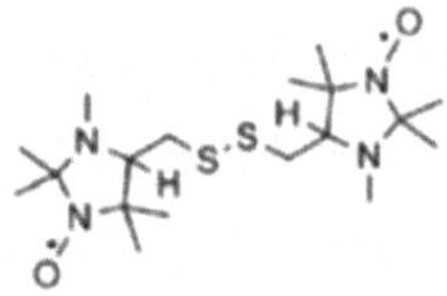

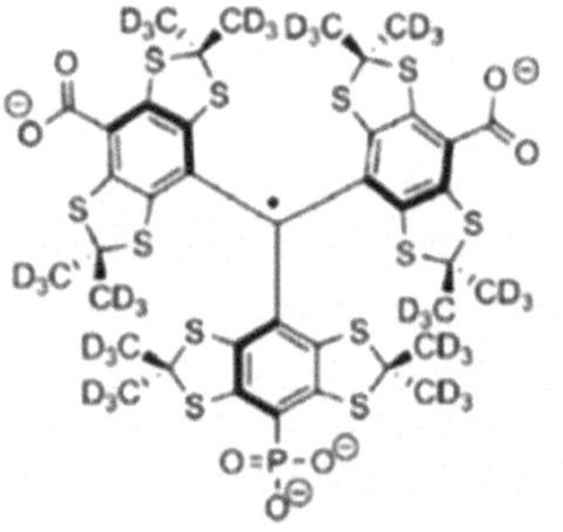

For the detection of spin probes, along with conventional versions of EPR spectroscopy, the method of field cycling proton electron double-resonance imaging (FC-PEDRI has recently been used [55, 56]. A new concept of variable field proton–electron double-resonance imaging (VF PEDRI), which allows for functional mapping using specifically designed paramagnetic probes (e.g., pH mapping) with MRI high-quality spatial resolution and short acquisition time, was proposed [57]. It was demonstrated that this method provides pH resolution of 0.1 pH units and a spatial resolution of 1.25 mm at 200 G field.

10.3.2 Spin Oximetry

Molecular oxygen plays a key role in many biochemical and physiological and pathophysiological processes such as mitochondrial respiration, synthetic and degradative reactions, oxidative damage, cell signaling, so forth. ESR oximetry is a technique that can make non-invasive sensitive and localized measurements of oxygen. Spin oximetry is a version of the method of spin label–spin probe in which molecular oxygen plays the role of a spin probe (Chap. 11). Nitroxides, stable aromatic radicals (derivatives of triarylmethyl radical, Indian ink), phthalocyanine, and micrococrystals (fusinite and carbohydrate chars) show narrow singlet ESR lines sensitive to O_2 concentration in water and cells, organs, small animals, and even parts of human body [55, 58, 59]. For example, the EPR linewidth of triarylmethyl radical can be served as spin probe for the measurement of pO_2 with accuracy 1 mm Hg and range, 1–100 mm Hg.

In pioneering work of James Hide group [60, 61] a method for measuring the oxygen diffusion-concentration product was invented. The method was based on the dependence of the spin–lattice relaxation time T_1 of the spin label on the bimolecular collision rate with oxygen [62]. Both time-domain and continuous-wave saturation methods showed strong Heisenberg exchange between spin label and oxygen which contributes directly to the T_1 of the spin label. Advantages in measurement of the oxygen transport parameter using saturation-recovery EPR have been clearly demonstrated in saturated (DMPC) and unsaturated (POPC) lipid bilayer membranes with the use of stearic acid (*n*-SASL) and phosphatidylcholine (*n*-PC) spin labels. Figure 10.21 illustrates a profile of the oxygen transport parameter across the POPC membrane obtained by X-, Q-, and W-band spectroscopy.

The effect of O_2 on the spin lattice relaxation time T_1, derived from saturation curves of the CW ESR spectra of nitroxide radicals, was examined for radicals incorporated into phosphatidylcholine (PC) liposomes and attached to lysozyme and

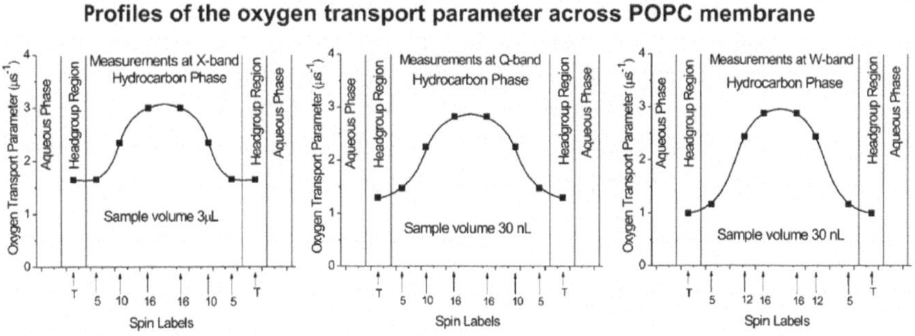

Fig. 10.21 A profile of oxygen transport (accessibility) obtained at X-, Q-, and W-band. Measurements and equilibrations with gas for X- and Q-band were performed at 29–30 °C. Measurements and equilibrations with gas for W-band were performed at room temperature [61]. Reprinted from [61], Copyright 2011 Elsevier

albumin [63]. O_2 increased the rates of spin–lattice and spin–spin relaxation by the factor $k_{ex}C$, where k_{ex} is a spin-exchange constant and C is the oxygen concentration. The experiments indicted that for the liposomes, $k_{ex}C$ increased as the spin label approached the center of the bilayer, while for the tagged proteins, $k_{ex}C$ was very close to that of the radicals in aqueous solution.

10.3.3 Spin pH Meter

A spin pH probes technique for the measurement of local acidity in biological and non-biological objects including organs and even living organisms was invented in work of Khramtsov, Weiner, Grigor'ev, and Volodarsky in 1982 and since then found wide application in biological systems, in vivo in particular [55, 64–68]. EPR spectra of stable nitroxides of the imidazoline and imidazolidine types

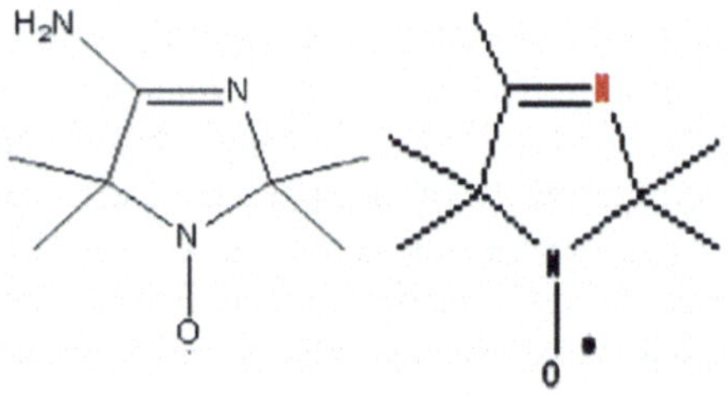

have been shown to be sensitive to pH and serve as spin pH probes.

Data on the synthesis of a wide set of pH-sensitive nitroxides is a pH spin probe (in the range from 6 to 8.0 with accuracy −0.05) of different sensitivity, stability to reduction, lipophilicity, covalent binding properties to macromolecules, and applications have been reported [65, 66, 68].

Typical pH-dependence of the observed hyperfine splitting constant, a_N, DNP spectra of the nitroxide NR1 acquired in acid, pH 4.98, and slightly alkaline, pH 7.62, and the effect of protonation of the radical heterocycle resulting in decreasing unpaired electron density at the nitrogen nucleus of the N–O fragment were presented. pH dependencies of the hyperfine splitting, a_N, for R_1 probe measured using L-band EPR spectroscopy are shown in Fig. 10.22.

10.3.4 Spin Redox Probe

Redox reactions play a key role in the fundamental chemical, photochemical, and biological processes and in the provision of energy of living organism, in particular. The spin redox probe techniques utilize ability of nitroxides and corresponding hydroxyl amines to proceed in following chemical reactions which include: (1) reduction of a

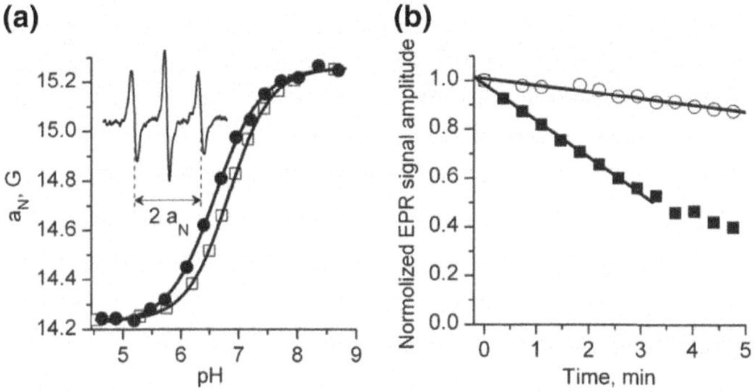

Fig. 10.22 pH dependencies of the hyperfine splitting, a_N, for R_I probe measured using L-band EPR spectroscopy. Temperature-dependent shift of the titration curve was observed with pK_a values being equal to 6.84 (□, 23 °C) and 6.60 (●, 37 °C). Insert: L-band EPR spectrum of the R_I probe measured in vivo after injection (10 μL, 10 mM) in mammary tumor tissue. The hyperfine splitting, a_N, was found to be equal to 14.72 G which corresponds to the value of $pH_e = 6.52$ assuming tumor tissue temperature 34 °C and $pK_a = 6.65$. **b** EPR signal decay of the probe R_I after injection (10 μL, 10 mM) in mammary glands (○) and mammary tumors (■). The analysis of the initial part of the kinetics yields the rates of the EPR signal reduction, k_{red}, in extracellular media of the tissues being equal to 0.5×10^{-3} s^{-1} and 2.5×10^{-3} s^{-1}, respectively [65]. Reprinted from reference [65], Copyright 2012 Wiley

nitroxide with reducing agent to corresponding hydroxyl amine, (2) oxidation of a nitroxide to oxoammonium cation and (3) oxidation of hydroxyl amine with an oxidant to correspondent nitroxide. All these reactions run as the simple one-electron processes and can be readily followed by the standard and advance ESR technique [5, 68–75]. The values of the nitroxide redox potential depend on their chemical structures (Chap. 3). For example, the redox potential of piperidine derivatives nitroxide is high enough to oxidize such biological compounds as ascorbic acid, semiquinones, and superoxide radical. As long ago as 1968, reduction of nitroxides was observed in electron transport in mitochondria [75]. Later, such processes were found to be typical of various biological systems.

A typical example of determination of in vivo pH and redox assessment of TME in mouse model of breast cancer using dual-function NR1 probe is described in Fig. 10.22 [65].

Real-time assessment of extracellular pH (pH(e)), redox, and intracellular glutathione were monitored in PyMT mice-bearing breast cancer tumors during treatment with granulocyte macrophage colony-stimulating factor using L-band EPR spectroscopy and nitroxide probe and disulfide nitroxide [76]. It was observed that (1) tumor pH(e) is about 0.4 pH units lower than that in normal mammary gland tissue, (2) treatment with granulocyte macrophage colony-stimulating factor decreased the value of pH(e) by 0.3 units compared with PBS control treatment, and (3) tumor tissue reducing capacity and intracellular glutathione were elevated compared with normal mammary gland tissue, and (4) granulocyte macrophage colony-stimulating factor treatment resulted in a decrease of the tumor tissue reducing capacity and intracellular glutathione content. In addition, pH(e) mapping was performed using recently proposed variable frequency proton–electron double-resonance imaging (VF PEDRI). The pH mapping superimposed with MRI image supports probe localization in mammary gland/tumor tissue showed high heterogeneity of tumor tissue pH(e) and a difference of about 0.4 pH units between average pH(e) values in tumor and normal mammary gland.

10.3.5 SH Moiety Assay

SH moiety assay is based on the use of RSSR probes (Fig. 10.21), in which two nitroxides are tethered with flexible bridge. The dynamic spin exchange in moment of close contact between the nitroxide leads to appearance new satellites in EPR spectra [55, 74, 77]. The reaction of thiol/disulfide exchange with a sulfide-bearing compound (glutathione, GSH for example) splits the RSSR disulfide bond resulting in formation of two monoradicals and cancelation of intramolecular spin-exchange between the monoradical fragments. Reaction of the RSSR probe with glutathione on the EPR spectra resulted in disappearance of the biradical spectral components and a corresponding increase of the intensity of the monoradical components. The rate of the increase of the amplitude of the monoradical component is proportional to the GSH concentration.

As an example, in the SH moiety assay, a set of RiSSRi biradicals (Fig. 10.23) was employed.

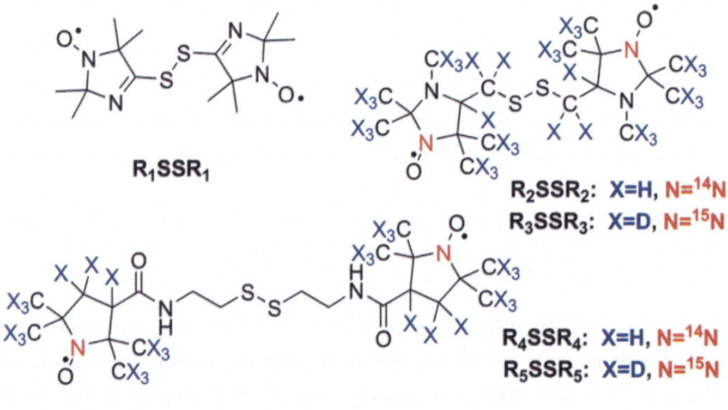

Fig. 10.23 Chemical structures of the imidazoline, R_1SSR_1, imidazolidine, R_2SSR_2 and R_3SSR_3, and pyrrolidine disulfide biradicals, R4SSR4 and R5SSR5 [77]. Reprinted from [77], Copyright 2017 American Chemical Society

10.3.6 Spin Imaging

The basic principles of using stable organic radicals involved in reversible exchange processes as functional paramagnetic probes were reviewed [77, 78]. It was demonstrated that these probes in combination with electron paramagnetic resonance (EPR)-based spectroscopy and imaging techniques provide analytical tools for quantitative mapping of critical parameters of local chemical microenvironment.

One way of field-cycled proton–electron double-resonance imaging (PEDRY) of free radicals designed in Lurie group and realized in collaboration with Khramtsov group was to use so-called phantoms [78, 79]. Images were acquired separately by changing the external magnetic field for EPR irradiation between ^{14}N and ^{15}N nuclei in FC-PEDRI. The center of the phantom consisted of several tubes of internal diameters 3–15 mm filled with a spin probe. For example, five tubes of internal diameters 15, 9, 5, 4, and 3 mm are filled with 2 mM TEMPOL solution. These were enclosed in a cylindrical container of diameter 4 cm that was filled with water doped with copper sulfate to give the same T_1 as that of the free radical solution (650 ms at 2.5 MHz). Fourteen sample tubes with internal diameters of 8 mm were attached around the outside of the cylinder, with alternate tubes being filled with 2 mM TEMPOL solution or copper sulfate-doped water. The overall diameter of the phantom was 6 cm, about the size of a small rat. The feasibility of redox imaging using nitroxyl radicals was validated by in vitro experiments showing the time-dependent redox reaction of both membrane-permeable $^{14}N/^{15}N$-3-methoxycarbonyl-2,2,5,5-tetramethyl-pyrrolidine-1-yloxy (MC-PROXYL) and membrane-impermeable $^{14}N/^{15}N$ carboxy-PROXYL in the presence of ascorbic acid. Images were acquired separately by changing the external magnetic field for EPR irradiation between ^{14}N and ^{15}N nuclei in FC-PEDRI. Seven phantom tubes comprising six outer tubes with liposomes encapsulated with 100 mM ascorbic acid were shown.

Simultaneous molecular imaging of redox reactions monitored by Overhauser enhanced MRI with ^{14}N- and ^{15}N-labeled nitroxyl radicals was reported in [80]. Time-dependent PEDRI of $^{15}N/^{14}N$-carboxy-PROXYL and $^{15}N/^{14}N$-MC-PROXYL in liposomes encapsulating ascorbic acid (100 mM) was detected in seven phantom tubes in the presence and the absence of liposomes. The images of the decay rates for ^{14}N- and ^{15}N-enhanced PEDRI images were calculated by assuming first-order kinetics for the time-dependent decrease of the contrast, and only MC-PROXYL showed decay images in both ^{14}N- and ^{15}N-enhanced PEDRI [80].

A pH map was extracted from two PEDRI acquisitions performed at EPR frequencies of protonated and unprotonated forms of a synthesized pH-sensitive probe [81]. Probe deuteration resulted in a narrow spectral line of 1.2 G compared to a non-deuterated analog line width of 2.1 G allowing for an increase of Overhauser enhancements and reduction in rf power deposition. Binding of the probe to the cell-impermeable tripeptide, glutathione (GSH), allowed for targeting to extracellular tissue space for monitoring extracellular tumor acidosis. In vivo VRF PEDRI was performed on Met-1 tumor-bearing mice (Fig. 10.24), and broad pH distribution with acidic mean pH_e (6.8 ± 0.1) in tumor tissue was observed.

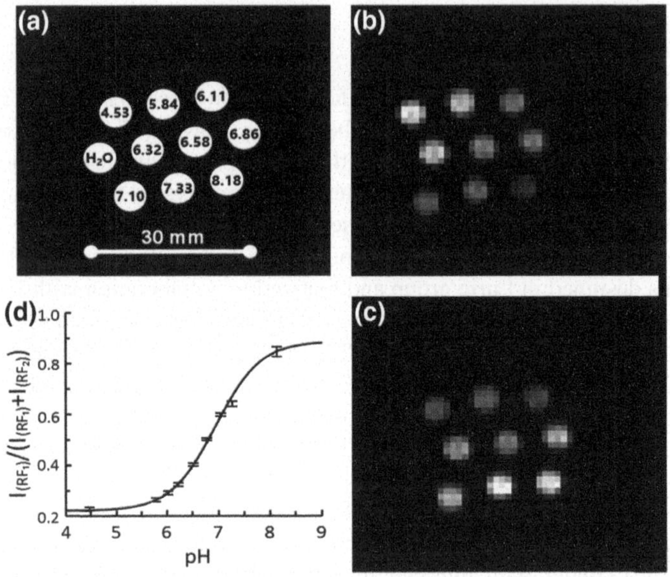

Fig. 10.24 VRF PEDRI pH calibration (**a–d**) and in vivo pH mapping (**e**, **f**) by using NR2 probe. Calibration has been performed by using a 10-tube phantom (**a**) at room temperature. PEDRI images were acquired at two EPR irradiations, $RF_2 = 559.3$ MHz (**b**) and $RF_1 = 562.1$ MHz (**c**), acquisition time, 8.4 s. The tube with water alone is not visible on either image due to lower intensity compared with the signal from tubes containing the paramagnetic probe. EPR off image (not shown) was subtracted from both EPR on images acquired at RF_1 and RF_2, yielding image intensities, $I_{(RF_1)}$ and $I_{(RF_2)}$. **d** pH dependence of the ratio $I_{(RF_1)}/(I_{(RF_1)}+I_{(RF_2)})$. Solid line is nonlinear least-squares fit of the data to a conventional titration equation, yielding pK_a value equal to 6.75–0.05. Error bars represent the standard deviation. In vivo PEDRI pH mapping was performed in the anesthetized breast tumor-bearing mouse (**e**, **f**). Paramagnetic NR2 probe was injected into the tumor (number 4 mammary gland, left) and normal mammary gland (number 9, right). **e** Qualitative visualization of the distribution of the NR2 probe, in vivo. The image is the average of two PEDRI images acquired at two EPR frequencies, $RF_1 = 562.1$ MHz and $RF_2 = 559.3$ MHz, NMR frequency, 784.9 kHz, matrix, 64×64; field of view, 80 mm. Irradiation time was 8.4 s for each acquisition. **f** pH map (in color) calculated from two PEDRI images superimposed with the MRI image (gray scale) showing the coronal view of the mouse [81]. Reprinted from [81], Copyright 2017 American Chemical Society

DNP-MRI imaging of free radical intermediates that are derived from endogenous species involved in metabolic processes was first reported [82]. Simultaneous images of free radical intermediates generated from the coenzyme Q10 (CoQ10), flavin mononucleotide (FMN), and flavin adenine dinucleotide (FAD) involved in the mitochondrial electron transport chain as well as carbamoyl-PROXYL and the radicals derived from vitamins E and K1 were visible. Each of the radical species can be distinguished in the spectroscopic images by the changing the frequency of irradiation. Figure 10.25 shows the EPR signals of the intermediate of various candidate probes.

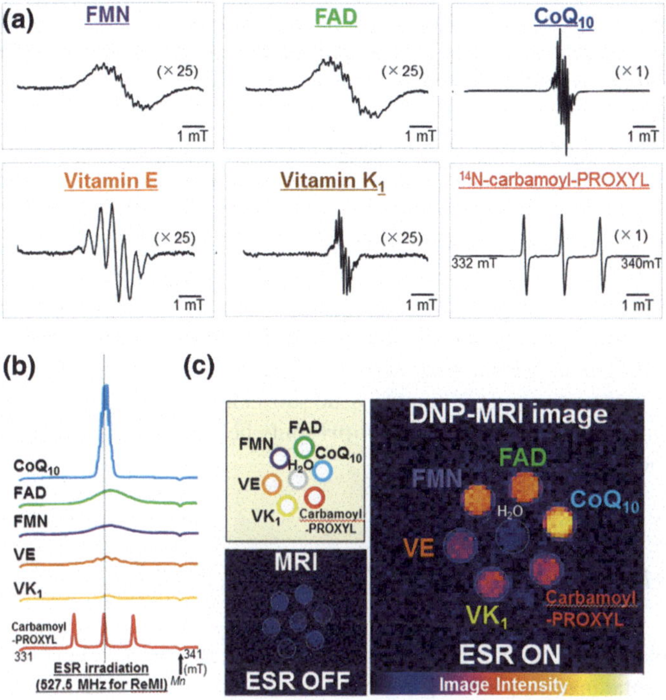

Fig. 10.25 Typical EPR spectra (**a**), the absorption EPR spectra (**b**), and simultaneous DNP-MRI image (**c**) of free radical intermediates from FMN, FAD, CoQ10, vitamin E, vitamin K1, and the synthetic stable radical ^{14}N carbamoyl-PROXYL. The vertical solid line in part **b** indicates the frequency of EPR irradiation for PEDRI (527.5 MHz), which is the resonance frequency of the central peak of carbamoyl-PROXYL. MRI spectra with and without EPR irradiation were obtained with a spin echo sequence at 850 kHz by using a homemade in vivo PEDRI system kept at room temperature [82]. Reprinted from [82], Copyright 2014 American Chemical Society

Two nitroxyl radicals, the cell membrane penetrable (carbamoyl-PROXYL) and the cell-impermeable (carboxy-PROXYL), were used in the in vivo PEDRI experiments to investigate the redox reactions which can occur in the intracellular or extracellular compartment [83]. Nitroxyl radical solution was injected to the same region of the mouse leg that had received BPVC pretreatment. In vivo DNP-MRI images and redox maps at 4 and 24 h after BPVC treatment in mice, taken after injection of 2.5 mM carbamoyl-PROXYL (50 lL) into both legs, were presented. Redox maps were calculated by using four pharmacokinetic in vivo DNP-MRI images, and decay rates were calculated for the DNP-MRI images of untreated and BPVC-treated legs. Differences between decay rates in BPVC-treated versus untreated (contralateral) legs in individual mice were shown. In vivo DNP-MRI images and redox map at 4 and 24 h after BPVC treatment in mice, taken after injection of 2.5 mM carboxy-PROXYL (50 lL) into both legs, were also indicated. Thus, the decay rate of the BPVC-treated leg was found to be faster than that of the control leg.

Real-time assessment of extracellular pH (pH(e)), redox, and intracellular glutathione was monitored in PyMT mice-bearing breast cancer tumors during treatment with granulocyte macrophage colony-stimulating factor using L-band EPR spectroscopy and nitroxide probe and disulfide nitroxide [76]. It was observed that (1) tumor pH(e) is about 0.4 pH units lower than that in normal mammary gland tissue, (2) treatment with granulocyte macrophage colony-stimulating factor decreased the value of pH(e) by 0.3 units compared with PBS control treatment, (3) tumor tissue reducing capacity and intracellular glutathione were elevated compared with normal mammary gland tissue, and (4) granulocyte macrophage colony-stimulating factor treatment resulted in a decrease of the tumor tissue reducing capacity and intracellular glutathione content. In addition, pH(e) mapping was performed using the VF PEDRI). The pH mapping superimposed with MRI image supports probe localization in mammary gland/tumor tissue shows high heterogeneity of tumor tissue pH(e) and a difference of about 0.4 pH units between average pH(e) values in tumor and normal mammary gland.

10.3.7 Molecular Dynamics. Spin Viscose Meter

The principles underlying the application of the nitroxide spin-label method as a tool for experimental investigation of protein molecular dynamics ("breathing") were formulated in the late 1960s and early 1970 by Likhtenshtein [62, 84, 85]. In parallel, complementary research on protein dynamics was performed by a proposed Moessbauer atom labeling [5, 86]. Transglobular conformational change in myoglobin and lysozyme as a consequence of a protein molecular dynamics was first demonstrated in [62].

A suggested approach for the investigation of protein dynamics relied on the well-known finding that the mobility of a molecule in a condensed phase is modulated to a great extent by the molecular dynamics of the surrounding molecules. Therefore, the particulars of nitroxide motion report on characteristics of the dynamics of the surrounding medium. An important feature of nitroxides is their structural anisotropy which causes anisotropy of magnetic interactions, such as the hyperfine splitting and the g-tensor. Radical motion leads to averaging of the hyperfine interaction and the g-tensor which causes significant changes to the ESR spectrum. Modern ESR techniques (Chap. 4) allow ones to access dynamic processes such as rotation and wobbling with a wide range of correlation time, $\tau_c = 10^2$–10^{-10} s and amplitude ending low-amplitude high-frequency vibration and the phonon processes in media. Essential knowledge about molecular dynamic state of system under investigation can be derived from the measurement of dynamic interactions between nitroxides and other paramagnetics at its encounters.

The following parameters of spin probes in liquid glasses, polymers, membranes, biomolecules, and proteins are sensitive to low-amplitude high-frequency and phonon dynamics: (1) line width of a radical pair (RP), (2) RP ESR signal splitting, (3) amplitude of RP ESR signal, (4) spin packet of a nitroxide probe $\Delta H_{1/2,}$

(5) nitroxide spin phase ($1/T_{1e} \sim \Delta H_{1/2}$) and spin–lattice relaxation rate ($1/T_{1e}$), (6) ESR spectra of nitroxides line width δ, and (7) nitroxides hyperfine splitting A_{aniso}. Atomic vibration and molecular motions of a nitroxide affect the spin relaxation parameters and line shapes of the nitroxide ESR spectra.

The first dynamic theory of the nitroxide ESR spectra for nitroxide rotation in the fast ($\tau_c = 10^{-9}$–10^{-10} s) and slow regions ($\tau_c = 10^{-7}$–10^{-8} s) was developed by Kivelson [86] and Freed [87], correspondingly. In the case of spin-labeled macromolecules (polymers, proteins, membranes, and nuclear acids), the situation appears to be more complicated when the nitroxide segment can be involved at least in two types of motions: high-amplitude low-frequency ($\tau_c \geq 10^{-8}$ s) motion or/and low-amplitude high-frequency wobbling ($\tau_c \leq 10^{-9}$ s) [87]. To distinguish from above-mentioned models, a method based on an analysis of correlation between the values of the A_{zz} shift and the line width induced by temperature and viscosity change was proposed [88]. Comparison of experimental data for a series of spin-labeled bioobjects, including biomembranes and proteins, indicated that the rotation of the nitroxyl fragments of the spin labels at an ambient temperature can be described by the model of slow anisotropic rotation with correlation time 10^{-7}–10^{-8} s under conditions where the rotation of the macromolecules and membranes is slow in the ESR time scale and can be neglected [88]. Pulsed multifrequency EPR was used to investigate orientational molecular motion of the nitroxide spin probe ($[(SO_3)_2NO^{\cdot}]$, Fremy's salt) in glycerol glass near the glass transition temperature [28, 29, 89, 90]. Measuring echo-detected EPR spectra at different pulse separation times at resonance frequencies of 3, 9.5, 95 and 180 GHz of Fremy's salt, in glycerol glass near the glass transition temperature allowed for estimating parameters of transverse relaxation tensor and thus for discriminating between different relaxation mechanisms and characterize the timescale of molecular reorientations (10^{-7}–10^{-10} s).

In work [91], dynamic behavior of di-tert-butyl nitroxide in aqueous solution has been investigated by means of an integrated computational approach, including Car–Parrinello molecular dynamics and quantum mechanical calculations involving a discrete–continuum embedding. Decoupling of the structural, dynamical, and environmental contributions to the nitroxide ESR spectra allowed one to elicit the role played by different dynamic effects and stress the importance of specific vibration modes. Data on motions in a protein detected by changes in the average distance and/or the shape and width of the distribution using DEER have been reported [92]. The following findings were described: (1) DEER detection of triggered conformational changes; (2) motion of a transmembrane helix during the transition from state A to state B alters the average distance (r_{av}) between spin labels; (3) the rotameric ensemble of each label generated from a rotamer library using the program MMM; (4) A and B are distinct conformers of different energies; (5) the conformational shift manifests primarily as a change in r_{av}, and (6) altering the biochemical conditions alters the contribution of each distinct conformation (dashed curves) to the distance distribution.

An extensive set of electron spin resonance spectra of the dynamics of spin-labeled T4 lysozyme was obtained over a wide range of frequencies (9, 95, 170, and 240 GHz) and temperatures (2–32 °C) [93]. The native side chain at solvent-exposed helical sites, 72 or 131, was labeled with nitroxide side chain (R1), or a methylated analog with hindered internal motion (R2). The spectra at all four frequencies were simultaneously fit with the slowly relaxing local structure (SRLS) model which suggests the global tumbling of the protein and the internal motion consisting of backbone fluctuations and side chain isomerizations. Quantitative spin labeling of SecB mutants containing a single cysteine per subunit or an exposed highly reactive new cysteine after removal of the nearby intrinsic cysteines was achieved with the methanethiosulfonate spin label (MTS) at positions C97 or E90C, respectively [94]. High-field (W-band) EPR measurements revealed that in labeled bovine pancreatic trypsin inhibitor (BPTI), the spin labels are exposed to a more polar/hydrophilic environment. Binding of BPTI led to a slight change in distances between labels at C97 but not at E90C.

General aspects of the dynamics of the tether linking methanethiosulfonate (MTSSL) spin probes to α-helices have been investigated with the purpose of rationalizing its effects on ESR line shapes [95]. Torsional profiles for the chain bonds have been calculated ab initio, and steric interactions with the α-helix and the neighboring residues have been introduced at the excluded-volume level. The implications for the ESR spectra of spin-labeled proteins were discussed, and suggestions for the introduction of realistic features of the spin probe dynamics into the line shape simulation were presented.

To investigate the single transmembrane (TM) domain conformational properties of integrin, β1a, 26 consecutive, single-cysteine mutants were generated for site-directed spin-labeling and continuous-wave electron paramagnetic resonance (CW-EPR) mobility and accessibility analyses [96]. The following results were presented: (1) the mobility analysis identified two integrin β1a-TM regions with different motional properties in micelles and an integrin β1a-TM helix with high immobility in liposomes; (2) the accessibility analysis verified the TM range (Val737-Lys752) of the integrin β1a-TMC in micelles; and (3) mobility and accessibility *comparisons* of the integrin β1a-TMC domains in micelles or liposomes identified a monomer embedded in detergent micelles and leucine-zipper-like homo-oligomeric clusters in liposomes were established. The phase diagram of 1,2-dipalmitoyl-*sn*-glycerophosphatidylcholin labeled (labeled at the end-chain 16-PC)–cholesterol binary mixtures versus temperature was studied using 2D-ELDOR at Ku band with the "full Sc" method [97]. The line shape changes and the homogeneous T_2's, extracted from the pure absorption spectra in the 2D-plus-mixing-time representation, allowed the characterization of the membrane phases with respect to their dynamic molecular structures and to determine the phase boundaries.

The dynamic behavior of spin labels with nitroxides located in different positions on the aliphatic chains in liposomes was first investigated with high-frequency (2-mm) ESR spectroscopy [98]. On the basis of temperature dependence of the spin probe spectra over the range temperatures, it was concluded that in the temperature range 220–260 K, the nitroxide rotation is essentially anisotropic with correlation time $\tau_c = 10^{-7}$–10^{8} s and about 10^{-9} s at physiological temperatures.

10.3.8 Spin Polarity Meter

The dependence of isotropic and anisotropic hyperfine splitting (A_{iso} and A_{aniso}) and g-factor for nitroxides on the spin density on nitrogen *atom and the energy of the* $n \rightarrow$ * *transition* in the N–O$^{\cdot}$ fragment makes this probes suitable for assessing local polarity and the availability of hydrogen bond donors to form H-bond with the N–O$^{\cdot}$ group [70, 99–102]. The physical reason for such a dependence is that the contribution of polarized resonance structure N–O$^{\cdot}$ $\leftrightarrow$ N^{+}–O^{-} increases in a polar media, causing an increase in the spin density on nitrogen atom observed as an increase in the A_{iso} value. At the same time, the A_{aniso} values are larger for the ionic structure.

The isotropic ^{14}N-hyperfine coupling, a_0^N, of nitroxyl spin labels depends linearly on the unpaired electron spin density on the nitrogen atom (ρ_π^N) and to a lesser extent on that on the oxygen atom (ρ_π^O)v [99]:

$$a_0^N = Q_N \rho_\pi^N + Q_{NO} \rho_\pi^O,$$

where $\rho_\pi^N + \rho_\pi^O \approx 1$ and the leading term is that involving Q_N ($\gg Q_{NO}$). The spin density distribution, and hence the hyperfine coupling, is perturbed linearly by the reaction field from the polar environment [22]. The polarity dependence of the principal z-element of the hyperfine tensor is determined by both the isotropic and anisotropic terms:

$$A_{zz} = a_0^N + 2\left|T_\perp^d\right|,$$

where $-\left|T_\perp^d\right|$ is the perpendicular element of the traceless hyperfine tensor that arises from the electron–nuclear dipolar interaction.

The g-factor values are also reported to be sensitive to the medium. A correlation between a decrease of g_X and an increase in A_z has been observed for nitroxides in simple liquids and labeled proteins [101]. A polarity effect (up to $\Delta A_{iso} = 0.13$ mT) was observed for the TEMPOL radical and for the formation of H-bond with N–O group of the nitroxides $\Delta A_{iso} = 0.15$ mT). ESR spectra of a nitroxide probe in solvent of various polarities can be used for estimation of local apparent dielectric constant in the vicinity of a nitroxide incorporated in system of interest.

The dependences of electronic paramagnetic resonance properties on environmental dielectric permittivity and proticity in biomolecular assemblies were reviewed [99].

10.4 Alternative Spin Labeling

The most commonly used SLs are nitroxyl radicals, but recently SL new approaches have been shown to be an attractive alternative for PD-EPR, particularly double electron–electron resonance (DEER). The first one, based on high-spin Gd^{3+} (S = 7/2) complexes, was designed and developed in Goldfarb group [103–109]. The long spin relaxation, narrow EPR signals, neutral chemical character, and stability of triarylmethyl radicals (trityls, TAMs) trityls at room temperature in liquid solutions make them a promising alternative for traditional nitroxides [15].

The following advantages of the first approach can be pointed out: (1) high sensitivity because high-field measurements at spectrometer frequencies higher than 30 GHz are associated with an increase in the absolute sensitivity; (2) Gd^{3+} complexes are stable toward oxidation or reduction, and therefore, it can be used in the reductive environment in cells; (3) neutral chemical character of this compounds, possessing neutral chemical character, does not alter the chemical properties of the parent molecule region to which they are attached; (4) and the long spin relaxation and narrow EPR signals.

The C7-Gd and C8-Gd tags, which are compact hydrophilic cyclen-based lanthanide tags for conjugation to cysteine residues in proteins, were loaded with Gd(III) on two mutants of the homodimeric ERp29 protein and investigated by EPR (Fig. 10.26) [105].

The W-band EPR spectra were found to differ between the tags in the free state and after conjugation to the protein. In addition, the spectra were sensitive to the labeling position, which may originate from an environment-dependent charge density on the Gd(III)-coordinating oxygens.

Two small and uncharged Gd(III) tags, propargyl-DO3A and C11 (Fig. 10.27), were attached to *E. coli* aspartate/glutamate-binding protein and the Zika virus NS2B–NS3 protease [106]. Distance measurements between the label on the proteins were performed by double electron–electron resonance experiments. Echo-detected EPR spectra and echo-decay spectra measurements were carried out at 10 K on a home-built pulse EPR spectrometer operating at W-band (94.9 GHz). Owning their

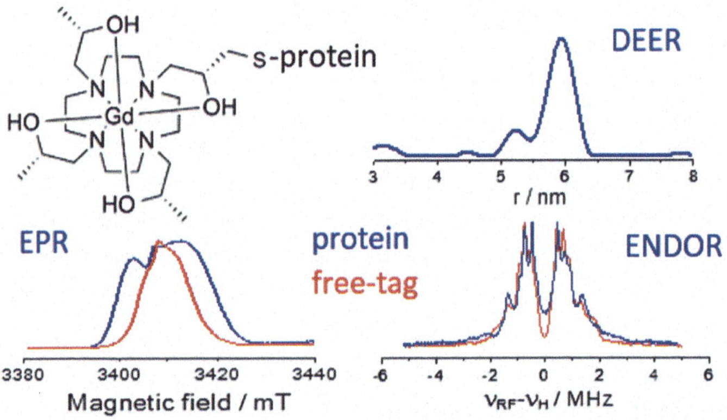

Fig. 10.26 Schematic presentation of Small Gd(III) Tags for Gd(III)–Gd(III) distance measurements in proteins by EPR spectroscopy [108]. Reprinted from [108], Copyright 2014 American Chemical Society

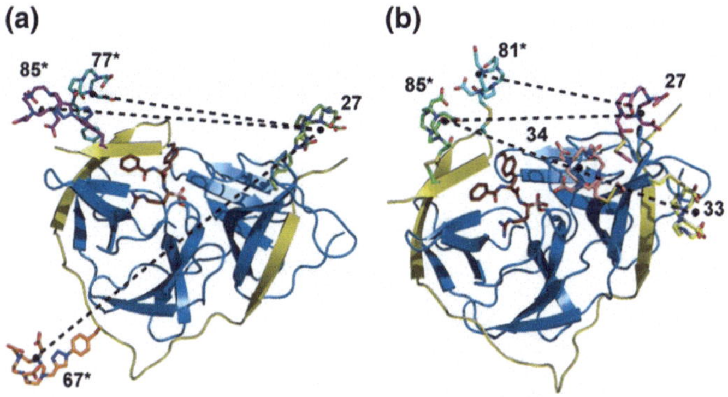

Fig. 10.27 Crystal structure of the Zika virus NS2B–NS3 protease (PDB: 5LC0) [34] with Gd(III) tags modeled at different sites. NS2B and NS3 are shown as yellow and blue ribbons, respectively. Tagged residues and the inhibitor cn-716 (indicated in orange) are shown in a line representation. Distances measured by DEER experiments are shown as dotted lines. **a** Structure with AzF residues tagged with propargyl-DO3A in positions 67*, 77*, 85* and 27. **b** Structure with cysteine residues tagged with C11 in positions 81*, 85*, 27, 33, and 34 [106]. Reprinted from [106], Copyright 2018 RCS

compact structure, the tags delivered narrower distance distributions. A closed conformation determined by X-ray crystallography in the presence of the high-affinity inhibitor cn-716 was confirmed in the EPR experiments, in which DEER measurements were performed in the presence as well as in the absence of this inhibitor. In this conformation, NS2Bc contributes to forming the substrate-binding site of the protease (Fig. 10.27).

The properties of the conformational landscape of a biomolecule, especially for proteins with disordered regions, are of capital importance to understand its function. To provide an upper bound of the statistical weight of each conformation of Ca^{2+}-bound calmodulin (CaM), Maximum Occurrence (MaxOcc) approach in combination with DEER was proposed [107]. For probing the interdomain distance distributions, three different mutants of calmodulin, N53C-T110C, N53C-A103C, and T34C-T117C, were labeled by Gd^{3+} DOTA-maleimide compound. One label was situated in the C-terminal domain and the other in the N-terminal domain. The reduction of the space that is covered by conformations with MaxOcc Z30% was appreciated with the viewpoint along the central helix, i.e., the fourth helix of the N-terminal domain. DEER data were highly complementary to NMR and small-angle X-ray scattering.

The spin labels C7-Gd

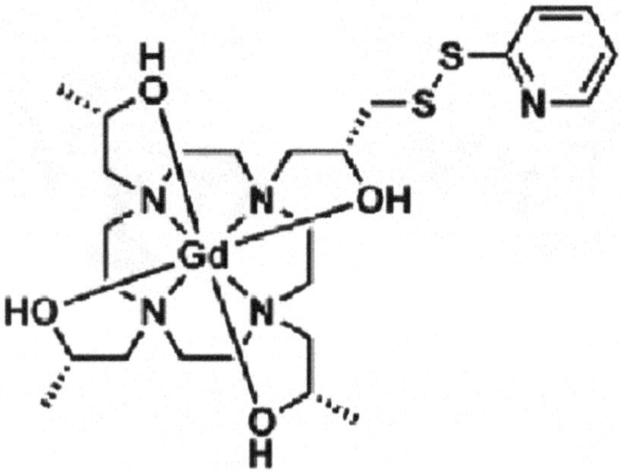

C7-Gd

and C8-Gd were loaded on two mutants of the homodimeric ERp29 protein (S114C/C157S and G147C/C157S) [108]. The labeled protein was investigated by an arsenal advance EPR techniques including the ^{1}H ENDOR, Mims ^{2}H-ENDOR, DEER, echo-detected EPR of 94.9 and 240 GHz. Due to their small size, short tether to the protein broad central EPR transition, narrow distance distribution, and sensitivity of EPR spectra to their nearby protein environment, the C7 and C8 tags were demonstrated superior for measurements of short (<4 nm) distances.

A combined method, employing NMR and EPR spectroscopies, was used to compare different types of nitroxide-based and Gd(III)-based spin labels attached to isolated RBDs of the polypyrimidine-tract-binding protein 1 (PTBP1) and to short RNA fragments (Fig. 10.28) [109]. DEER and continuous-wave EPR spectroscopy performance, resulting distance distributions, and their consistency with the predictions from the spin-label rotamers analysis were described in detail. As an example, the assumption of incomplete RBD/RNA complex formation was tested by DEER measurements in three-spin systems with two Gd(III)-based or nitroxide-based spin labels attached to the RBD2 and one IAP spin label at the U15 site of the RNA stem loop (Fig. 10.29).

In the last three decades, triarylmethyl (TAM)

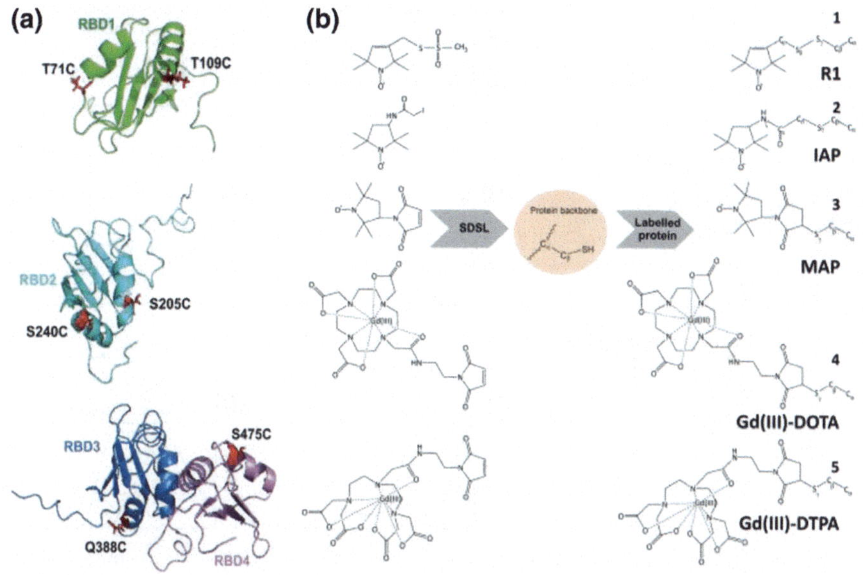

Fig. 10.28 Solution-NMR structures of the individual RBDs showing the selected mutation *EXIST* positions (red) in the a-helices of each domain (**a**). RBD34 is treated as one inter-domain construct, i.e., as a single rigid body, because of its hydrophobic interactions. SDSL with the respective spin labels, nitroxide radicals (1–3), and Gd(III)-based spin labels (4, 5), is shown in (**b**). The MTSSL-based side chain (R1) is formed by an S–S bond formation which is reversible under reducing conditions (1). Iodo-acetamido proxyl (IAP) and succinimide-containing labels (MAP, Gd(III)–DOTA, Gd(III)–DTPA) form a C–S–C thioether group (2–5), including the sulfur and the Cb atom of the cysteine residue, which makes the spin labels irreversible also under reducing conditions [109]. Reprinted from [109], Copyright 2017 RCS

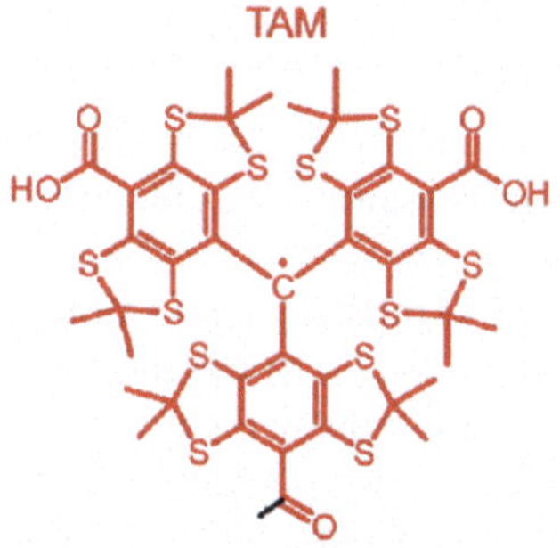

radicals have been widely employed as spin probes for oxymetry in EPR and EPR tomography owing to the narrow EPR linewidth and high stability in living systems. Nowadays, TAM successively used as spin labels for studies on the structure of proteins and nucleic acids utilizing site-directed spin labeling (SDSL) and PD-EPR spectroscopy [15, 110, 111]. The propeller-shaped structure of a TAM drawn

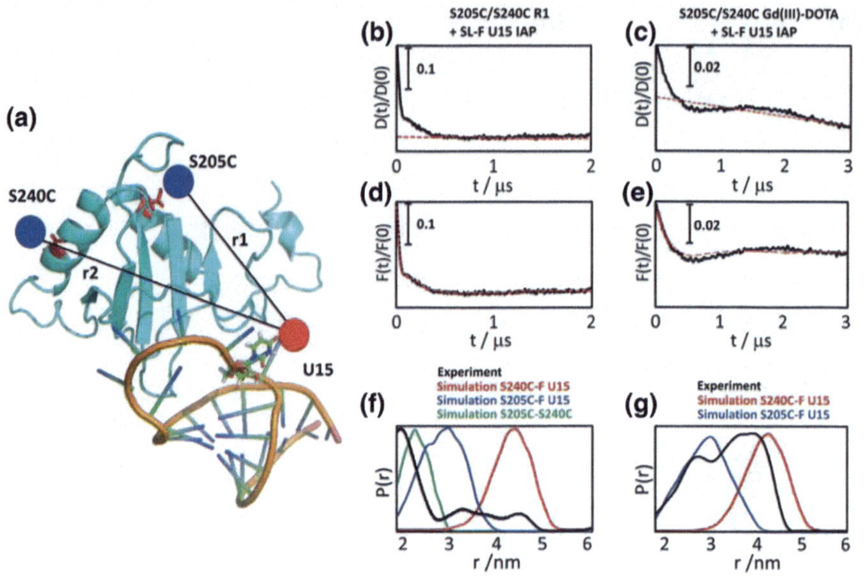

Fig. 10.29 Spectroscopically orthogonal two-label system and control with three non-orthogonal labels. **a** Representation of the RBD2–SLF complex with labeled positions on the protein and on RNA (blue and red dots). **b**, **c** Experimental data (black) and background fits (red, dashed line) for three nitroxide labels (**b**) and two Gd(III) labels combined with one nitroxide label (**c**). **d**, **e** Background-corrected DEER data (black) and form factor fit (red, dashed line) for three nitroxide labels (**d**) and two Gd(III) labels combined with one nitroxide label (**e**). **f** and **g** Experimental distance distributions (black) and simulated distance distributions (colored lines) for three nitroxide labels (**f**) and two Gd(III) labels combined with one nitroxide label (**g**) [109] Reprinted from [109], Copyright 2017 RCS

using EPR, ENDOR, and quantum-chemical analysis of two simple symmetric trityl radicals was established [110].

An idea of use of carbon-centered triarylmethyl (trityl) radicals instead of nitroxides for nanometer-distance measurements was first introduced and realized in 2012 [112].

Specifically, the tetrathiatriaryl methyl gave an EPR spectrum with one line only and has a transverse relaxation time T_m in the microsecond regime, even at room temperature in the liquid state. To evaluate the potential of trityls as spin labels for nanometer-distance measurements, two derivatives of Poly(para phenyleneethynylene)s (PolyPPEs) were prepared. Compound 1 contains one trityl and a typical nitroxide, whereas compound 2 has two trityl groups. PELDOR and double-quantum coherence (DQC), both in combination with DEER analysis showed the spin–spin distance values of about 35 and 50 Å for compounds 1 and 2, respectively, in excellent agreement with their structure. Triarylmethyl radicals were attached via disulfide linkages to substituted cysteine residues at positions 65 and 80 or 65 and 76 in T4 lysozyme immobilized on Sepharose. Interspin distances determined using double-quantum coherence (DQC) in solution were found to be close to those expected from

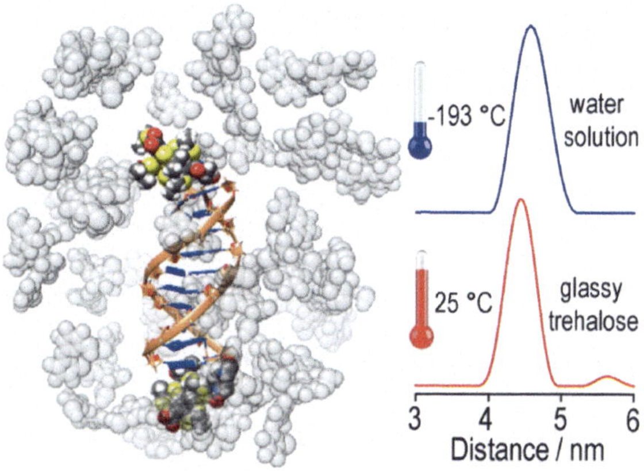

Fig. 10.30 Schematic representation of TAM-labeled DNA duplex in trehalose distance distribution studied by X-band DQC and DEER [114]. Reprinted from [114], Copyright 2014 American Chemistry Society

models [113]. The experiments revealed the narrow distance distribution in each case which indicates that the TAM-based spin label is relatively localized.

The first detailed investigation of trehalose as a prospective immobilizing agent of spin-labeled nucleic acids for pulse EPR distance measurements at room temperatures was reported [114].

Two other saccharides of close structure, sucrose and glucose, and compare their performance with trehalose were also investigated. The 10-mer DNA duplex doubly spin-labeled with TAM radical was used a model system for pulse EPR distance measurements. Schematic representation of TAM-labeled DNA duplex in trehalose distance distribution studied by X-band DQC and DEER is shown in Fig. 10.30.

In work [115], the broad range of triarylmethyl radicals (TAMs) varying polarity, number, and nature of substituents attached to carboxyl moieties located in para-positions of TA aryl rings, and structure of the TAM core were synthesized, and their relaxation times at room temperature in liquids were measured. Pulsed and low-temperature CW-EPR experiments were carried out at the X-band (9 GHz) and Q-band (34 GHz). Measurements of electron spin relaxation ($T_m \sim T_2$, T_1) were performed at the X- and Q-bands at $T = 300$ K. T_m was measured using a two-pulse electron spin echo (ESE) sequence; T_1 was measured using inversion–recovery technique with inversion π-pulse and detecting two-pulse ESE sequence. Data on Electron Spin Dephasing Time, T_m, and Electron Spin–Lattice Relaxation Time, T_1, for Investigated Radicals at 300 K at the X- and Q-Bands were tabulated and discussed in detail.

An approach that keeps structural conformation and unity of immobilized double-stranded DNA labeled with TAM (Fig. 10.31) using saccharides trehalose, sucrose, and glucose as immobilizing media, which provides conditions suitable for room-temperature EPR distance measurements, was proposed [116]. Room-temperature electron spin dephasing time of triarylmethyl-labeled DNA in trehalose was found to be noticeably longer compared to previously used immobilizers, thus providing a broader range of available distances.

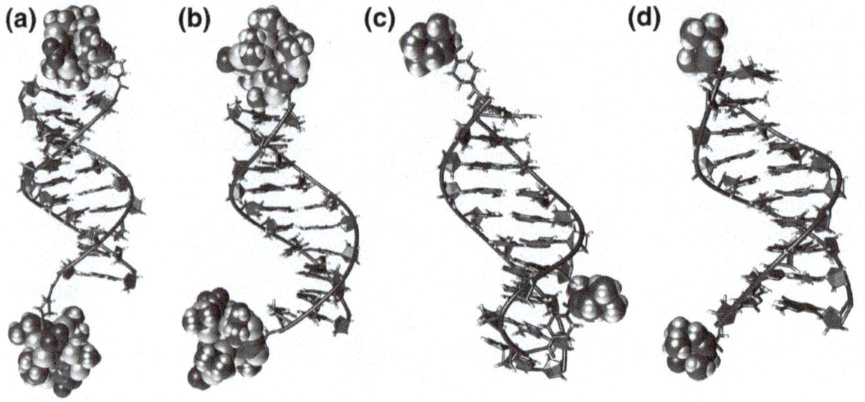

Fig. 10.31 Typical positions of spin labels relative to DNA duplexes: **a** I (D1-Pip-TAM/TAM); **b** VI (D1-NH-TAM/TAM); **c** IV (D1-Pip-NIT/NIT); **d** V (D1-NH-NIT/NIT) [116]. Reprinted from [116], Copyright 2010, American Chemical Society

The results of a nucleic acid study obtained using TAM were compared with data on nitroxyl spin labels (Fig. 10.32). In order to find structure/property relations in a series synthesized nitroxides with different substituents adjacent to NO-moiety (spirocyclohexane, spirocyclopentane, tetraethyl and tetramethyl groups), electron spin relaxation times (T_1, $T_m \approx 700$ ns) of these radicals immobilized in trehalose were measured at room temperature at X- and Q-bands (9/34 GHz) [117]. In addition, a comparison was made with the corresponding relaxation times in nitroxide-labeled DNA immobilized in trehalose. The room-temperature value of T_m in trehalose is weakly dependent on the structure of substituents adjacent to NO-moiety of nitroxide.

A new site-directed 2,5-bis(spirocyclohexane)-substituted spin label, based on the attachment of a linker containing an aliphatic amino group to the target nucleotide

Fig. 10.32 Nitroxide-labeled DNA immobilized in trehalose [117]. Reprinted from [117], Copyright 2016 Elsevier

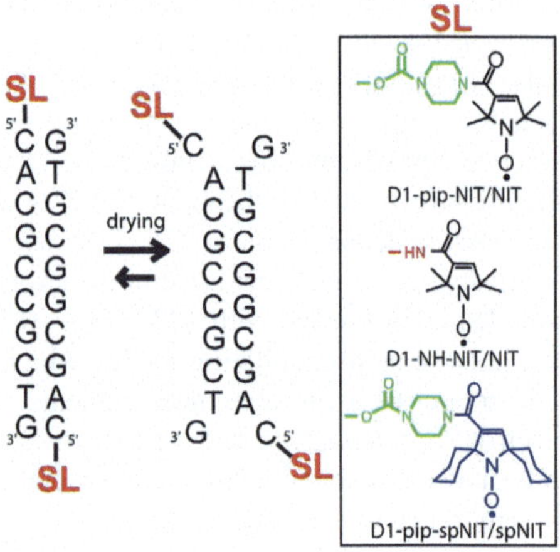

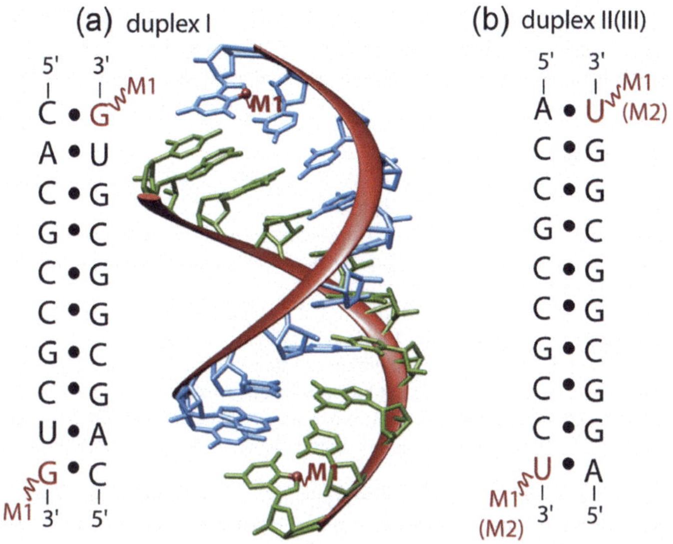

Fig. 10.33 Schematic presentation of RNA labeling by nitroxide [111]. Reprinted from [111], Copyright 2014 RCS

residue followed by selective coupling of a spin label to this amino group, was utilized [111]. This label with advanced stability and relaxation properties was tethered to the desired RNA residue via a sequence-specific reaction with the derivatives of oligodeoxyribonucleotides (Fig. 10.33). It was found that the attached label distance distribution measured employing Q-band (34 GHz) pulsed double electron–electron resonance corresponds well to values expected from model RNA duplex with known structure and distance between corresponding residues. The obtained results were validated by studying a similar RNA duplex, where the linker with the aliphatic amino group was introduced via solid-phase synthesis.

The features of PD-EPR and RE at ambient temperatures, requirements on electron spin phase memory time, ways of immobilization of biomolecules, the influence of a linker between the spin probe and biomolecule, and future opportunities were comprehensively reviewed [15]. Various aspects of spin labeling have been discussed in [119–121].

References

1. T.J. Stone, T. Buckman, P.L. Nordio, H.M. McConnell, Spin-labeled biomolecules. Proc. Natl. Acad. Sci. **54**, 1010–1017 (1965)
2. E. Fremi, Ann. Chim. Phys. Ser. 3 **15**, 408–488 (1845)
3. M.B. Neiman, É.G. Rozantzev, Y.G. Mamedova, Free radical reactions involving no unpaired electrons. Nature **196**, 472 (1962)

4. G.I. Likhtenshtein, Determination of the topography of proteins groups using specific paramagnetic labels. Mol. Biol. (Moscow) **2**, 234–240 (1968)
5. G.I. Likhtenshtein, *Spin Labeling Method in Molecular Biology* (Wiley, New York, 1976)
6. J.C. Taylor, J.S. Leigh, M. Cohn, The effect of dipole–dipole interaction between nitroxide radical and a paramagnetic ion on the line shape of the ESR spectra of radical. Proc. Natl. Acad. Sci. USA **64**, 219–206 (1969)
7. G.I. Likhtenshtein, Study on the proteins microstructure by method of spin-label paramagnetic probe. Mol. Biol. (Moscow) **4**, 782–789 (1970)
8. A.I. Kulikov, G.I. Likhtenshtein, E.G. Rozantzev, V. Suskina, A.V. Shapiro, Nitroxide bi- and polyradicals as standard models for distance estimation between the nitroxide moities. Biofizika **17**, 42–49 (1972)
9. A.I. Kokorin, K.I. Zamaraev, G.L. Grigoryan, V.P. Ivanov, E.G. Rozantsev, Distance estimation between nitroxyl radicals. Biofizika **17**, 34–41 (1972)
10. A.V. Kulikov, G.I. Likhtenstein, Application of saturation curves for evaluating distances in biological objects by the method of double spin-labels. Biofisika **19**, 420–424 (1974)
11. G.I. Likhtenshtein, Depth of immersion of paramagnetic centers, in *Magnetic Resonance in Biology*, ed. by L. Berliner, S. Eaton, G. Eaton (Kluwer Academic Publishers, Dordrecht, 2000), pp. 1–36
12. G.D. Case, J.S. Leigh Jr., Intramitochondrial position of cytochrome haem groups determined by dipolar interaction with paramagnetic cations. Biochem. J. **160**, 769–783 (1976)
13. A.V. Kulikov, G.I. Likhtenshtein, The use of spin-relaxation phenomena in the investigation of the structure of model and biological systems by method of spin labels. Adv. Mol. Relax. Interact. Process. **10**, 47–78 (1977)
14. G.I. Likhtenshtein, *Electron Spin in Chemistry and Biology: Fundamentals, Methods, Reactions Mechanisms, Magnetic Phenomena, Structure Investigation* (Springer, 1976)
15. O. Krumkacheva, E. Bagryanskaya, EPR-based distance measurements at ambient temperature. J. Magn. Reson. **280**, 117–126 (2017)
16. L.A. Syrtsova, L.A. Levchenko, E.N. Frolov, G.I. Likhtenshtein, N.N. Pisarscaya, L.V. Vorob'ev, V.A. Gromoglasova, Structure and function of the nitrogenase components from *Azotobacter vinelandii*. Mol. Biol. (Moscow) **6**, 62 (1972)
17. E.N. Frolov, G.I. Likhtenshtein, L.A. Syrtsova, Study of nonheme iron proteins. Dokl. A.N. SSSR **196**, 1149 (1971)
18. A.D. Milov, R.I. Samoilova, Y.D. Tsvetkov, M. De Zotti, C. Toniolo, J. Raap, PELDOR conformational analysis of bis-labeled Alamethicin aggregated in phospholipid vesicles. J. Phys. Chem. B **112**, 13469–13472 (2008)
19. G. Jeschke, DEER distance measurements on proteins. Annu. Rev. Phys. Chem. **63**, 419–446 (2012)
20. J.H. Freed, New technologies in electron spin resonance. Annu. Rev. Phys. Chem. **51**, 655–689 (2000)
21. G.R. Eaton, S.S. Eaton, D.P. Barr, R.T. Weber, *Quantitative EPR* (Springer, 2010)
22. A.V. Kulikov, Determination of distance between the nitroxide label and a paramagnetic center in spin-labeled proteins from the parameters of the saturation curve of the ESR spectrum of the label at 77K. Mol. Biol. (Moscow) **10**, 109–116 (1976)
23. C.L. Motion, J.E. Lovett, S. Bell, S.L. Cassidy, P.A.S. Cruickshank, D.R. Bolton, R.I. Hunter, H. El Mkami, S. Van Doorslaer, G.M. Smith, DEER sensitivity between iron centers and nitroxides in heme-containing proteins improves dramatically using broadband, high-field EPR. J. Phys. Chem. Lett. **7**, 1411–1415 (2016)
24. B. Selmke, P.P. Borbat, C. Nickolaus, R. Varadarajan, J.H. Freed, W.E. Trommer, Open and closed form of maltose binding protein in its native and molten globule state as studied by electron paramagnetic resonance spectroscopy. Biochemistry **57**(38), 5507–5512 (2018)
25. A.L. Lai, E.M. Clerico, M.E. Blackburn, N.A. Patel, C.V. Robinson, P.P. Borbat, J.H. Freed, L.V. Gierasch, Key features of an Hsp70 chaperone allosteric landscape revealed by ion-mobility native mass spectrometry and double electron-electron resonance. J. Biol. Chem. **292**(21), 8773–8785 (2017)

26. M.V. Airola, D. Huh, N. Sukomon, J. Widom, R. Sircar, P.P. Borbat, J.H. Freed, K.J. Watts, B.R. Crane, Architecture of the soluble receptor Aer2 indicates an in-line mechanism for PAS and HAMP domain signaling. J. Mol. Biol. **425**(5), 886–901 (2013)
27. G.T. Merz, P.P. Borbat, A.R. Muok, M. Srivastava, D.N. Bunck, J.H. Freed, B.R. Crane, Site-specific incorporation of a Cu^{2+} spin label into proteins for measuring distances by pulsed dipolar electron spin resonance spectroscopy. J. Phys. Chem. B **122**(41), 9443–9451 (2018)
28. B.J. Wylie, B.G. Dzikovski, S. Pawsey, M. Caporini, M. Rosay, J.H. Freed, A.E. McDermott, Dynamic nuclear polarization of membrane proteins: covalently bound spin-labels at protein-protein interfaces. J. Biomol. NMR **61**(3–4), 361–367 (2015)
29. Y. Yang, F. Yang, Y.-J. Gong, T. Bahrenberg, A. Feintuch, X.-C. Su, D. Goldfarb, High sensitivity in-cell epr distance measurements on proteins using an optimized Gd(III) spin label. J. Phys. Chem. Lett. **9**, 6119–6123 (2018)
30. S. Dunkel, L.P. Pulagam, H.-J. Steinhoff, J.P. Klare, In vivo EPR on spin labeled colicin A reveals an oligomeric assembly of the pore-forming domain in *E. coli* membranes. Phys. Chem. Chem. Phys. **17**(7), 4875–4878 (2015)
31. L.G.M. Basso, L.F.S. Mendes, A.J. Costa-Filho, The two sides of a lipid–protein story. Biophys. Rev. **8**, 179–191 (2016)
32. R.D. Nielsen, K. Che, M.H. Gelb, B.H. Robinson, A ruler for determining the position of proteins in membranes. J. Am. Chem. Soc. **127**, 6430–6442 (2005)
33. D. Snead, A.L. Lai, R.T. Wragg, D.A. Parisotto, T.F. Ramlall, J.S. Dittman, J.H. Freed, D. Eliezer, Unique structural features of membrane-bound C-terminal domain motifs modulate complexin inhibitory function. Front. Mol. Neurosci. **10**, 1–17 (2017)
34. R.S. Cooper, E.R. Georgieva, P.P. Borbat, J.H. Freed, E.E. Heldwein, Structural basis for membrane anchoring and fusion regulation of the herpes simplex virus fusogen gB. Nat. Struct. Mol. Biol. **25**, 416–424 (2018)
35. A.L. Lai, J.H. Freed, HIV gp41 fusion peptide increases membrane ordering in a cholesterol-dependent fashion. Biophys. J. **106**(1), 172–181 (2014)
36. P. Lueders, H. Jäger, M.A. Hemminga, G. Jeschke, M. Yulikov, Distance measurements on orthogonally spin-labeled membrane spanning WALP23 polypeptides. J. Phys. Chem. B **117**, 2061–2068 (2013)
37. T.F. Segawa, M. Doppelbauer, L. Garbuio, A. Doll, Y.O. Polyhach, G. Jeschke, Water accessibility in a membrane-inserting peptide comparing Overhauser DNP and pulse EPR methods. J. Chem. Phys. **144**(19), 194201/1–194201/12 (2016)
38. I.D. Sahu, G.A. Lorigan, Site-directed spin labeling EPR for studying membrane proteins. Biomed. Res. Int. **2018**, 3248289 (2018)
39. E.R. Georgieva, Nanoscale lipid membrane mimetics in spin-labeling and electron paramagnetic resonance spectroscopy studies of protein structure and function. Nanotechnol. Rev. **6**(1) (2016). https://doi.org/10.1515/ntrev-2016-0080
40. D. Marsh, T. Pali, The protein–lipid interface: perspectives from magnetic resonance and crystal structures. Biochim. Biophys. Acta: Biomembr. **1666**(1–2), 118–141 (2004)
41. M.A. Hemminga, L.J. Berliner, *ESR spectroscopy in membrane biophysics* (Springer, New York, 2007)
42. E.S. Karp, J.J. Inbaraj, M.L. Laryukhin, G.A. Lorigan, Electron paramagnetic resonance studies of an integral membrane peptide inserted into aligned phospholipid bilayer nanotube arrays. J. Am. Chem. Soc. **128**, 12070–12071 (2006)
43. A.N. Smith, U.T. Twahir, T. Dubroca, G.E. Fanucci, J.R. Long, Molecular rationale for improved dynamic nuclear polarization of biomembranes. J. Phys. Chem. B **120**, 7880–7888 (2016)
44. S.Y. Liao, M. Lee, T. Wang, I.V. Sergeyev, M. Hong, Efficient DNP NMR of membrane proteins: sample preparation protocols, sensitivity, and radical location. J. Biomol. NMR **64**, 223–237 (2016)
45. L.G. Mansor Basso, L.F. Santos Mendes, A.J. Costa-Filho, The two sides of a lipid–protein story. Biophys, Rev **8**, 179–191 (2016)

46. J.M. Franck, S. Chandrasekaran, B. Dzikovski, C.R. Dunnam, J.H. Freed, Focus: two-dimensional electron–electron double resonance and motions: the challenge of higher frequencies. J. Chem. Phys. **142**, 212302 (2015). (Review)
47. S.Th. Sigurdsson, Nitroxides and nucleic acids: chemistry and electron paramagnetic resonance (EPR) spectroscopy. Pure Appl. Chem. **83**, 677–686 (2011)
48. Y. Sun, P.P. Borbat, V.M. Grigoryants, W.K. Myers, J.H. Freed, C.P. Scholes, Pulse dipolar ESR of doubly labeled mini TAR DNA and its annealing to mini TAR RNA. Biophys. J. **108**, 893–902 (2015)
49. G.W. Reginsson, O. Schiemann, Spin labeling of DNA and RNA, in *Encyclopedia of Biophysics*, ed. by G.C.K. Roberts (Springer, Berlin, Heidelberg, 2013)
50. G. Masliah, C. Maris, L.B. König, M. Yulikov, F. Aeschimann, A.L. Malinows, J. Mabille, J. Weiler, A. Holla, J. Hunziker, N. Meisner-Kober, B. Schuler, G. Jeschk, F. H-T Allain, Structural basis of siRNA recognition by TRBP double-stranded RNA binding domains. EMBO J. **37**(6) (2018)
51. E.S. Babaylova, A.A. Malygin, A.A. Lomzov, D.V. Pyshnyi, M. Yulikov, G. Jeschke, O.A. Krumkacheva, M.V. Fedin, E.G. Bagryanskaya, Complementary-addressed site-directed spin labeling of long natural RNAs. Nucleic Acids Res. **44**(16), 7935–7943 (2016)
52. O. Duss, M. Yulikov, F.H.T. Allain, G. Jeschke, Combining NMR and EPR to determine structures of large RNAs and protein–RNA complexes in solution. Methods Enzymol. **558**(Structures of Large RNA Molecules and Their Complexes), 279–331 (2015)
53. J.M. Esquiaqui, E.M. Sherman, J.D. Ye, G.E. Fanucci, Conformational flexibility and dynamics of the internal loop and helical regions of the kink-turn motif in the glycine riboswitch by site-directed spin-labeling. Biochemistry **55**(31), 4295–305 (2016)
54. G.I. Likhtenshtein, *Electron Spin in Chemistry and Biology: Fundamentals, Methods, Reactions Mechanisms, Magnetic Phenomena, Structure Investigation* (Springer, 2016)
55. V.V. Khramtsov, In vivo molecular electron paramagnetic resonance-based spectroscopy and imaging of tumor microenvironment and redox using functional paramagnetic probes. Antioxid. Redox Signal. **28**, 1365–1377 (2018)
56. D.J. Lurie, M.A. Foster, D. Yeung, J.M.S. Hutchison, Design, construction and use of a large-sample field-cycled PEDRI imager. Phys. Med. Biol. **43**, 1877–1886 (1998)
57. V.V. Khramtsov, G.L. Caia, K. Shet, E. Kesselring, S. Petryakov, J.L. Zweier, A. Samouilov, Variable field proton–electron double-resonance imaging: application to pH mapping of aqueous samples. J. Magn. Reson. **202**, 267–273 (2010)
58. G. Ilangovan, J.L.Zweier, P. Kuppusamy, Microximetry: simultaneous determination of oxygen consumption and free radical production using electron paramagnetic resonance spectroscopy. Methods Enzymol. **381**(Oxygen Sensing), 747–762 (2004)
59. H. Swartz, Seeing is believing—visualizing drug delivery *in vitro* and in *vivo*. Adv. Drug Deliv. Rev. **57**, 1085–1086 (2005)
60. W.K. Subczynski, J.S. Hyde, The diffusion-concentration product of oxygen in lipid bilayers using the spin-label T_1 method. Biochim. Bioph. Acta **43**, 283–291 (1986)
61. W.K. Subczynski, L. Mainali, T.G. Camenisch, W. Froncisz, J.S. Hyde, Spin-label oximetry at Q- and W-band. J. Magn. Reson. **209**, 142–148 (2011)
62. G.I. Likhtenshtein, Yu.D. Akhmedov, L.V. Ivanov, L.A. Krinitskaya, Yu.V. Kokhanov, Investigation of the lysozyme macromolecule by a spin-labeling method. Mol. Biol. (Moscow) **8**, 40–48 (1974)
63. E.I. Yudanova, A.V. Kulikov, Determination of spin-exchange frequency of nitroxide radicals and oxygen by continuous saturation of ESR spectra. Biofizika **29**, 925–929 (1984)
64. V.V. Khramtsov, L.M. Weiner, I.A. Grigor'ev, L.B. Volodarsky, Proton exchange in stable nitroxyl radicals. ESR study of the pH of aqueous solutions. Chem. Phys. Lett. **91**, 69–72 (1982)
65. I.A. Grigor'ev, A. Samouilov, V.V. Khramtsov, In vivo monitoring of pH, redox status, and glutathione using L-band EPR for assessment of therapeutic effectiveness in solid tumors. Magn. Reson. Med. **67**, 1827–1836 (2012)

66. V.V. Khramtsov, In vivo spectroscopy and imaging of nitroxide probes, in *Nitroxides: Theory, Experiment and Applications*, ed. by A.I. Kokorin (2012), pp. 317–346
67. I.A. Kirilyuk, A.A. Bobko, V.V. Khramtsov, I.A. Grigor'ev, Nitroxides with two pK values—useful spin probes for pH monitoring within a broad range. Org. Biomol. Chem. **3**, 1269–1274 (2005)
68. L.B. Volodarsky (ed.), *Imidazolin Radicals* (CRC Press, Bosa Raton, 1988)
69. D. Meisel, G. Czapski, One-electron transfer equilibriums and redox potentials of radicals studied by pulse radiolysis. J. Phys. Chem. **79**, 1503–1509 (1975)
70. G.I. Likhtenshtein, *Biophysical Labeling Methods in Molecular Biology* (Cambridge University Press, Cambridge, New York, 1993)
71. A.P. Jagtap, I. Krstic, N.C. Kunjir, R. Hänsel, T.F. Prisner, S.T. Sigurdsson, Sterically shielded spin labels for in-cell EPR spectroscopy: analysis of stability in reducing environment. Free Radic. Res. **49**(1), 78–85 (2015)
72. N. Kocherginsky, H.M. Swarts, *Nitroxide Spin Labels. Reactions in Biology and Chemistry* (CRC Press, 1995)
73. L.B. Volodarsky, I.A. Grigor'ev, S.A. Dikanov, V.A. Reznikov, *Imidazoline Nitroxide Radicals* (Nauka (Siberian Branch), Novosibirsk, 1988), pp. 188–193)
74. V.V. Khramtsov, V.I. Yelinova (Popova), L.M. Weiner, T.A. Berezina, V.V. Martin, L.B. Volodarsky, Quantitative determination of SH groups in low- and high-molecular-weight compounds by an electron spin resonance method. Anal. Biochem. **182**, 58–63 (1989)
75. L. Ya, L. Gendel, M.A. Goldfeld, V.K. Koltover, E.G. Rozanzev, V.I. Suskina, Biofizika **13**, 1114–1115 (1968)
76. A.A. Bobko, T.D. Eubank, J.L. Voorhees, O.V. Efimova, I.A. Kirilyuk, S. Petryakov, D.G. Trofimiov, C.B. Marsh, J.L. Zweier, I.A. Grigor'ev, A. Samouilov, V.V. Khramtsov, *In vivo* monitoring of pH, redox status, and glutathione using L-band EPR for assessment of therapeutic effectiveness in solid tumors. Magn. Reson. Med. **67**, 1827–1836 (2012)
77. V.V. Khramtsov, A.A. Bobko, M. Tseytlin, B. Driesschaert, Exchange phenomena in the electron paramagnetic resonance spectra of the nitroxyl and trityl radicals: multifunctional spectroscopy and imaging of local chemical microenvironment. Anal. Chem. **89**, 4758–4771 (2017)
78. S. Kishimoto, M.C. Krishna,V.V., Khramtsov, H. Utsumi, D.J. Lurie, In Vivo Application of Proton-Electron Double-Resonance Imaging. ANTIOXIDANTS & REDOX SIGNALING 28, 1345–1363 (2018)
79. D.J. Lurie, J.M. Hutchison, L.N. Bell, I. Nicholson, D.M. Bussell, J.R. Mallard, Field-cycled proton-electron double resonance imaging of free radicals in large aqueous samples. J. Magn. Reson. **84**, 431–437 (1989)
80. H. Utsumi, K-I. Yamada, K. Ichikawa, K. Sakai, Y. Kinoshita, S. Matsumoto, M. Nagai, Simultaneous molecular imaging of redox reactions monitored by Overhauser enhanced MRI with ^{14}N- and ^{15}N-labeled nitroxyl radicals. Proc. Natl. Acad. Sci. USA **103**, 1463–1468 (2006)
81. A. Samouilov, O.V. Efimova, A.A. Bobko, Z. Sun, S. Petryakov, T.B. Eubank, D.G. Trofimov, I.A. Kirilyuk, I.A. Grigor'ev, W. Takahashi, I.L. Zweier, V.V. Khramtsov, In vivo proton–electron double-resonance imaging of extracellular tumor pH using an advanced nitroxide probe. Anal. Chem. **86**, 1045–1052 (2014)
82. F. Hyodo, S. Ito, K. Yasukawa, R. Kobayashi, H. Utsumi, Simultaneous and spectroscopic redox molecular imaging of multiple free radical intermediates using dynamic nuclear polarization-magnetic resonance imaging. Anal. Chem. **86**, 7234–7238 (2014)
83. H. Eto, F. Hyodo, N. Kosem, R. Kobayashi, K. Yasukawa, M. Nakao, M. Kiniwa, H. Utsumi, Redox imaging of skeletal muscle using in vivo DNP-MRI and its application to an animal model of local inflammation. Free Radic. Biol. Med. **89**, 1097–1104 (2015)
84. G.I. Likhtenshtein, A.P. Pivovarov, P.Kh. Bobodzhanov, E.G. Rozantsev, N.B. Smolina, Biofizika **13**, 396–400 (1968)
85. G.I. Likhtenshtein, Yu.B. Grebenshchikov, T.V. Avilova, Mol. Biol. (Moscow) **6**, 52–60 (1972)

86. D. Kivelson, Theory of EPR [electron paramagnetic resonance] line widths of free radicals. J. Chem. Phys. **33**, 1094–1106 (1960)
87. J.H. Freed, Theory of the ESR spectra of nitroxides, in *Spin Labeling. Theory and Applications*, vol. 1, ed. by L. Berliner (Academic Press, New York, 1976)
88. L.I. Antsiferova, O.V. Belonogova, V.V. Kochetkov, G.I. Likhtenshtein, Izvestiya Akademii Nauk SSSR, Seriya Biologicheskaya 494–501 (1989)
89. M.A. Voinov, D.B. Good, M.E. Ward, S. Milikisiyants, A. Marek, M.A. Caporini, M. Rosay, R.A. Munro, M. Ljumovic, L.S. Brown, V. Ladizhansky, A.I. Smirnov, Cysteine-specific labeling of proteins with a nitroxide biradical for dynamic nuclear polarization NMR. J. Phys. Chem. B **119**, 10180–10190 (2015)
90. E.P. Kirilina, I.A. Grigoriev, S.A. Dzuba, Oriental motion of nitroxide in molecular glasses. J. Chem. Phys. **121**, 12465–12471 (2004)
91. M. Pavone, P. Cimino, F. De Angelis, V. Barone, J. Am. Chem. Soc. **128**, 4338–4347 (2006)
92. H.S. Mchaourab, P.R. Steed, K. Kazmier, Toward the fourth dimension of membrane protein structure: insight into dynamics from spin-labeling EPR spectroscopy. Structure **19**, 1549–1561 (2011)
93. Z. Zhang, M.R. Fleissner, D.S. Tipikin, Z. Liang, J.K. Moscicki, K.A. Earle, W.L. Hubbell, J.H. Freed, Multifrequency electron spin resonance study of the dynamics of spin labeled T4 lysozyme. J. Phys. Chem. B **11**, 5503–5521 (2010)
94. M.M. Haimann, Y. Akdogan, R. Philipp, R. Varadarajan, D. Hinderberger, W.E. Trommer, Conformational changes of the chaperone SecB upon binding to a model substrate—bovine pancreatic trypsin inhibitor (BPTI). Biol. Chem. **392**, 849–858 (2011)
95. F. Tombolato, A. Ferrarin, J.H. Freed, Dynamics of the nitroxide side chain in spin-labeled proteins. J. Phys. Chem. B **110**, 26248–26259 (2006)
96. L. Yu, W. Wang, S. Ling, S. Liu, L. Xiao, Y. Xin, C. Lai, Y. Xiong, L. Zhang, C. Tian, CW-EPR studies revealed different motional properties and oligomeric states of the integrin β1a transmembrane domain in detergent micelles or liposomes. Sci Rep. **5**, 7848 (2015)
97. Y.-W. Chiang, A.J. Costa-Filho, J.H. Freed, Dynamic molecular structure and phase diagram of DPPC-cholesterol binary mixtures: a 2D-ELDOR study. J. Phys. Chem. B **111**, 11260–11270 (2007)
98. V.I. Krinichnyi, O.Ya. Grinberg, E.I. Judanova, M.L. Borin, Ya.S. Lebedev, G.I. Likhtenshtein, Study of molecular mobility in biological membranes by two millimeter band ESR spectroscopy. Biofizika **32**, 59–65(1988)
99. D. March, Spin-label EPR for determining polarity and proticity in biomolecular assemblies: transmembrane profiles. Appl. Magn. Reson. **37**, 435–454 (2010)
100. T.I. Smirnova, T.G. Chadwick, M.A. Voinov, O. Poluektov, J. van Tol, A. Ozarowski, G. Schaaf, M.M. Ryan, V.A. Bankaitis, Local polarity and hydrogen bonding inside the Sec14p phospholipid-binding cavity. Biophys. J. **92**, 3686–3695 (2007)
101. V.I. Krinichnyi, *2-mm Wave Band EPR Spectroscopy of Condensed Systems* (CRC Press, Boca Raton, IL, 1995)
102. G.I. Likhtenshtein, J. Yamauchi, S. Nakatsuji, R. Smirnov, *Tamura Nitroxides: Application in Chemistry, Biomedicine, and Materials Science* (Wiley, Weinhem, 2008)
103. D. Goldfarb, Gd^{3+} spin labeling for measuring. distances in biomacromolecules: why and how? Methods Enzymol. **563**, 416–457 (2015)
104. A. Feintuch, G. Otting, D. Goldfarb, Gd^{3+} spin labeling for measuring. distances in biomacromolecules: why and how? Methods Enzymol. **563**, 416–457 (2015)
105. G. Prokopiou, M.D. Lee, A. Collauto, E.H. Abdelkader, T. Bahrenberg, A. Feintuch, M. Ramirez-Cohen, J. Clayton, J.D. Swarbrick, B. Graham, G. Otting, D. Goldfarb, Measurements in proteins by EPR spectroscopy. Inorg. Chem. **57**(9), 5048–5059 (2018)
106. M.C. Mahawaththa, M.D. Lee, A. Giannoulis, A. Adams, A. Feintuch, J.D. Swarbrick, B. Graham, C. Nitsche, D. Goldfarb, G. Ottgorin, Small neutral Gd(III) tags for distance measurements in proteins by double electron–electron resonance experiments. Phys.Chem.Chem. Phys. **20**, 23535 (2018)

107. L. Gigli, W. Andrałojć, A. Dalaloyan, G. Parigi, E. Ravera, D. Goldfarb, C. Luchinat, Assessing protein conformational landscapes: integration of DEER data in maximum occurrence analysis. Phys. Chem. Chem. Phys. **20**, 27429 (2018)
108. G. Prokopiou, M.D. Lee, A. Collauto, E.H. Abdelkader, T. Bahrenberg, A. Feintuch, M. Ramirez-Cohen, J. Clayton, J.D. Swarbrick, B. Graham, G. Otting, D. Goldfarb, Small Gd(III) tags for Gd(III)–Gd(III) distance measurements in proteins by EPR spectroscopy. Inorg. Chem. **57**(9), 5048–5059 (2018)
109. C. Gmeiner, G. Dorn, F.H.T. Allain, G. Jeschke, M. Yulikov, Spin labelling for integrative structure modelling: a case study of the polypyrimidine-tract binding protein 1 domains in complexes with short RNAs. Phys. Chem. Chem. Phys. **19**, 28360 (2017)
110. M.K. Bowman, C. Mailer, H.J. Halpern, The solution conformation of triarylmethyl radicals. J. Magn. Reson. **172**, 254 (2005)
111. E.S. Babaylova, A.V. Ivanov, A.A. Malygin, M.A. Vorobjeva, A.G. Venyaminova, Yu.F. Polienko, I.A. Kirilyuk, O.A. Krumkacheva, M.V. Fedin, G.G. Karpova, E.G. Bagryanskaya, A versatile approach for site-directed spin labeling and structural EPR studies of RNAs. Org. Biomol. Chem. **12**, 3129 (2014)
112. G.W. Reginsson, N.C. Kunjir, STh Sigurdsson, O. Schiemann, Trityl radicals: spin labels for nanometer-distance measurements. Chem. Eur. J. **8**, 13580–13584 (2012)
113. Z. Yang, Y. Liu, P. Borbat, J.L. Zweier, J.H. Freed, W.L. Hubbell, Pulsed ESR dipolar spectroscopy for distance measurements in immobilized spin labeled proteins in liquid solution. J. Am. Chem. Soc. **134**, 9950–9952 (2012)
114. G.Y. Shevelev, O.A. Krumkacheva, A.A. Lomzov, A.A. Kuzhelev, O.Y. Rogozhnikova, D.V. Trukhin, T.I. Troitskaya, V.M. Tormyshev, M. Fedin, D. Pyshnyi, E.G. Bagryanskaya, Physiological-temperature distance measurement in nucleic acid using triarylmethyl-based spin labels and pulsed dipolar EPR spectroscopy. J. Am. Chem. Soc. **136**, 9874–9877 (2014)
115. A.A. Kuzhelev, D.V. Trukhin, O.A. Krumkacheva, R.K. Strizhakov, O.Y. Rogozhnikova, T.I. Troitskaya, M.V. Fedin, V.M. Tormyshev, E.G. Bagryanskaya, Room-temperature electron spin relaxation of triarylmethyl radicals at the X- and Q-bands. J. Phys. Chem. B **119**, 13630–13640 (2015)
116. A.A. Kuzhelev, G.Yu. Shevelev, O.A. Krumkacheva, V.M. Tormyshev, D.V. Pyshnyi, M.V. Fedin, E.G. Bagryanskaya, Saccharides as prospective immobilizers of nucleic acids for room-temperature structural EPR studies. J. Phys. Chem. Lett. **7**, 2544–2548 (2016)
117. A.A. Kuzhelev, R.K. Strizhakov, O.A. Krumkacheva, Y.F. Polienko, D.A. Morozov, GYu. Shevelev, D.V. Pyshnyi, I.A. Kirilyuk, M.V. Fedin, E.G. Bagryanskaya, Room-temperature electron spin relaxation of nitroxides immobilized in trehalose: Effect of substituents adjacent to NO-group. J. Magn. Reson. **266**, 1–7 (2016)
118. P. Roser, M.J. Schmidt, M. Drescher, D. Summerer, Site-directed spin labeling of proteins for distance measurements *in vitro* and in cells. Org. Biomol. Chem. **14**, 5468–5476 (2016)
119. N. Alonso-Garcia, I. Garcia-Rubio, J.A. Manso, R.M. Buey, R.H. Urien, A. Sonnenberg, G. Jeschke, J.M. de Pereda, Combination of X-ray crystallography, SAXS and DEER to obtain the structure of the FnIII-3,4 domains of integrin α6β4. Acta Crystallogr. D: Biol. Crystallogr. **71**(4), 969–985 (2015)
120. C. Gmeiner, D. Klose, E. Mileo, V. Belle, S.R.A. Marque, G. Dorn, F.H.T. Allain, B. Guigliarelli, G. Jeschke, M. Yulikov, Orthogonal tyrosine and cysteine site-directed spin labeling for dipolar pulse EPR spectroscopy on proteins. J. Phys. Chem. Lett. **8**(19), 4852–4857 (2017)
121. E.A.W. van der Cruijsen, E.J. Koers, C. Sauvee, R.E. Hulse, M. Weingarth, O. Ouari, E. Perozo, P. Tordo, M. Baldus, Biomolecular DNP-supported NMR spectroscopy using site-directed spin labeling. Chem. Eur. J. **21**, 12971–12977 (2015)

Chapter 11
Nitroxides Miscellaneous Application

11.1 Multispin Nitroxides

In the last two decades, superparamagnetic organic compounds including polymer magnets with stability at ambient temperature and/or higher magnetic ordering have been attracting attention as models of multispin systems and potential magnetic devices [1–6].

As an example, in a recent paper [6], the three-spin electron–electron–nucleus (e–e–n) cross-effect (CE) was investigated. The influence of compounds bearing nitroxides on the nuclear longitudinal relaxation time constant T_{1n}, the build-up time constants of nuclear magnetic resonance (NMR) signal, T_{DNP} and DNP-enhancement of NMR signal in a series of radicals with a systematic increase in their local electron spin concentration (mono-radical, bi-radicals, triradical, and a dendrimer with 9 nitroxide moieties (Fig. 11.1) were examined. All static DNP, NMR, and EPR

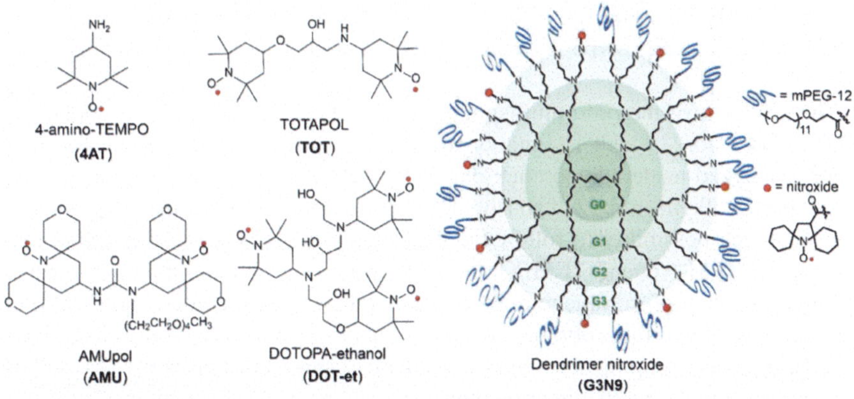

Fig. 11.1 Chemical structures of the mono-, bi-, tri-nitroxides, and dendrimer nitroxides [6]. Reprinted from [6], Copyright 2018 RCS

G. I. Likhtenshtein, *Nitroxides*, Springer Series in Materials Science 292,
https://doi.org/10.1007/978-3-030-34822-9_11

experiments were conducted at 7 T and at 4 K. Magic angle spinning MAS-DNP experiments employing a Bruker 400 MHz DNP-NMR spectrometer were performed at 92 K. The T_{1n} and T_{DNP} values for the five radicals were found to decrease significantly from mono-, bi-, triradical to the dendrimer having 37-60-fold smaller T_{1n} and T_{DNP} than the monomer, respectively, increasing e–e dipolar coupling. T_{1e} moderately increased, and T_M decreased with increasing local electron spin concentration. No correlation was found between T_M and T_{1n}.

Data on electron relaxation times and spectral diffusion rates for different radicals with 10 mM global electron spin concentration at 7 T magnetic field and 4 K temperature were tabulated. From the experimental observations and theoretical analysis made in this study, it was concluded that: (1) an increase in electron spin clustering led to an increased electron spin depolarization, (2) a dramatically shortened T_{1n} and T_{DNP} time constants under static and magic angle spinning (MAS) conditions takes place, (3) e–e–n polarization transfer rates are smaller or comparable to the nuclear spin diffusion rates, (4) polarization efficiency of nuclear spins tens to hundreds of nm away from the polarizing agent can result from a combination of direct e–e–n polarization transfer and nuclear spin diffusion processes, (5) DNP oversaturation is due to electron spin depolarization beyond optimal conditions with the dendrimer sample and attributed to the presence of strong e–e dipolar couplings, (6) strong e–e interactions, caused by electron spin clustering, increase the cross-effect rate, (7) fast cross-effect rates will benefit DNP at liquid helium temperatures, or at higher magnetic fields and pulsed DNP, and (8) slow e–e–n polarization transfer rate can be a key bottleneck to achieving maximal DNP performance.

In work [7], it was found that the cross-effect CE polarization transfer was truncated at the EPR resonant frequencies. Specifically, this effect was observed when doping Trityl-OX063 with a pyrroline nitroxide radical that possesses electron-withdrawing tetracarboxylate substituents (tetracarboxylate-ester-pyrroline or TCP) in vitrified water/glycerol at 6.9 T and at 3.3–85 occurs. The CE mechanism was replicated by quantum mechanical simulations with a mixed radical system, where the two radicals maximize the EPR spectral overlap to fulfill the cross-effect conditions, while one radical type is easily saturated but not the other due to very short T_{1e}.

The synthesis of the rigid multispin molecule based on tetrakis (4-azidophenyl) methane as a core, which is functionalized with nitroxide moieties via fourfold copper-catalyzed azide alkyne click chemistry between the azide functions of the core **1** and the alkyne moiety of an isoindoline nitroxide, was performed [8]. In order to evaluate the possibility to follow the exchange process between the isoindoline nitroxide of TPM-NO and TEMPO, the EPR spectra of the individual compounds and mixtures thereof in toluene at room temperature were recorded. The TEMPO nitroxide has a hyperfine coupling constant of about 15.5 Gauss, while TPM-NO 3 shows a hyperfine coupling constant of about 14.1 Gauss, most probably as result of spin exchange in the multispin compounds.

Diradicals 2–7 with $S = 1$ ground-state and $S = 3/2$ ground-state triradical 8 (Fig. 11.2) were described [9]. Theoretically, estimates' difference between the singlet and triplet states (ΔEST) of 2, 4, and 5 was found to be ~1.8 kcal mol^{-1},

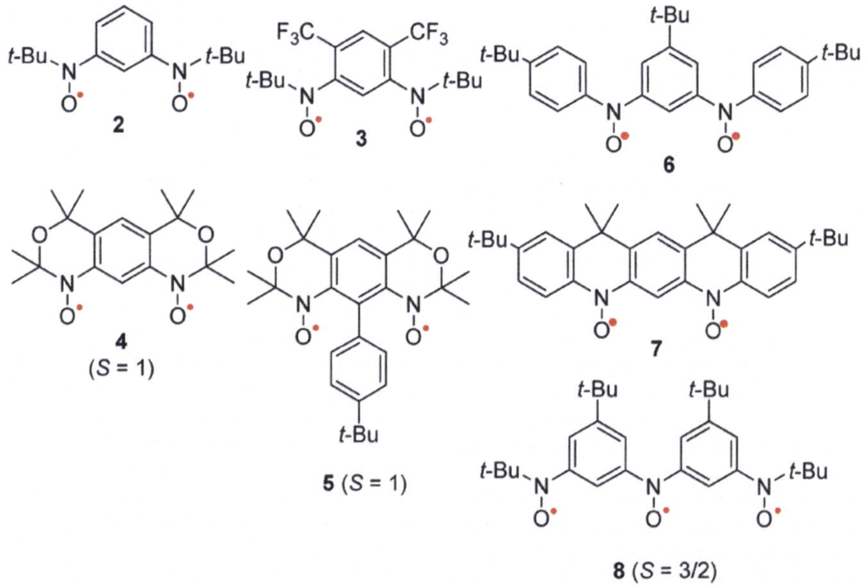

Fig. 11.2 High-spin molecules incorporating nitroxide spin centers [1]. Reprinted from [1], Copyright 2015 American Chemical Society

3.3 kcal mol^{-1}, and 2.2 kcal mol^{-1}, respectively. For triradical 8, a low energy gap between the quartet and excited doublet states, $\Delta E_{DQ} \approx 0.5$ kcal mol^{-1}, was measured by magnetic susceptibility.

11.1.1 Spin Trapping

The pioneering works, in which the principle possibility of effectiveness of the spin trapping was demonstrated, came to the light in 1968 [9]. Since then it is customary to use this technique for investigation radical processes in chemistry and biology, and many chemical, instrumental, and biochemical efforts have being undertaken to achieve "an ideal" structure [9–18].

^{1}H NMR and ESR titrations were used to determine the association constants of the complexes of R-phenyl-*N*-*tert*-butylnitrone (PBN) analogs and their superoxide spin adducts, respectively, both incorporated in methylated *â*-cyclodextrin, 2,6-di-*O*-methyl-*â*-cyclodextrin (Fig. 11.3) [13]. After the superoxide radical spin trapping reaction, EPR titrations (Fig. 11.4) afforded the association constants of the corresponding cyclodextrin-nitroxide complexes. Two-dimensional EPR simulations indicated a bimodal inclusion of the nitroxide free radical spin adducts into the cyclodextrins. For all the nitrone–cyclodextrin and nitroxide–cyclodextrin complexes, the association constants were always higher for the nitroxide complexes than

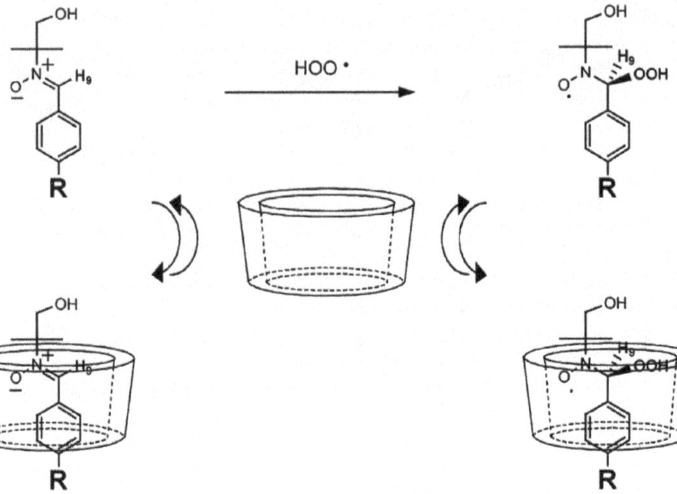

Fig. 11.3 Schematic presentation of double superoxide trapping by nitron and dextrin [13]. Reprinted from [13], Copyright 2005 American Chemical Society

for the nitrone complexes. The efficiency of the cyclodextrin inclusion technique to trap superoxide and to resist bioreduction by sodium L-ascorbate was also evaluated.

A series of new 11 hybrid 2-(diethoxyphosphoryl)-*N*-(benzylidene)propan-2-amine oxide derivatives with different aromatic substitution (PPNs) (Fig. 11.5) were synthesized, and their spin trapping potential was evaluated by EPR [14]. The experimental screening of these hydroxyphenyl- and methoxyphenyl-substituted PPNs and biocompatible nitrones 4d, 4g, and 4i deriving from caffeic, gallic, ferulic, and sinapic acids, which combined improved EPR probing of ROS formation, vasorelaxant action, and antioxidant potency was carried out. The PPNs antioxidant activity NO-donation properties in vitro, cytotoxicity and vasoprotective effect on precontracted rat aortic rings were also investigated. Figure 11.6 illustrates the author's approach to the trapping of hydroxyl and methyl radicals formatted in the supernatant of rat aortic rings stimulated by NADPH.

Data on synthesis of series of tetramethylpyrazine (TMP) derivatives and on investigation of their abilities for scavenging free radicals and preventing against oxidative stress-induced neuronal damage in vitro were presented [15]. Compounds, consisted of TMP, caffeic acid, and a nitrone group showed variety of activities: (1) radical scavenging, (2) broad neuroprotective effects, (3) rescuing iodoacetic acid-induced neuronal loss, (4) preventing from *tert*-butylhydroperoxide (*t*-BHP)-induced neuronal injury, and (5) exerting against *t*-BHP injury via activation of the phosphatidyl inositol 3-kinase (PI3K)/Akt signaling pathway. In addition, in a rat model of permanent middle cerebral artery occlusion, this compound significantly improved neurological deficits, and alleviated the infarct area and brain edema.

Two α-phenyl-*N*-tert-butylnitrone (PBN) derivatives were designed by adding a cholesterol and sugar moieties to parent nitrone to increase its lipophilicity and

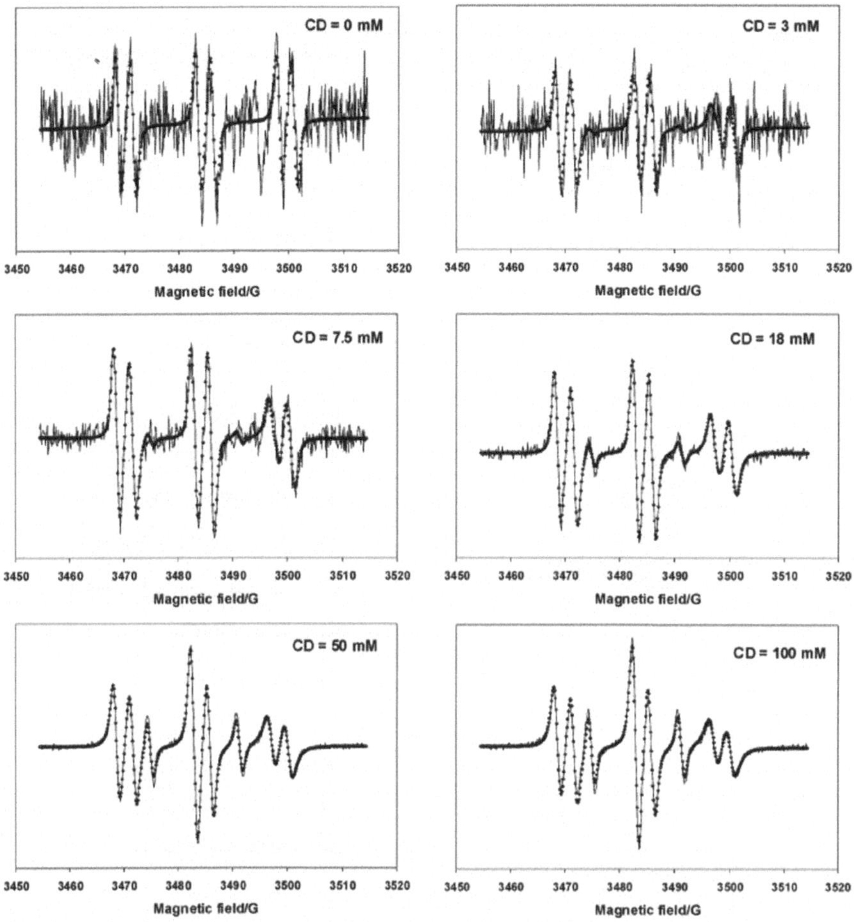

Fig. 11.4 ESR spectra after superoxide trapping by nitron and α-cyclodextrin (CD) [13]. Reprinted from [13], Copyright 2005 American Chemical Society

hydrophilic properties in the same time [16]. It was found that the new compound is able to form stable monolayers at the air/water interface and for the two derivatives bearing a sugar group, repulsive interactions with 1,2-dilinoleoyl-sn-glycero-3-phosphocholine (DLPC) in liposome occurred. The protective effect against lipid oxidation was tested using the AAPH thermolysis which generates peroxyl radicals and the Fenton reagent. The formation of conjugated dienes was quantified spectrophotometrically.

An approach using EPR and a cocktail of spin traps was designed [17]. The rationale for using this cocktail was that it would cover a wide range of biologically relevant free radicals and have a large range of hydrophilicity and lipophilicity in order to trap free radicals produced in different cellular compartments. Use of a

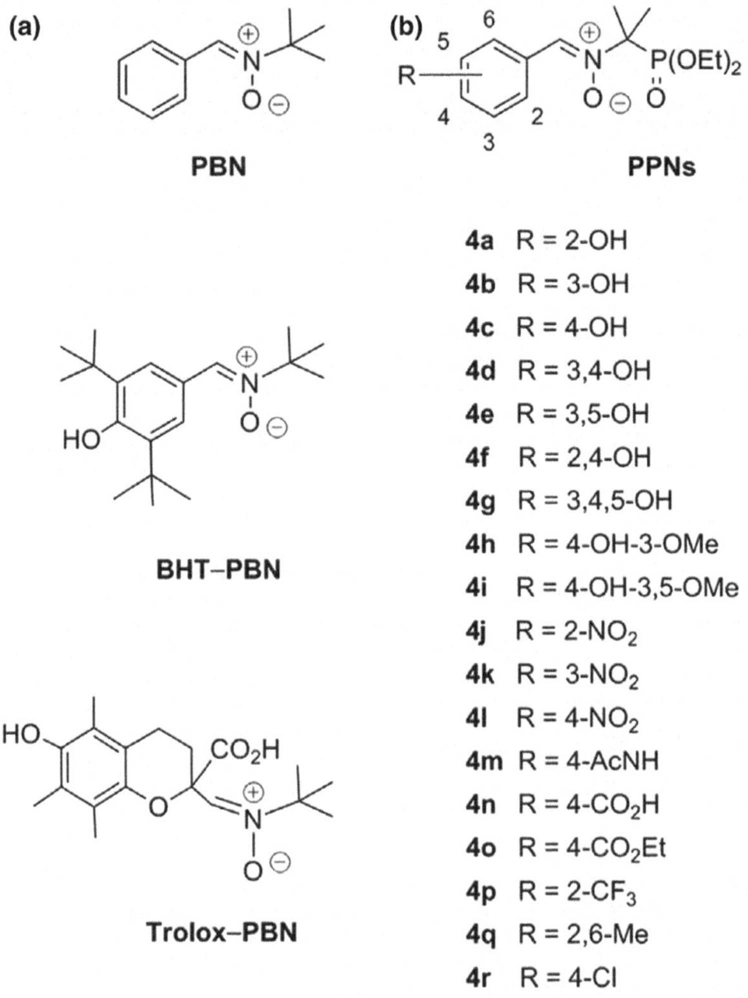

Fig. 11.5 Structures of **a** PBN and its phenol-based derivatives BHT_PBN Trolox_PBN successfully tested in in vitro and animal models of oxidative stress, and **b** phenolic (4a_4i) and non-phenolic (4j, 4q) derivatives of PPN (R ¼ H) [14]. Reprinted from [14], Copyright 2016 Elsevier

cocktail of spin traps. The ability of the technique to unambiguously detect free radical (O-, N-, C-, or S-centered) production in cells was well documented.

A novel glycine-nitronyl nitroxide (GNN) conjugate, which showed a synergetic protection against renal ischemia/reperfusion (I/R) injury, was designed, and its action mechanism was established using a hypoxia/reoxygenation (H/R) injury model of human umbilical vein endothelial cells (HUVECs) [18]. It was found that the GNN conjugate significantly elevated the cell viability via reducing the apoptosis rate in H/R-treated HUVECs and conjugate-attenuated H/R-induced mitochondrial fragmentation, mitochondrial membrane potential reduction, cytochrome *c* release,

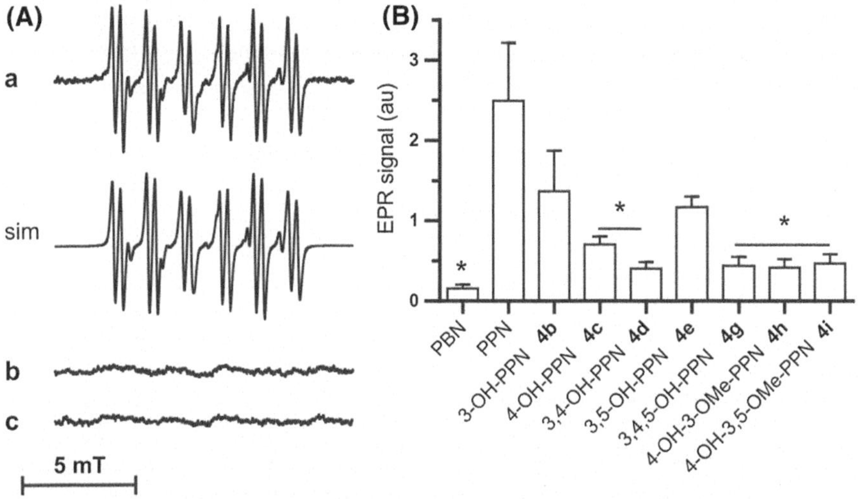

Fig. 11.6 Indirect spin trapping evidence for hydroxyl radical formation in the supernatant of rat aortic rings stimulated by NADPH (1 mM) for 80 min. After 60-min NADPH stimulation in KH buffer, the spin trap (15 mM) diluted in DMSO (1% final concentration) was added to the medium and incubated with rings for 20 min (**A**) EPR signal recorded in (**a**) the complete incubation system containing NADPH, DMSO, DTC (30 mM), and 3-OH-PPN 4b [14]. Reprinted from [14], Copyright 2016 Elsevier

and autophagy. On the example of an in vivo hind limb I/R model, the extensive applicability of GNN conjugate in different I/R models and its effect in remote organs were established.

11.2 Spin Label-Spin Probe Method

11.2.1 Fundamentals

A first version the spin label-spin probe (SLSP) approach designed and developed by Likhtenshtein group in 1970th [19–23] was based on the dynamic spin exchange interaction between a stable radical (i.e., nitroxide) attached to a molecular object of interest and chemically inert paramagnetic complex capable of diffusing freely in solution.

According to Salikhov et al. [20], the rate constant of the exchange interaction at the paramagnetic species encounters

$$k_{\mathrm{ex}} = P_{\mathrm{ex}} k_d = \frac{f_{\mathrm{g}} f_{\mathrm{ns}} k_d J^2 \tau_{\mathrm{c}}^2}{(1 + J^2 \tau_{\mathrm{c}}^2)} \tag{11.1}$$

where k_d is the rate constant of encounters in solution, P_{ex} is the probability of spin exchange in the course of lifetime (τ_c) of the encounter complex, f_g is the geometric steric factor, f_{ns} is the nuclear statistical factor, and J is the exchange integral of interaction in the encounter complex at the direct contact between the particles

If $J^2\tau_c^2 \gg 1$, k_{ex} is not dependent on J (strong exchange) and

$$k_{ex} = f_g f_{ns} k_d \tag{11.2}$$

In the case of weak exchange $J^2\tau_c^2 \ll 1$ and

$$k_{ex} = f_g f_{ns} k_d J^2 \tau_c^2 \tag{11.3}$$

For a particular pair of paramagnetic species, the value of k_{ex} depends on the microstructure, microviscosity, steric hindrances, and local charges in the region of encounters (Fig. 11.7). Hence, the methods of examining k_{ex} can be used for experimental study of these factors in biological and other objects.

A variant of the SLSP method based on direct measurement of electron spin–lattice relaxation times was developed in Hide group [21].

The principle of the SLSP was used in the following versions (1) label is a paramagnetic metal complex (e.g., iron in heme proteins) and a probe is nitroxide, (2) label

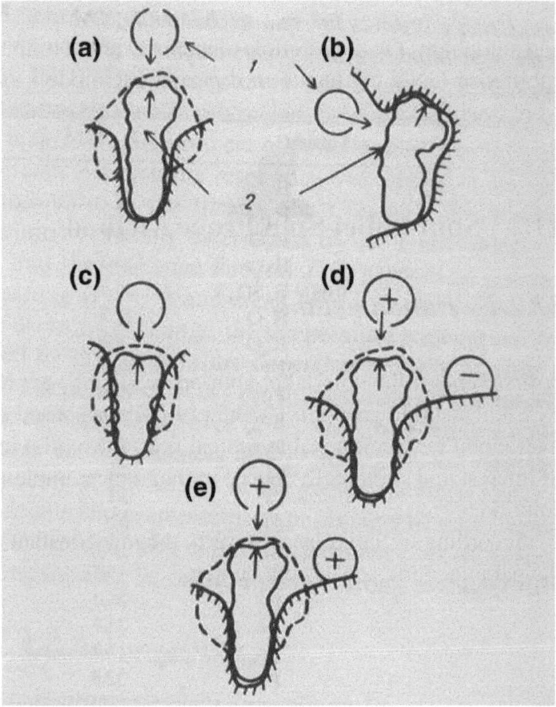

Fig. 11.7 Schematic illustration of spin label–spin probe method [19]

is a chromophore in the excited triplet state, and a probe is paramagnetic species (e.g., nitroxide), and (3) label is a molecular bearing paramagnetic nucleus (e.g., proton) and probe is a paramagnetic species (e.g., nitroxide).

11.2.2 Nitroxide in Investigation of Electrostatic Effects

11.2.2.1 Local Charges in Proteins and DNA

Electrostatic interactions are known to play one of the major roles in the myriad of cellular and molecular biology processes, ranging from enzyme catalysis, electron transfer, protein folding to insertion of proteins, toxins, and viruses into membranes as well as other more complex events.

The idea of using charged nitroxide radicals for the experimental study of local charges and electrolytic potential was first put forward and implemented in the Likhtenshtein group in 1970th as part of the spin label-spin probe method [19, 22, 23]. Subsequently, this approach was widely used to study electrostatic effects in biological molecular systems [21–30].

Methods of determination of the electrostatic potential and local charges in the vicinity of groups bearing spin are based on the sensitivity of spin relaxation rates of radicals and nuclei to encounters with charged paramagnetic species, stable radicals, or paramagnetic complexes in solution. This approach can be applied to three types of problems [26–28]. It can be used to investigate the electrostatic fields in the vicinity of an attached spin label using a second paramagnetic species with a different charge, ferricyanide anion, or diphenylchromium cation, for example [19, 22, 23]. The second approach involves monitoring the effect of a paramagnetic species, such as a complex of paramagnetic ion with a protein or the active site of a metalloenzyme, on the spin relaxation parameters of nitroxide spin probes of different charges freely diffusing in solution [28]. Measurements of the effect of paramagnetics (i.e., nitroxides) with different charges on the spin relaxation rates of the nuclei allow quantitative characterization of electrostatic effects [26–28].

Calculations of local charge Z_x in the vicinity of a paramagnetic particle (such as the active site of metalloprotein or a spin label) colliding with a nitroxide or metallocomplex with known charge Z_p can be carried out with the use of the Debye equation (11.4):

$$\frac{k^+}{k^0} \text{ or } \frac{k^-}{k^0} \frac{Z_p Z_x \alpha}{\exp(Z_p Z_x \alpha) - 1} \tag{11.4}$$

where k^+, k^-, and k^0 are the rate constants of encounters for positively charged, negatively charged, and neutral uncharged particles, respectively; and $\alpha = e^2/k_B T \varepsilon_0 r$, where e is the charge of an electron; k_B is the Boltzmann constant; temperature T = 293 K, ε_0 is the dielectric constant of water, and r is the distance between the charges in the encounter complex. The values of (k^+/k^0) and (k^-/k^0) are determined by measurement of the rate constants of spin exchange which describes the effect of paramagnetic species on spin-phase and spin–lattice relaxation rates of the radical.

A diagram (Fig. 11.8) allows one to calculate the Z_pZ_x product using the appropriate experimentally measured $\log(k^+/k^0)$ or $\log(k^-/k^0)$ value and an r value estimated for the encounter complex. From the value of Z_xZ_p, one can ascertain the Z_x charge in the vicinity of the region of the paramagnetic complex encountered by the charged nitroxide spin probe since the Z_p charge of the nitroxide spin probe is known.

Electron-carrier horse cytochrome c and dioxygen-carrier sperm-whale myoglobin were served as models for determination of local electrostatic charges in the vicinity of paramagnetic active sites of metalloenzymes and metalloproteins, using nitroxide radicals [28]

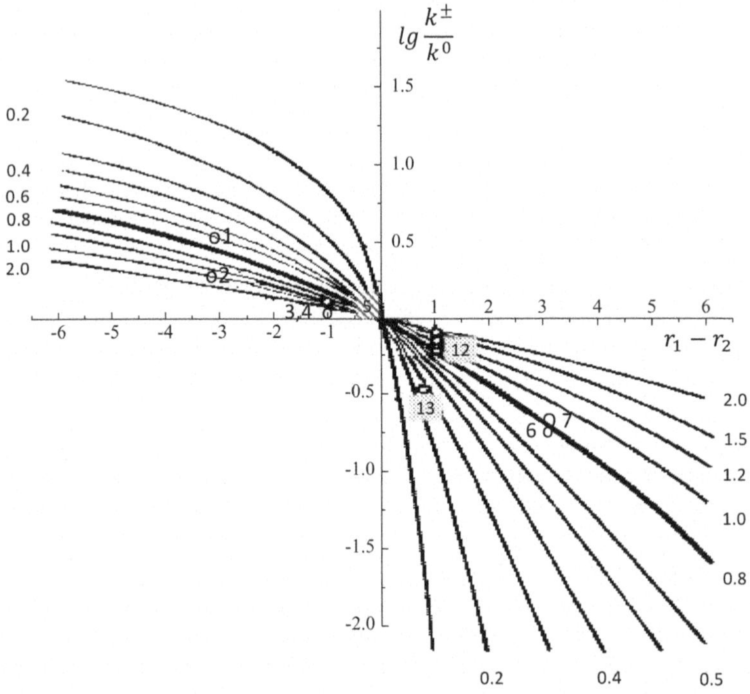

Fig. 11.8 A diagram for calculating the Z_pZ_x product [23]

Table 11.1 Apparent spin-exchange rate constants ($k+$, $k-$, and $k0$, in $M^{-1}S^{-1}$) of the positively charged (**III**), negatively charged (**II**), and neutral (**I**) nitroxide spin probes with hemin, ferricyanide anion, and heme proteins in aqueous solutions and apparent local charges (Z_H) in the vicinity of the heme groups [28]

Paramagnetic species	pH	$k^{+}10^{-8}$	$k^{-} \times 10^{-8}$ ($M^{-1}s^{-1}$)	$k^{0} \times 10^{-8}$ ($M^{-1}s^{-1}$)	Z_H (± 0.1)
Hemin[b]		c	c	24	
$K_3Fe(CN)_6$[e]	6.9	38	2	6	c
Cytochrome c (Fe^{3+})[g]	10.7	c	0.974	0.72	+0.5
Cytochrome c (Fe^{3+})[e]	6.9	c	0.841	0.71[f]	+0.3
Cytochrome c (Fe^{2+})[e]	6.9			~0	
Myoglobin[g]	7.0	c	0.561	0.77	−0.5

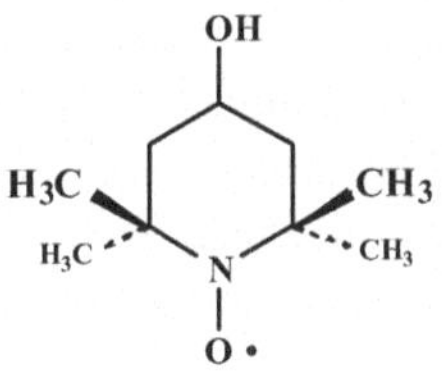

I

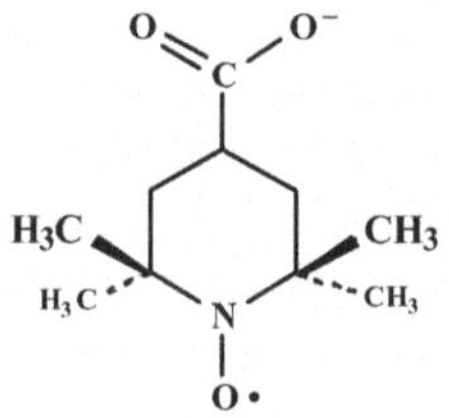

II

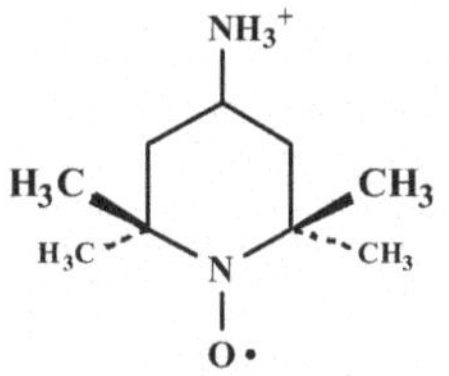

III

as spin probes. Calculations of local charge Z_x in the vicinity of a paramagnetic particle can be performed using the diagram based on 11.4 for encounters between two charged particles in solution (Fig. 11.8). The values of k^{i} and the resulting charges, Z_H calculated with 11.4, are presented in Table 11.1.

The ability to measure local charges and the potential with the help of the above-described tag probe was confirmed by several subsequent studies [24–32].

SLSP approach was employed in Hecht et al. [24] for determination of the electrostatic potential near the surface of calf thymus DNA. Spin–spin interaction between an ^{14}N-nitroxide derivative of 9-aminoacridine attached to DNA and free ^{15}N-labeled nitroxides of different charges was monitored by electron–electron double resonance (ELDOR). An approach developed in [30] also uses fast-relaxing charged and neutral paramagnetic relaxation agents (PRAs) to increase nitroxide spin-label relaxation rate solely through collisional spin exchange. With this purpose, T PRAs were calibrated in experiments on small nitroxides of known structure and charge to account for differences in their relaxation efficiency. Nitroxide longitudinal [R(1)] and transverse [R(2)] relaxation rates were measured by applying lineshape analysis to progressive saturation spectra. The ratio of measured R(1) increases for each pair of charged and neutral PRAs according to electrostatic potential. Voltage at the spin label is then calculated using the Boltzmann and Debye–Hückel equations. Voltage for spin-labeled myosin fragment S1 also agrees with calculation based on the pK shift of the reacted.

11.2.2.2 Effect of Charge on Dipolar Interactions Between Protons and a Paramagnetic Species

Proton spin–lattice relaxation is affected by paramagnetic ions and spin probes. According to theory of Hwang and Freed [31], the spin–lattice relaxation rate of proton nuclei, $1/T_{1p}$, upon encounter with a "radical" free electron, are dominated by dipolar interaction because of low electron spin density at the van der Waals distance. Slopes, $k = d(1/T_{1p})/d[R\cdot]$ of the dependence of proton spin–lattice relaxation rate on the concentration of the nitroxide probes [R•] can characterize the dipole–dipole interaction that dominates in an encounter between a proton and paramagnetic species To explore the effect of electrostatic interaction on the dipole–dipole relaxation rate, the ratios of experimental values k^+/k^0 or k^-/k^0 were used to estimate the electrostatic potential, $U(R_0)$, in the vicinity of a proton [26–28, 32].

The simple organic molecules and amino acids provide well-defined molecules to study the factors that influence electrostatic fields in more complex biologically important molecules [26–28, 32]. A molecular modeling calculation of local electrostatic potential was applied to a number of small charged (ethylamine, propionate anion, and imidazole) and uncharged (ethanol) molecules with an algorithm developed in [29]. The sign and magnitudes of the experimental and calculated parameters $U(R_0)_{exp}$ and $U(R_0)_{calc}$ (Table 11.2) correspond to expected values from simple electrostatic considerations.

Three charged hexosamines D-glucosamine·HCl, D-galactosamine·HCl, D-mannosamine·HCl, a glucuronide metabolite of paracetamol [p-acetamidophenyl-β-D-glucuronide sodium salt, and a mononucleotide [thymidine-3′-phosphate sodium salt (8)], were investigated in [32]. Ratios of these apparent relaxation rate constants

Table 11.2 Rate constants of spin–lattice relaxation of protons in small molecules under the effect of nitroxides **I-III** of different charge [27]

Molecule	Proton	$U(R_0)^{*,H}_{calcd}$	$U(R_0)^{*}_{calcd}$	$U(R_0)^{*,\#,\S}_{exptl}$
		$\epsilon = 1, R = 1.2$ Å	$\epsilon = 80, R = 8$ Å	
Ethanol	H(α)	15(1)	+0.03	
Ethanol	H(β)	10(1)	+0.02	
EtNH3]$^+$Cl$^-$	H(α)	126(2)	+0.25	+0.29(6)
EtHN3]$^+$Cl$^-$	H(β)	109(4)	+0.22	+0.26(8)
Imidazolium$^+$Cl$^-$	H(2)	151(3)	+0.30	+0.30(8)
Imidazolium$^+$Cl$^-$	H(4,5)	130(3)	+0.26	+0.30(4)
EtC0$_2$-Na$^+$	H(α)	−80(4)	−0.16	−0.29(3)
EtC0$_2$-Na$^+$	H(β)	−71(3)	−0.14	−0.26(4)
Aspartate-Na$^+$	H(α)			−0.30(5)
Aspartate-Na$^+$	H(β)			−0.31(6)

The k^+/k^0 and k^-/k^0 ratios (in this Table) are markedly different for protons of positively charged histidine and negatively charged aspartate

($k_+/k_- > 1$ or $k_-/k_+ > 1$.) provided a measurement of the sign and magnitude of the local electrostatic fields surrounding a particular proton in the target.

Biophysical methods to probe local electrostatic potentials of proteins and lipid bilayer systems that are based on an observation of reversible protonation of nitroxides by EPR were described in [33, 34]. The review [33] includes the following sections: (1) EPR characterization of pH-sensitive thiol-specific nitroxide labels, (2) mapping of local pH and electrostatics of peptide and protein systems with methanethiosulfonate derivatives of ionizable nitroxides, (3) pH-sensitive spin-labeled lipids for measuring surface electrostatics of lipid bilayers by EPR, (4) preparation of spin-labeled lipid vesicles, (5) EPR titration experiments with lipid bilayers, (6) analysis of EPR spectra in titration experiments, (7) interfacial pKa and the choice of an electrically neutral reference interface, and (8) surface charge and potential calculation using the Gouy–Chapman theory are outlined.

The ionizable nitroxide labels and the nitroxide-labeled phospholipids described so far cover an exceptionally wide range of ca. 2.5–7.0 pH units, making them suitable to study a broad range of biophysical phenomena, especially at the negatively charged lipid bilayer surfaces. Chemical structures of phospholipid-based nitroxide electrostatic EPR probes, IMTSL-PTE ((S)-2,3-bis(palmitoyloxy)propyl 2-(((1-oxyl-2,2,3,5,5-pentamethylimidazolidin- 4-yl)methyl)disulfanyl)ethyl phosphate) and IKMTSL-PTE ((S)-2,3-bis (palmitoyloxy)propyl 2-((4-(4-(dimethylamino)-2-ethyl-1-oxyl-5,5-dimethyl-2,5-dihydro-1H-imidazol-2-yl)benzyl)disulfanyl)ethyl phosphate) were compared with non-ionizable analog of IMTSL-PTE–MTSL-PTE ((S)-2,3-bis(palmitoyloxy)propyl 2-(((1-oxyl-2,2,5,5-tetramethyl-2,5-dihydro-1H-pyrrol-3-yl)methyl)disulfanyl)ethyl phosphate), and of the adduct of IKMTSL with 2-mercaptoethanol (IKMTSL-2me).

11.3 Immersion Depth of Nitroxide Fragment

At interpretation of data on spin-label dynamics, it is important to know the immersion death in molecular object of interest (r_{im}). A method was developed for determining the nearest distance (r_{min}) between a stable radical (R•) and an ion of paramagnetic metal, an ion-relaxator (IR), which has effects on the spin–lattice relaxation time of R (T_{1e}) and is randomly distributed in the bulk of the vitrified sample [28, 35–37]. In fact, the r_{min} value is equal to the radical immersion depth. The contribution of the dipole interaction of IR to the R• spin relaxation rate is expressed by the equation:

$$\Delta(1/T_{1e}) = \frac{A_d \mu^2 \gamma^2 \tau_{1e} C}{r_{min}^3} \ (2) \quad (11.5)$$

where $1/T_{1e}$ is the nitroxide spin–lattice relaxation rate, which is derived by analysis of ESR spectra saturation curves, C is the IR concentration, μ and τ_{1e} are magnetic moment and the spin relaxation rate of the IR, respectively, and A_d is a factor that

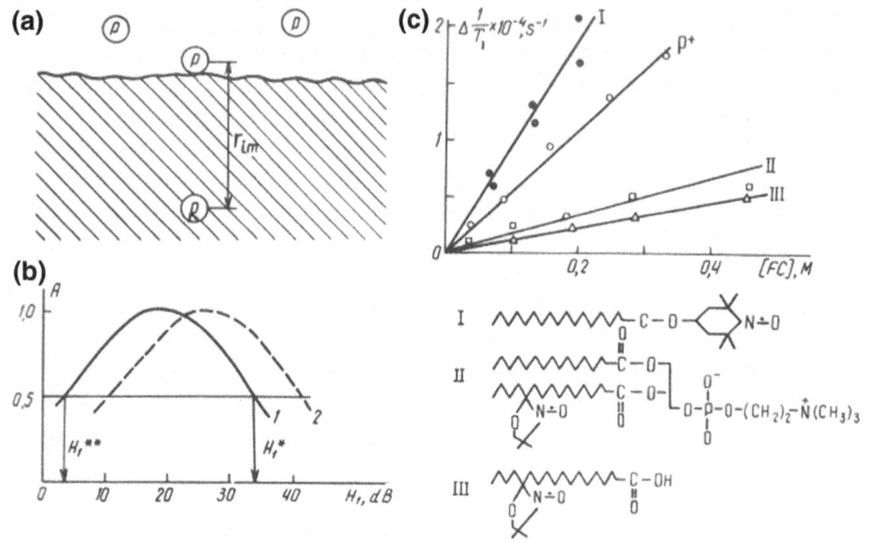

Fig. 11.9 Schematic illustration method of the immersion depth of nitroxide fragment measurement

depends on the geometry of the surface. For example, if the surface is flat, A_d = 0.2. The sensitivity of the ESR spectra saturation curves for spin-labeled cotton and cellulose fibers to the nature of the object and presence of ion-relaxator ferricyanide were schematically illustrated on examples of nitroxide spin probes incorporated in biomembranes, cotton fibers, and cellulose (Fig. 11.9).

A method has been developed for measuring depth of immersion of a fluorescent chromophore in biological matrices, such as biomembranes [38]. The method is based on dynamic quenching of chromophore fluorescence by a nitroxide probe freely diffused in solution. By measuring the quenching rate constant and making use of theoretical consideration and experimental data relating to exchange processes, it is possible to estimate the depth of immersion of chromophore. The ratio between the quenching rate constants for a chromophore–quencher pair in solution (diffusion limit) and after ducking in a matrix (kinetic limit) is given by:

$$k_q^k / k_q^d = \tau_c^2 10^a \exp 2[-\beta(R_0 - r_v)] \quad (11.6)$$

and the depth of immersion of the center under investigation by:

$$(R_0 - r_v) = 0.5 \times \beta^{-1}[\ln(k_q^d / k_q^k) + \ln(\tau_c^2 10^a)] \quad (11.7)$$

where k^d and k^k are the rate constant of encounters in solution, and after immersion, correspondingly; τ_c is the lifetime of the encounter complex; β ∇ 2 and 1.3 A and a ∇ 28 and 26 for the intersystem crossing (ICHA) and electron transfer (ET) mechanisms, respectively. Equation (11.7) can be used to estimate the depth of immersion of a fluorescent chromophore in a "non-conductive" matrix by experimental measurement of ratio *k*/*k* with a reasonable choice of the equation parameters. A fluorescence–photochrome probe, 4.4′-dimethylaminocyanostilbene (DMACS) was incorporated in lecithin liposomes and membranes from *Bacillus subtilis* grown in the absence and presence of chloramphenicol. The quenching of the probe fluorescence by a water-soluble nitroxide radical (TEMPOL) was investigated.

To elucidate the location of a phosphorescent label (erythrosine) covalently incorporated in cotton fiber, which was a source of information about the fiber's local dynamics, a method for determining the depth of the immersion of a luminescence chromophore (r_{im}) was developed [39]. The method is based on experimental measurements of rate constants of quenching luminescence in conditions of free access between chromophore and quencher (k_{qd}) and between immersed chromophore and free quencher (k_{qk}). For the measurement of r_{im}, the 11.7 was employed. Values of immersion depth of spin label I in fibers (r_{min}, in nm) were found to be 1.0 and 1.1 nm for α-cellulose and cotton fiber cotton fiber, respectively. These values are close to the immersion depth of phosphorescence triplet label of about 0.8 nm [43]. Therefore, the labels' mobility reflects the molecular dynamics of the polymers chains in the vicinity of the label location.

11.4 Nitroxides in Cotton and Cellulose Physicochemistry and Technology

Cotton and cellulose are ones of the essential materials for the human civilization. The role of cotton fibers and cellulose in industry, medicine, and human beings' everyday life is impossible to overestimate. Since pioneering works of Marupov and Likhtenshtein groups [40, 41], nitroxides are widely used in solving a number of challenging problems of structure and functions of cotton fibers and cellulose [42–54]. This section is intended to provide the physical principles of methods based on the use of the covalently bound and adsorbed nitroxides (spin labels and spin probe, correspondingly) and its applications in the investigation of cotton fibers and cellulose microstructure and molecular dynamics and it connection with these materials durability. An arsenal of modern physical methods including electron spin resonance (ESR), Messbauer and luminescence techniques is involved in investigation of the labeled samples [43]. Involving nitroxide in chemical processes in this area is also considered.

The covalent modification of an OH group of cotton and cellulose fibers performed with correspondent nucleophilic reactions using nitroxide spin label

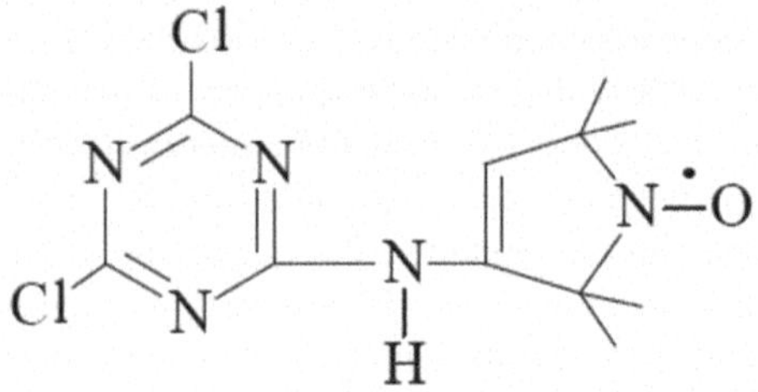

was described in detail in [40, 41]. Samples of cellulose labeled with stable nitroxyl radical

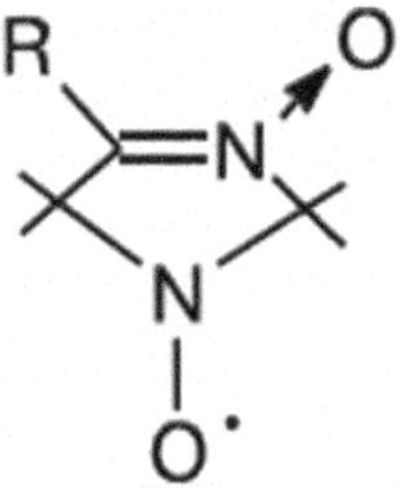

1—3

R = Me (**1**), COOH (**2**), CH_2Br (**3**)

were prepared through mechanochemical synthesis [42]. The samples were studied by IR and EPR spectroscopy.

Modern ESR techniques allow ones to access dynamic processes that are characterized by a wide range of correlation time, $\tau_c = 10^2$–10^{-10} s (Chap. 4). Experiments show (Fig. 11.10) [43] that the investigated cotton fibers consist of two fractions: major (component h, $\tau_c \leq 10^{-8}$) and minor (component h′, $\tau_c \geq 10^{-8}$ s). Apparently, the latter is attributed to loose microscopic defects of the fibers. The contribution of these fractions in different samples is 94–98 and 6–2%, respectively. Contribution of the loose fraction enhances with increase of temperature.

The dynamic behavior of spin labels with nitroxides located in samples of cotton fiber was investigated with the use of high resolution, high-frequency (2-mm) ESR spectroscopy (Fig. 11.11) [43–45]. The studies confirmed conclusions formulated on the base of the 3-cm band (V_1) and also indicated strong anisotropic of the label nitroxide fragment

Notably, the nitroxide fragment dynamics in this microsecond temporal region and its nanosecond dynamics occur in the same temperature region (273–373 K). These findings suggest distribution of the label dynamic parameters modulated by the fibers dynamics. This conclusion was confirmed in study of molecular dynamics

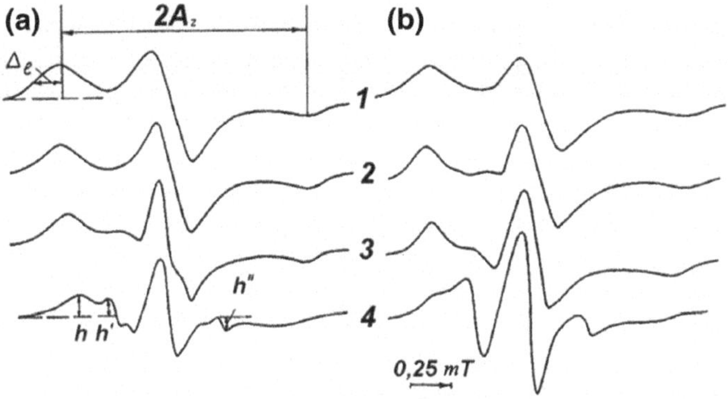

Fig. 11.10 First harmonic ESR spectra (V_1) for spin-labeled cotton fibers **a** «5595-B», **b** "Tashkent 1" Temperature (1) 123 K, (2) 213 K, (3) 323 K, (4) 373 K. Relative humidity $P/P_0 = 0.96$ [43]

Fig. 11.11 High-resolution (2 mm) ESR spectrum of spin-labeled cotton fiber «5595-B» at relative humidity $P/P_s = 0.04$ and temperature 150 K (**a**); dependence of parameters of g and A tensors on temperature (**b**) [44]

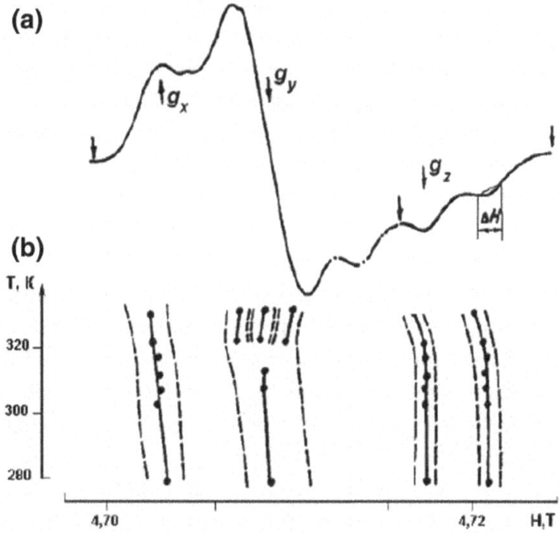

of cotton fibers by methods of fluorescence and phosphorescence labels [43]. The free activation energy distribution of mobility of the chromophore label was found to be 10.5 kJ/mole (data on fluorescence) and 6.7 kJ/mole (data on phosphorescence).

Values of immersion depth of spin label I in fibers (r_{min}, in nm) were found to be 1.0 and 1.1 for α-cellulose and cotton fiber cotton fiber, respectively [43]. These values are close to the immersion depth of phosphorescence triplet label of about 0.8 nm.

The role of the microstructure of fibers in its stability under tensile stress is a basic challenge and applied problem. The design, fabrication, testing, or an operation stage of a polymer product's life in particular in a great extend related to its durability. A

Table 11.3 Parameters of the Zhurkov 11.8. Loose-phase (defects) fracture of cotton fibers (n) derived by the spin labeling and Zhurkov parameters for the samples at $T = 300$ K; $\xi = d(h'/h)/dT$ is the rate of increase of parameter $'h/h$ (Fig. 11.12), which characterizes a loosing of the polymer structure at a temperature increase above 333 K [43]

Fiber	n-defects fracture (%)	ξ (K^{-1})	σ_p (MPa)	U_0 (kJ/MOLE)	$\gamma.10^{-4}$ (m^3/MOLE)
«5595-B»	2 ± 0.5	0.009	280 ± 12	142 ± 7	2.6
«Taskent-1A»	6 ± 2	0.030	180 ± 12	141 ± 7	4.9
«*Tashkent-1B*»	2.8 ± 2	0.070	150 ± 12	141 ± 7	6.0

connection between durability of various samples of cotton fibers and its molecular properties was established using method of nitroxide spin labels (Table 11.3) [43].

A connection between durability of various samples of cotton fibers and its molecular properties was established using method of nitroxide spin labels and a Zhurkov model [46] (Table 11.3). Widely accepted Zhurkov model, assuming the kinetic nature of bond destruction through the thermo-fluctuation mechanism, enables one to obtain a simple, easy-to-use, and physically meaningful formula for the evaluation of the durability of a material after the given time in operation at the given temperature and under the given stress. According to this model, a sample durability (r) can be described with the following equation

$$r = \tau_0 e^{\frac{U_0 - \sigma\gamma}{kT}} \quad (11.8)$$

where U_0 is the energy activation of mechanic destruction, τ_0 is the time of atomic vibration, σ is the applied tension, and γ is the activation volume sensitive to the sample structure. The experimental parameters of 11.8 are presented in Table 11.3.

As it is shown in Table 11.3, the energy activation of mechanic destruction is practically the same for all samples, which indicates that the process occurs as a break in chemical bonds. The difference in the samples' duration is caused by differences in fracture structural defects (n). The larger n is, the smaller ξ and σ_p are, and the higher γ is—this equates to lower sample durability. This finding opens the way to predict the durability of products made from cotton fibers or cellulose using EPR spectroscopy even in the processes of its production.

A series of microstructure studies of cotton fiber and cellulose was performed using non-covalently bound probes [47–54]. These studies confirmed the main results of earlier works on the one hand and revealed important details on the other.

Spin probe ESR method was used to study local polarity and heterogeneous viscosity of two types of unprocessed cotton fibers, *G. hirsutum* and *G. barbadense* [51]. These fibers were loaded with two nitroxide probes that differ in polarity—TEMPO and its more hydrophilic derivative TEMPOL. A series of polar and non-polar solvents was utilized. Results indicated coexistence of cellulose nanodomains with different physicochemical properties, such as polarity and microviscosity, that are affected by solvents and temperature. EPR studies of cellulose microstructure employing the nitroxide spin-labeled albumin, as spin probe entrapped in cellulose triacetate fibers, were carried out [52]. The experiments show that within the fiber, two phases are present: medium viscosity inside microcavities and more rigid phase.

The effect of water content on the physicochemical properties of the amorphous regions in cotton was studied by measuring the electron paramagnetic resonance of TEMPOL [47]. Mobility of nitroxide radicals, deposited in cotton at different loadings, as a function of the relative humidity (RH) and temperature, was investigated. The method of nitroxide spin labels of different charges developed earlier in [22, 23] was employed to investigate the effect of charge on spin probe interactions and dynamics in the nanopores of cotton, using 4-amino-TEMPO (T–NH_2) and 4-carboxy-TEMPO (T–COOH) deposited on dry cotton [48].

Forty samples of cotton filaments from naturel and mutant plants spin-labeled by a nitroxide 2,2,5,5-tetramethyl-3-aminopyrrolidine-(3,5-dichlorotriazine)-1-oxyl (I), chemically bound by the cellulose hydroxyl group, were prepared [49]. Experiments revealed different conformational changes of cotton cellulose grown from seeds exposed to different doses of gamma-irradiation. The spin-label investigation of flax shive cellulose structural and molecular dynamics properties in the process of its nitration using the spin label I revealed that the macromolecules' packing density in the less ordered nitrate fields is higher than that of natural flax cellulose. A correlation between the "order" and "disorder" relation in natural cellulose and in finite nitrogen products was established. The plasticizing capacity of the solvents at 0° decreases through the EtOH, H_2O, and $CHCl_3$ series. A facile grafting to protocol based on the generation of radicals at the surface of cellulose by mild UV irradiation ($\lambda_{max} \sim 311$ nm) of an immobilized photoinitiator was conducted [50]. The radicals photoinduced generated in cellulose were trapped with a nitroxide-functionalized polymer. Spin capturing was reported, using nitrones that after a first radical reaction generate a nitroxide which are able to undergo a second radical coupling. A nitroxide amphiphilic block copolymer of poly(ethylene glycol)-b-poly(2,2,6,6-tetramethylpiperidinyloxy-4-yl-methacrylate) was used as a recoverable catalyst for selective catalytic oxidation of cellulose [53].

In conclusion, covalent modification cotton fibers and cellulose by the labels and the use electron spin resonance are allowed to establish the label location and molecular motion the labeled samples which in turn is modulated by the polymers local molecular dynamics and its distribution within 10^{-4}–10^{-9} s range of correlation times. Data on dependencies of the fibers' molecular dynamics on origin, temperature, water, and other plasticizing agents, nutrition, period of maturing, virus infection, and radiation of seeds have been presented. Combination of ESR experiments and Zhurkov technique have revealed a strong dependence of fibers' resistance to stress (durability) on microscopic structural defects. Employing non-covalently bound nitroxides as spin probes provided a way for studies of effect of humidity on the supramolecular structure of cotton, local polarity, and heterogeneous viscosity of two types of unprocessed cotton fibers, cellulose modification by nitroxide-mediated graft polymerization, composites of poly(lactic) acid reinforced with TEMPO-oxidized fibrillated cellulose, and photoinduced macromolecular functionalization of cellulose vianitroxide spin trapping.

11.5 Spin Cascade

Measurements of active encounters between molecules in native membranes containing ingredients, including proteins, are of prime importance. To estimate rare encounters in a high range of rate constants (rate coefficients) in membranes, a spin cascade method (SCM) was invented and developed in Likhtenshtein group [54–57]. The SCM is based on four well-known fundamental physical phenomena: (1) singlet triplet transition in an excited chromophore with the formation of a paramagnetic state with a relative long lifetime (triplet probe, TP), (2) quenching of excited triplet state of TP by a paramagnetic species (spin probe, SP), (3) *cis-trans*-isomerization of a photochrome compound (fluorescence–photochrome probe, FPP), and (4) triplet–triplet energy transfer between TP and FPP.

As was shown in the Likhtenshtein group, paramagnetism and the relative long lifetime of chromophores in the excited triplet state provide the possibility for these compounds to be used as phosphorescence (triplet) labels for solving a number of structural and dynamics problems [57, 58]. Dexter triplet–triplet TT energy transfer is a process of exchanging both spin and energy between a pair of molecules or molecular fragments [59]. Nitroxides are effective quenchers of excited states by the exchange mechanisms [57].

Within the spin cascade method, specifically, a spin probe is nitroxide (5-doxyl stearic acid), triplet probe is erythrosin B in exited state, and fluorescence–photochrome probe (FPP) is (4-dimethylamino-4-aminostilbene) [56].

Stilbenes exhibit a diverse photochemical behavior in solution such as reversible *cis/trans*-isomerization cyclization of cis-stilbene which involves rearrangement along a double bond [54, 60, 61]. Due to the fact that *trans*-stilbene exhibits fluorescence and *cis*-stilbene is fluorescence silent, the photoisomerization process can be readily monitored by the highly sensitive fluorescent techniques. The *trans–cis* photoisomerization of stilbenes can proceed by two in-principle mechanisms: (i) the direct process involving exited single state and (ii) the triplet-sensitized photoreaction of stilbenes that caused *trans–cis* photoisomerization.

The spin cascade method was employed to estimate rare encounters in a high range of rate constants (rate coefficients) in multilamellar liposomes (Fig. 11.12) [55].

Experimental rate constant of the triplet probe phosphorescence quenching by nitroxide (k_q) and the rate constant of the triplet–triplet energy transfer (k_T) evaluated in 2D terms were obtained as $k_q = 1.05 \times 10^{15}$ cm^2/(mol s) and $k_T = 1.26 \times 10^{12}$ cm^2/(mol s), respectively. The values of diffusion rate constants in model lipid membranes together with similar data obtained from other methods, that cover characteristic times over eight orders of magnitude, were found to be in good agreement with the advanced theory of diffusion-controlled reactions in two dimensions (Fig. 11.13) [62]. In addition, using the spin cascade method, the following dynamic parameters of the cascade system components can be experimentally measured: (1) the spin-label rotation correlation time and spin relaxation parameters, (2) the fluorescence and phosphorescence polarization correlation times, (3) the rate constants

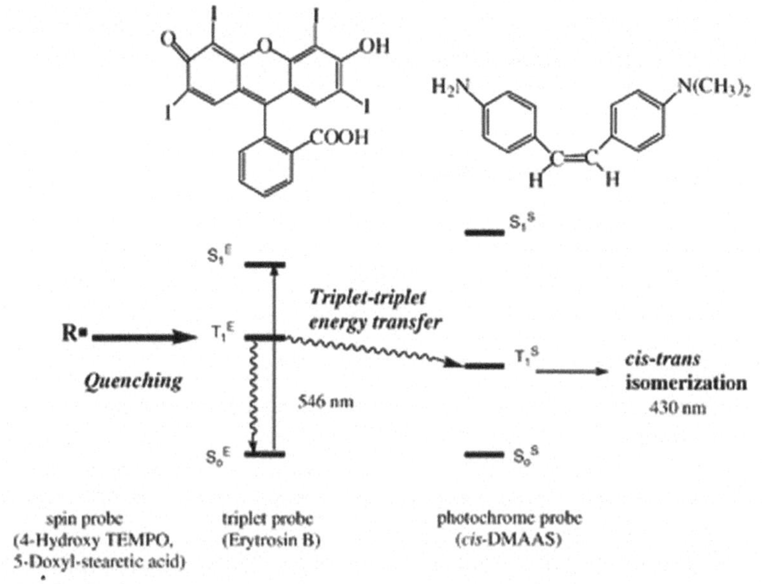

Fig. 11.12 Representation of energy levels of cascade reactants and competition between the $T_1^E \to T_1^S$ and $T_1^E \to S_0^S$ processes [56]

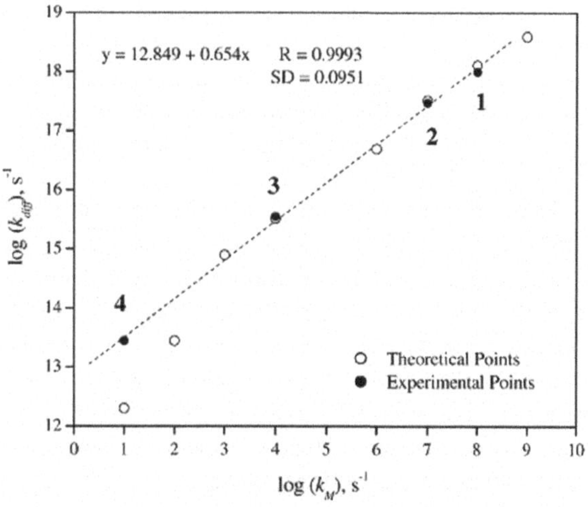

Fig. 11.13 Values of experimental and theoretical diffusion rate constants in model lipid membranes versus corresponding characteristic rate coefficient [56]

of the chromophore singlet- and triplet-state quenching by nitroxides, and (4) the rate constant of photoisomerization. This set of parameters is a cumulative characteristic of the dynamic state of biomembranes in the wide range of the probes amplitude and characteristic time. Proficiency of the method can be expanded by a choice of the cascade components with the higher efficiency of triplet–triplet energy transfer, higher sensitizer lifetime, and by an increase of the time of integration of experimental data on a photochrome photoisomerization.

11.6 Nitroxides in Matrices

Multifrequency continuous wave (CW) and echo-detected (ED) electron paramagnetic resonance (EPR) were employed to study the mobility of 4-methoxy-2,2,6,6-tetramethylpiperidine-*N*-oxyl (MT) and *N*-(2-methylpropyl)-*N*-(1-diethylphosphono-2,2-dimethylpropyl)-aminoxyl (DEPN) confined in nanocapsules of *p*-hexanoyl calix[4]arene, using W-band and 360 GHz cw [63]. Experiments indicated that between 90 and 200 K, the caged nitroxide undergoes isotropic small-angle fluctuations (librations), whereas at higher temperatures restricted rotations of the radical with correlation times of 0.75×10^{-10} s and 1.2×10^{-10} s dominate at 325 and 300 K, respectively. The activation energy of the rotational motion of encapsulated MT radicals was evaluated as $E_a = 1.0$ kcal mol^{-1}.

New covalently linked spin-labeled cyclodextrin (CD) was prepared via the attachment of nitronyl nitroxide 2-(4-hydroxyphenyl)-4,4,5,5-tetramethyl-4,5-dihydro-1H-imidazole-3-oxide-1-oxyl to permethylated β-cyclodextrin I, and its molecular dynamics were investigated employing ESEEM/CW EPR approaches [64].

The experiments indicated that: (1) nitronyl nitroxide exhibits dynamic equilibrium between conformations with radical fragment outside the cavity and radical capping the cavity of CD, (2) covalent attachment of nitronyl nitroxide to CD leads to a corresponding increase of the rotational correlation time by an order of magnitude; (3) nitronyl nitroxide covalently attached to CD retains its sensitivity to nitric oxide, and (4) the reduction rate constant of nitroxide fragment in I is close to that of corresponding free radical.

Rotation of methyl groups which present in spin labels including MTSL contributes in spin echo dephasing [65]. These effects were removed in iodoacetamide azaadamantyl spin label in fluid solution and in 9:1 trehalose/sucrose glass. For example, in 9:1 toluene/CH_2Cl_2 solution at 293 K, the isotropic nitrogen hyperfine coupling is 19.2G, T_1 is 0.37 ms, and T_m is 0.30–0.35 ms. Although complexation of the azaadamantyl spin label with β-cyclodextrin slows tumbling in aqueous solution by about a factor of 10, it does not affect significantly on $1/T_1$ or $1/T_m$ in 9:1 trehalose:sucrose between 80 and 293 K. It was suggested that relaxation is dominated by Raman and local mode processes.

Interfacial electrostatic properties of hydrated mesoporous and nanostructured alumina powders are determining factors for the use of these materials in heterogeneous catalysis and as a sorption media for filtration and chromatographic applications. The purpose of work [66] was to measure Stern potential in cylindrical nanosized hydrated channels of the mesoporous molecular sieves (MMS) MCM-41 and SBA-15 with channel of different diameters (2.3–8.1 nm) utilizing pH-sensitive nitroxide radicals (NR).

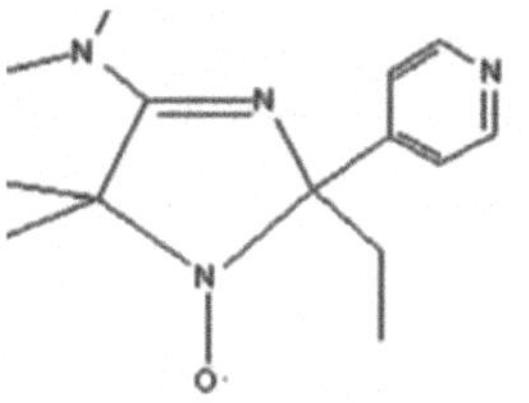

R1

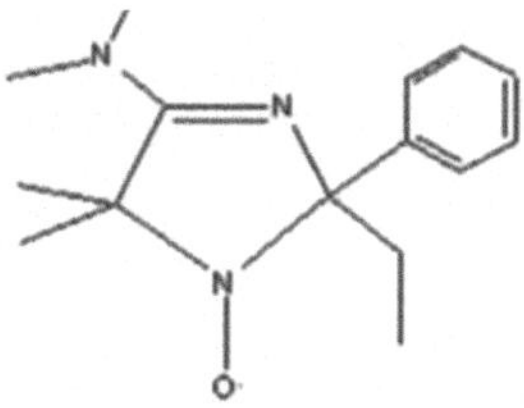

R2

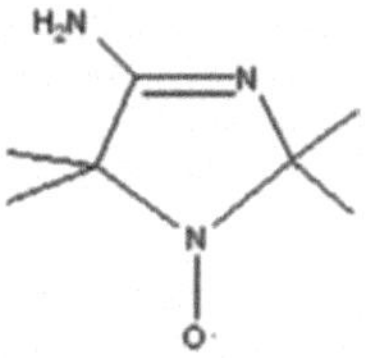

R3

Stern potential is an electric potential on the external boundary of the Stern layer versus the bulk electrolyte. The Stern layer accounts for ions' finite size and consequently an ion's closest approach to the electrode. The layer is on the order of the ionic radius. EPR spectroscopy was employed to study rotational dynamics of pH-sensitive NRs adsorbed in the water-filled nanochannels. The following experimental findings were reported: (1) Two populations of the probe molecules in the sieve channels, one undergoing a fast rotational diffusion and the other one being significantly slower motion of the probe, were observed; (2) the solution acidity (pKel) inside the channels and near-surface electrical potential was estimated using the pH spin probes; (3) negative and positive values of Stern potential for the positively and negatively charged surfaces of MMS channels were found; and (4) the ranges of pH of external solution for the near-zero charge, boost-charging of the channel surface, and for dissociation of functional groups in the MMS studied were determined. Values of near-surface potential of the positively (4–19 mv) and negatively (from -110 to -179 mv) charged MMS were shown in a table.

EPR spectroscopy [X-band (9 GHz)] of new pH-sensitive nitroxides (4-dimethylamino-5,5-dimethyl-2-(4-(chloromethyl) phenyl)-2-ethyl-2,5-dihydro-1H-imidazol-1-oxyl hydrochloride semihydrate (nitroxide R1) was employed to evaluate the surface charge and interfacial acid–base equilibria at the pore surface of mesoporous powders of α-Al_2O_3, γ-Al_2O_3, $Al_2\ O_3 \times nH_2O$, and basic γ-Al_2O_3 and nanostructured γ-Al_2O_3 [67]. Mesoporous γ-Al_2O_3 was synthesized by heating aluminum hydroxide. The materials in the form of pristine were modified with aluminum-tri-sec-butoxide, hydroxyaluminum glycerate, phospholipids, soy phosphatidylcholine, cholesterol, negative composition of soy lecithin Epikuron 200, phosphatidylcholine, and cetyl pyridinium chloride. It was found that for basic γ-Al_2O_3 and $Al_2O_3 \times nH_2O$, the values of nitroxide R1 pKa were shifted by ΔpKa $\approx +0.6$ and up to $\approx$+1.2 pH units, respectively, while the shift for γ-Al_2O_3 was found as ΔpKa $= +3.5$. The order of descending magnitude of the effective surface electrostatic potential Ψ is mesoporous γ-Al_2O_3 > Al_2O_3 $\times$ basic γ-Al_2O_3 > α-Al_2O_3.

EPR spectroscopy of two pH-sensitive NRs of the imidazoline type, 4-dimethylamine-2-ethyl-5,5-dimethyl-2-pyridine- 4-yl-2,5-dihydro-1H-imidazole-1-oxyl (R1) and 4-dimethylamine- 2-ethyl-5,5-dimethyl-2-phenyl-4-yl-2,5-dihydro-1H-imidazole-1-oxyl (R2), and of the imidazole type, 4-amino-2,2,5,5-tetramethyl-2,5-dihydro-1H- imidazole- 1-oxyl (R3) was employed to examine the conditions for

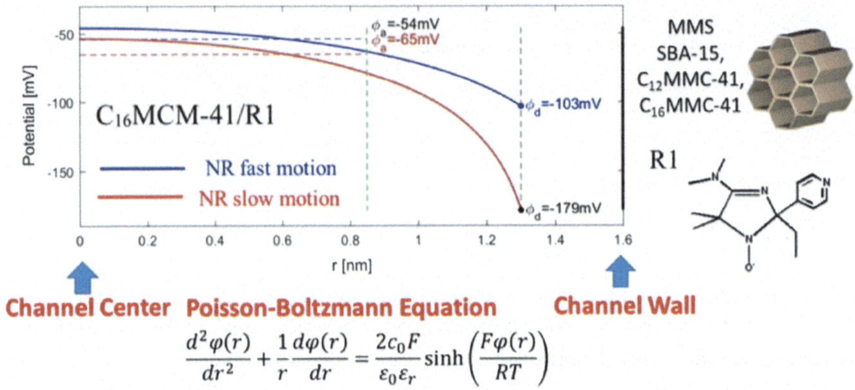

Fig. 11.14 Radial profiles of electrostatic potential $\varphi(r)$ obtained from numerical solutions of the PB equation using different boundary conditions, channel center Poisson–Boltzmann equation and the nanochannels of silica-based mesoporous molecular sieves mode [68]. Reprinted from [68], Copyright 2018 American Chemical Society

the efficient electrostatic surface potential screening inside the nanochannels of silica-based mesoporous molecular sieves (MMS) filled with water at ambient conditions and a ionic strength of 0.1 M [68]. All-silica MCM-41 materials were synthesized using dodecyltrimethylammonium bromide (C12MCM-41) and hexadecyl trimethylammonium bromide (C16MCM-41) and exhibited average channel diameters of 2.3 and 3.2 nm, respectively. CW X-band (9 GHz) CW EPR spectra of NRs in bulk aqueous solution and upon incorporation into the channels of solid MMS samples were recorded at temperature 293 K. EPR experiments with water-soluble ionizable pH-sensitive nitroxides provided the means to determine an average effective local pH_{loc}. The experiments, using water-soluble ionizable pH-sensitive nitroxides, showed that the nitroxide molecules can diffuse in aqueous volume inside the nanopores of the MMS. In SBA-15 channels filled with aqueous solutions, when the pore radius is about fourfold larger than the Debye radius, pH_{loc} averaged over the central section of the nanochannel was found to be higher by 0.2–0.3 pH units than pH_{ext}. Data on the average electrostatic potential over the central volume of the nanochannels measured by "mobile" EPR probes and the surface potential at the channel surface determined from EPR titrations of immobilized nitroxides were in a good agreement with results of the classical Poisson–Boltzmann theory for the channels down to 3.2 $\pm$ 0.1 nm in diameter. The main results of the work are schematically illustrated in Fig. 11.14.

References

1. N.M. Gallagher, A. Olankitwanit, A. Rajca, High-spin organic molecules. J. Org. Chem. **80**(3), 1291–1298 (2015)

2. X. Meng, W. Shi, P. Cheng, Magnetism in one-dimensional metal nitronyl nitroxide radical system. Coord. Chem. Rev. **378**, 134–150 (2018). https://doi.org/10.1016/j.ccr.2018.02.002
3. S. Kumar, Y. Kumar, S.K. Keshri, P. Mukhopadhyay, Recent advances in organic radicals and their magnetism. Magnetochemistry **2**, 42 (2016). https://doi.org/10.3390/magnetochemistry2040042
4. K.-A. Hansen, J.P. Blinco, Nitroxide radical polymers a versatile material class for high-tech applications. Polym. Chem. **9**, 1479–1516 (2018)
5. Y. Xie, K. Zhang, M.J. Monteiro, Z. Jia, Conjugated nitroxide radical polymers: synthesis and application in flexible energy storage devices. ACS Appl. Mater. Interfaces. **11**(7), 7096–7103 (2019). https://doi.org/10.1021/acsami
6. A. Leavesley, S. Jain, I. Kamniker, H.H. Zhang, S. Rajca, A. Rajca, S. Han, Maximizing NMR signal per unit time by facilitating the e-e-n cross effect DNP rate. Phys. Chem. Chem. Phys. **20**(43), 27646–27657 (2018)
7. A. Equbal, Y. Li, A. Leavesley, S. Huang, S. Rajca, A. Rajca, S. Han, Truncated cross effect dynamic nuclear polarization: an overhauser effect doppelganger. J. Phys. Chem. Lett. **9**(9), 2175–2180 (2018)
8. A. Bihlmeier, G. Jeschke, S. Bräse, M. Tsotsalas, Radical exchange reaction of multi-spin isoindoline nitroxides followed by EPR spectroscopy. RSC Adv. **6**, 55715–55719 (2016)
9. E.G. Janzen, B.J. Blackburn, Detection and identification of short-lived free radicals by an electron spin resonance trapping technique. J. Am. Chem. Soc. **90**, 5909–5910 (1968)
10. A. Nawab, A. Nichols, R. Klug, J.I. Shapiro, K. Sodhi, Spin trapping: a review for the study of obesity related oxidative stress and Na +/K + -ATPase. J. Clin. Cell Immunol. **8**(3), 505 (2017)
11. M. Heinenberg, B. Menges, S. Mittler, H. Ritter, Polymeric nitrons. 2. Synthesis, irradiation and waveguide mode spectroscopy of polymeric nitrons derived from polymeric Benzaldehydes and N Isopropylhydroxylamine. Macromolecules **35**(9), 3448–3455 (2002)
12. P.L. Zamora, F.A. Villamena, Theoretical and experimental studies of the spin trapping of inorganic radicals by 5,5-Dimethyl-1-pyrroline *N*-Oxide (DMPO). J. Phys. Chem. A **116**(26), 7210–7218 (2012)
13. D. Bardelang, A. Rockenbauer, H. Karoui, J.P. Finet, P. Tordo, Inclusion complexes of PBN-type nitrone spin traps and their superoxide spin adducts with cyclodextrin derivatives: parallel determination of the association constants by NMR titrations and 2D-EPR simulations. J. Phys. Chem B **109**, 10521–10530 (2005)
14. M. Cassien, C. Petrocchi, S. Tetiot-Laurent, M. Robin, E. Ricquebourg, C. Kandouli, A. Asteian, A. Rockenbauer, A. Mercier, M. Culcasi, S. Pietri, On the vasoprotective mechanisms underlying novel phosphorylated nitrones: focus on free radical characterization, scavenging and NO-donation in a biological model of oxidative stress. Eur. J. Med. Chem. **119**, 197–217 (2016)
15. H. Chen, G. Tan, J. Cao, G. Zhang, P. Yi, Y. Pei, Y. Sun, Z. Zhang, Y. Wang, Design, synthesis, and biological evaluation of novel tetramethylpyrazine derivatives as potential neuroprotective agents. Chem. Pharm. Bull. **65**, 56–65 (2017)
16. L. Socrier, M. Rosselin, F. Choteau, G. Durand, S. Morandat, Cholesterol-nitrone conjugates as protective agents against lipid oxidation: A model membrane study. Biochim. Biophys. Acta Biomembr. **1859**(12), 2495–2504 (2017)
17. V. Marchand, N. Charlier, J. Verrax, P.P. Buc-Calderon, P. Levêque, B. Gallez, Use of a cocktail of spin traps for fingerprinting large range of free radicals in biological systems. PLoS ONE **12**(3), e0172998 (2017)
18. X. Gao, Y. Bi, K. Chi, Y. Liu, T. Yuan, X. Li, W. Bi, Glycine-nitronyl nitroxide conjugate protects human umbilical vein endothelial cells againsthypoxia/reoxygenation injury via multiple mechanisms and ameliorates hind limb ischemia/reperfusion injury in rats. Biochem. Biophys. Res. Commun. **488**(1), 239–246 (2017)
19. G.I. Liktenshtein, Y.B. Grebenshcikov, P.K. Bobodzhanov, Y.V. Kokhanov, Investigation of the macrostructure of proteins the paramagnetic label-paramagnetic probe technique. Moleculyarnaya Biologiya **4**, 682–691 (1970)

20. K.M. Salikhov, A.B. Doctorov, Y.N. Molin, K.I. Zamaraev, Spin relaxation of radicals and complexes upon encounters in solution. J. Magn. Reson. **5**, 189–196 (1971)
21. J.S. Hyde, H.M. Swartz, W.E. Antholine, The spin probe-spin label methods, in *Spin Labeling. Theory and Application*, vol. 2, ed. by L. Berliner (Academic Press, New York, 1976), pp. 72–113
22. G.I. Likhtenstein, Y.B. Grebentchikov, P.K. Bobodzhanov, Y.V. Kokhanov, Study on the proteins microstructure by method of spin-label paramagnetic probe. Mol. Biol. (Moscow) **4**, 782–789 (1970)
23. G.I. Likhtenstein, Y.B. Grebentchikov, E.G. Rosantev, V.P. Ivanov, Study on the electrostatic charges in proteins by method of paramagnetic probes. Mol. Biol. (Moscow) **6**, 498–507 (1972)
24. J.L. Hecht, B. Honig, Y.K. Shin, W.L. Hubbell, Electrostatic potentials nearthe- surface of DNA—Comparing theory and experiment. J. Phys. Chem. **99**(19), 7782–7786 (1995)
25. Y.K. Shin, W.L. Hubbell, Determination of electrostatic potentials at biological interfaces using electron electron double resonance [Article]. Biophys. J. **61**(6), 1443–1453 (1992)
26. G.I. Likhtenshtein, I. Vaisbuch, I. Adin, A. Shames, R. Glaser, Distribution of electrostatic-field around biological molecules studied by methods of spin-probes and NMR. Biophys. J. **72**, A129 (1997)
27. G.I. Likhtenshtein, I. Adin, A. Krasnoselsky, I. Vaisbuch, A. Shames, Glaser, NMR and ESR studies of electric field distribution around biologically important molecules. Biophys. J. **77**, 443–454 (1999)
28. G.I. Likhtenshtein, Depth of immersion of paramagnatic centers, in *Magnetic Resonance in Biology*, ed. by L. Berliner, S. Eaton, G. Eaton (Kluwer Academic Publishers, Dordrecht, 2000), pp. 1–36
29. B.J. Depmeier, A.J. Driessen, W.J. Hehre, J. Johnson, A.C. Peng, L. Lou, J. Yu, *MacSpartan version 1.0.2* (Wavefunction, Irvine, CA, 1996)
30. J.T. Surek, D.D. Thomas, A paramagnetic molecular voltmeter. J. Magn. Reson. **190**(1), 7–25 (2008)
31. L.P. Hwang, J.H. Freed, Dynamics effect of pair correlation function on spin-relaxation by translational diffusion in liquids. J. Chem. Phys. **63**, 4017–4025 (1975)
32. R. Glaser, A. Novoselsky, A. Shames, G.I. Likhtenshtein, NMR studies of electrostatic fields around charged monosaccharides and related molecules. Isr. J. Chem. Special Lemieux Issue **40**, 263–269 (2000)
33. M.A. Voinov, A.I. Smirnov, Ionizable nitroxides for studying local electrostatic properties of lipid bilayers and protein systems by EPR, in *Methods in Enzymology*, vol. 564 (Electron Paramagnetic Resonance Investigations of Biological Systems by Using Spin Labels, Spin Probes, and Intrinsic Metal Ions, Part B) (2015), pp. 191–192
34. M.A. Voinov, A. Ruuge, V.A. Reznikov, I.A. Grigor'ev, A.I. Smirnov, Mapping local protein electrostatics by EPR of pH-sensitive thiol-specific nitroxide. Biochemistry **47**(20), 5626–5637 (2008)
35. A.V. Kulikov, Determination of distance between the nitroxide label and a paramagnetic center in spin-labeled proteins from the parameters of the saturation curve of the ESR spectrum of the label at 77K. Mol. Biol. (Moscow) **10**, 109–116 (1976)
36. A.V. Kulikov, G.I. Likhtenstein, Application of saturation curves for evaluating distances in biological objects by the method of double spin-labels. Biofisika **19**, 420–424 (1974)
37. A.V. Kulikov, G.I. Likhtenshtein, The use of spin-relaxation phenomena in the investigation of the structure of model and biological systems by method of spin labels. Adv. Molecul. Relax. Proc. **10**, 47–78 (1977)
38. N.V. Strashnikova, N. Medvedeva, G.I. Likhtenshtein, Depth of immersion of fluorescent chromophores in biomembranes studied by quenching with nitroxide radical. J. Biochem. Biophys. Methods **48**, 43–60 (2001)
39. I.K. Yusupov, G.I. Likhtenshtein, Phosphorescence quenching as an approach for estimating localization of triplet label in cotton fibers. Biofizika **57**(2), 286–291 (2012)
40. R.M. Marupov, P.K. Bobodzhanov, N.V. Kostina, A.B. Shapiro, Spin label study of the structure and conformational properties of cotton filament grown from γ-irradiated seeds. Biofizika **21**, 825–828 (1976)

41. R.M. Marupov, P.K. Bobodzhanov, I.K. Usupov, E.N. Frolov, G.I. Likhtenshtein, Study of temperature stability of cotton fibers by spin labeling. Biofizika **24**, 519–523 (1979)
42. A.V. Dushkin, I.B. Troitskaya, V.V. Boldyrev, I.A. Grigor'ev, Mechanochemical method for introduction of a spin marker in cellulose. Russ. Chem. Bull. **54**, 1155–1159 (2005)
43. I.K. Ysupov, G.I. Likhtenshtein, Study of microstructure and molecular dynamics of cotton and cellulose fibers by methods of physical labels. Int. Res. J. Pure Appl. Chem. **6**(3), 105–119 (2015)
44. V.I. Krinichnyi, O.Y. Grinberg, I.K. Yusupov, R.M. Marupov, P.K. Bobodzhanov, G.I. Likhtenshtein, Y.S. Lebedev, Two-millimeter band ESR study of spin-labeled cotton fiber. Biofizika **31**, 482–485 (1986)
45. V.I. Krinichnyi, *2-mm Wave Band EPR Spectroscopy of Condensed Systems* (CRC Press, Boca Raton, Florida, 1995)
46. S.N. Zhurkov, E.A. Egorov, Effect of tensile stress on the molecular mobility in oriented polymers. Dokl. Akad. Nauk SSSR **152**, 1155–1158 (1963)
47. S. Frantz, G.A. Hübner, O. Wendland, E. Roduner, C. Mariani, M.F. Ottaviani, S.N. Batchelor, Effect of humidity on the supramolecular structure of cotton, studied by quantitative spin probing. J. Phys. Chem. B. **109**(23), 11572–11579 (2005)
48. S. Frantz, O. Wendland, E. Roduner, C.J. Whiteoak, S.N. Batchelor, Effect of charge on spin probe interaction and dynamics in the nanopores of cotton. J. Phys. Chem. C **111**(39), 14514–14520 (2007)
49. S. Islomov, R. Marupov, R.G. Zhbankov, PKh Bobodzhanov, L.V. Zabelin, G.N. Marchenko, Spin-label study of structural properties of nitrates based on flax shive cellulose. Zh. Prikl. Spektrosk. **45**, 633–638 (1986)
50. G. Dietrich, M.J.P. Blinco, A. Hirschbiel, M. Bruns, L. Barner, C. Barner-Kowollik, Photo-induced macromolecular functionalization of cellulose via nitroxide spin trapping. Biomacromol **13**(5), 1700–1705 (2012)
51. A. Marek, M.A. Voinov, A.I. Smirnov, Spin probe multi-frequency EPR study of unprocessed cotton fibers. Cell Biochem. Biophys. **75**(2), 211–226 (2017)
52. S. Gulinelli, E. Mantovani, A. Zanobi, EPR characterization of cellulose triacetate fibers used for enzyme immobilization. Appl. Biochem. Biotechnol. **6**, 129–141 (1981)
53. S. Liu, Y. Xing, J. Han, E. Tang, Catalytic oxidation of cellulose with a novel amphiphilic nitroxide block copolymer as a recoverable catalyst. Cellulose (Dordrecht, Netherlands) **24**(9), 3635–3644 (2017)
54. G.I. Likhtenshtein, *Stilbenes: Application in Chemistry, Life Science and Material Science* (WILEY-VCH, Weinhem, 2009)
55. V. Papper, N. Medvedeva, I. Fishov, G.I. Likhtenshein, Quenching of cascade reaction between triplet and photochrome probes with nitroxide radicals: a novel labeling method in study of membranes and surface systems. Appl. Biochem. Biotechnol. **89**, 231–248 (2000)
56. N. Medvedeva, V. Papper, G.I. Likhtenshten, Study of rare encounters in a membrane using quenching of cascade reaction between triplet and photochrome probes with nitroxide radicals. Phys. Chem. Chem. Phys. **7**, 3368–3374 (2005)
57. G.I. Likhtenshtein, *Electron Spin in Chemistry and Biology: Fundamentals, Methods, Reactions Mechanisms, Magnetic Phenomena, Structure Investigation* (Springer, Berlin, 2016)
58. V.M. Mekler, A.I. Kotel'nikov, G.I. Likhtenstein, Study of model and biological membranes by probes emitting annihilated delayed fluorescence. Biofizika **28**, 503–504 (1983)
59. D.L. Dexter, A theory of sensitized luminescence in solids. J. Chem. Phys. **21**, 836–850 (1953)
60. G.S. Hammond, J. Saltiel, Photosensitized *Cis-Trans* Isomerization of the Stilbenes. J. Am. Chem. Soc. **84**, 4983–4984 (1962)
61. D.H. Waldeck, Photoisomerization dynamics of stilbenes. Chem. Rev. **91**, 415–436 (1991)
62. K.R. Naqvi, J. Martins, E. Melo, Recipes For analyzing diffusion-controlled reactions in two dimensions: time-resolved and steady-state measurements. J. Phys. Chem. B **104**, 12035–12038 (2000)
63. E.G. Bagryanskaya, D.N. Polovyanenko, M.V. Fedin, L. Kulik, A. Schnegg, A. Savitsky, K. Mobius, A.W. Coleman, G.S. Ananchenko, J.A. Ripmeester, Multifrequency EPR study of the

mobility of nitroxides in solid-state calixarene nanocapsules. Phys. Chem. Chem. Phys. **11**(31), 6700–6707 (2009)

64. R.K. Strizhakov, E.V. Tretyakov, A.S. Medvedeva, V.V. Novokshonov, V.G. Vasiliev, V.I. Ovcharenko, O.A. Krumkacheva, M.V. Fedin, E.G. Bagryanskaya, Permethyl-b-cyclodextrin spin-labeled with nitronyl nitroxide: synthesis and EPR study. Appl. Magn. Reson. **45**, 1087–1098 (2014)
65. S.S. Eaton, A. Rajca, Z. Yang, G.R. Eaton, Azaadamantyl nitroxide spin label: complexation with β-cyclodextrin and electron spin relaxation. Free Radical Res. **52**(3), 319–326 (2018)
66. E.G. Kovaleva, L.S. Molochnikov, E.L. Golovkina, M. Hartmann, I.A. Kirilyuk, I.A. Grigoriev, Electrical potential near hydrated surface of ordered mesoporous molecular sieves assessed by EPR of molecular pH-probes. Microporous Mesoporous Mater. **203**, 1–7 (2015)
67. E.G. Kovaleva, L.S. Molochnikov, D.P. Stepanova, A.V. Pestov, D.G. Trofimov, I.A. Kirilyuk, A.I. Smirnov, Interfacial electrostatic properties of hydrated mesoporous and nanostructured alumina powders by spin labeling EPR. Cell Biochem. Biophys. **75**, 159–170 (2017)
68. E.G. Kovaleva, L.S. Molochnikov, D.O. Antonov, D.P. Tambasova Stepanova, M. Hartmann, A.N. Tsmokalyuk, A. Marek, A.I. Smirnov, Proton activity in nanochannels revealed by electron paramagnetic resonance of ionizable nitroxides: a test of the Poisson-Boltzmann double layer theory. J. Phys. Chem. C **122**, 20527–20538 (2018)

Index

G. I. Likhtenshtein, *Nitroxides*, Springer Series in Materials Science 292,
https://doi.org/10.1007/978-3-030-34822-9

O

P